CMOS

IEEE Press
445 Hoes Lane
Piscataway, NJ 08854

IEEE Solid-State Circuits Society, Sponsor

CMOS

Mixed-Signal Circuit Design

Second Edition

R. Jacob Baker

IEEE Press Series on Microelectronic Systems
Stuart K. Tewksbury and Joe E. Brewer, *Series Editors*

IEEE PRESS

A JOHN WILEY & SONS, INC., PUBLICATION

Published by John Wiley & Sons, Inc., Hoboken, New Jersey.
Published simultaneously in Canada.

For general information on our other products and services or for technical support, please contact our Customer Care Department within the United States at (800) 762-2974, outside the United States at (317) 572-3993 or fax (317) 572-4002.

Wiley also publishes its books in a variety of electronic formats. Some content that appears in print may not be available in electronic format. For information about Wiley products, visit our web site at www.wiley.com.

Library of Congress Cataloging-in-Publication Data is available.

ISBN 978-0-470-29026-2

Printed in the United States of America.

10

To Julie (again)

Brief Contents

Contents

Preface

Designs that combine analog circuits with digital signal processing, DSP, are called mixed-signal designs, MSDs. Designs that use both digital and analog circuits but no DSP, like a 555 timer or a pipeline ADC, are not, by this definition, MSDs. These mixed-mode or mixed analog/digital circuits aren't as robust as circuits designed using MSD techniques because they require precise components and often calibrations or tuning. The use of DSP in a mixed-signal circuit relaxes the requirements placed on the analog components (**important**) by overcoming the shortcomings, like low transistor gain and poor matching, found in nanometer CMOS.

This book provides a tutorial introduction to MSD techniques. The content is suitable for use in a senior/graduate electrical engineering course or as a reference for a working engineer doing MSD. The assumed background of the reader is a course in signals/systems, and courses in digital and analog integrated circuit design in CMOS technology.

Chapter 1 covers basic signals, filters, and tools. The reader may be inclined to skip this material; however, the author would recommend against it unless the reader can: 1) Give an example of an imaginary signal (answer using words and without math or equations). 2) Explain why imaginary numbers are used. 3) Explain why *I/Q* signals are used. 4) Describe the differences between delaying and non-delaying integrators. While we could keep listing questions, even the most seasoned systems person will likely benefit from reading and thinking about the material in Ch. 1. **Don't skip it!** Make an attempt to understand what is going on in the discussions and not just to understand the math. Once the reader feels they have mastered this material they should try to provide physical metaphors to describe a concept or equation (for example, use a cup, water, and a bucket to describe Eq. [1.56]). Remember that math is easy. Understanding what's going on is tough.

Simulation examples are used throughout the book to provide an additional avenue towards understanding the book's content. In Ch. 1, for example, simulations are available at **CMOSedu.com** that aren't present or discussed in the book. An important part of learning is modifying a simulation, thinking about what should happen and why, then running the simulation to verify your understanding.

Chapter 2 covers sampling and aliasing. These are key topics because sampling is required in any MSD system. Further, decimation (down-sampling) and interpolation (up-sampling) are commonly used in digital signal processing and thus MSD. In addition, circuits used for sampling are presented. Chapters 3 and 4 cover analog and digital filtering. The focus is on practical and useful circuits that can be used in MSD. Chapter 5 covers noise and signal-to-noise ratio. The remaining chapters in the book describe the design of data converters using MSD techniques and the associated trade-offs.

An attempt has been made to present circuits and information with the goal of answering the reader's questions and provoking thought. Hopefully, this will lead the reader towards creative solutions to their circuit design problems. For example, the delta-sigma data converters presented in Ch. 7 use an active integrator. Why? Why can't a passive integrator be used? Chapter 6 develops data converters using passive elements. Both the benefits and problems associated with passive topologies are discussed leading up to answering why an active integrator is used in most delta-sigma data converters.

Finally, one of the main (perceived) limitations of MSD techniques is speed. Often, in a MSD, time is traded-off for precision. The results are circuits that are precise, but slow. The pipeline ADC mentioned earlier is an example of a fast circuit that needs to be precise. Since the pipeline ADC doesn't use MSD techniques, its design can be more than challenging, especially for a production-worthy design, requiring special layout attention or extensive calibrations. The last chapter in the book, Ch. 9, presents a high-speed topology, the K-Delta-1-Sigma topology, that uses MSD techniques that may prove useful in ultimately replacing the pipeline ADC in nanometer CMOS technology nodes. The design procedures used in this topology also provide a good summary of the MSD techniques presented in the book.

Acknowledgments

I would like to thank and acknowledge the reviewers, students, colleagues, and friends that have helped to make this book a possibility: Jenn Ambrose, Hemanth Ande, Jeanne Audino, Kyri Baker, Mahesh Balasubramanian, Amine Bermak, Bertan Bakkaloglu, Joe Brewer, Prashanth Busa, Kris Campbell, Mike Engelhardt, Gilda Garretón, Shantanu Gupta, Bob Hay, Bahar Jalali-Farahani, Kaijun Li, Richard G. Lyons, Pui-In Mak, Brittany Rotert, Steven Rubin, Vishal Saxena, Stu Tewksbury, Donna Welch, Thad Welch, and Aruna Vadla.

R. Jacob (Jake) Baker

Chapter

1

Signals, Filters, and Tools

Mixed-signal circuit design requires a fundamental knowledge of signals, signal processing, and circuit design. In this chapter we provide an overview of signals, filtering, and the mathematical tools. The chapter may be a review for the reader; however, we use it to ensure a good foundation to build on in the coming chapters and to provide a quick reference for the mathematical formulas we'll use throughout the book.

1.1 Sinusoidal Signals

Let's take a fundamental look at the sinewave. While there are many ways (equations and formulas) of representing a sinewave, we must remember it is an empirically determined function. Naturally occurring signals, shapes, or constants are determined or described through empirical measurements or observations. For example, π is determined by dividing the circumference of a circle by its diameter

$$\pi = \frac{\text{circumference}}{\text{diameter}} \tag{1.1}$$

The goal of this section is to provide intuitive discussions that will help create a deeper understanding of what's going on in a circuit or system.

1.1.1 The Pendulum Analogy

Consider the (ideal, that is, lossless) moving pendulum seen in Fig. 1.1a. In this figure the pendulum is moving back and forth between Points 1 and 3 repeatedly over time. As the pendulum leaves Point 1 it starts out slow, gaining maximum speed as it passes Point 2, and finally reaching Point 3. At Point 3 it stops and reverses direction. The time it takes to make this complete journey back to the starting point, Point 1 in this discussion, is the period, T. In Fig. 1.1b we plot the movement of the pendulum along the arched path. We record the position to define a function, $f(t)$, that indicates the pendulum's position at a specific time

$$Position = f(t) = f(t+nT), \text{ where } n \text{ is an integer} \tag{1.2}$$

This signal, we should all recognize, is a sinusoid or sinewave which repeats its position with a frequency, f_o, of $1/T$.

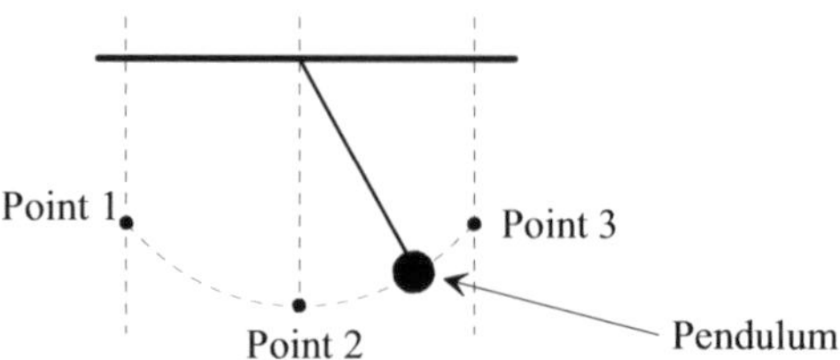

(a) Ideal (never stops or deviates from same path) swinging pendulum in motion. The time it takes to swing from Point 1 to Point 3 and back to Point 1 is the period T.

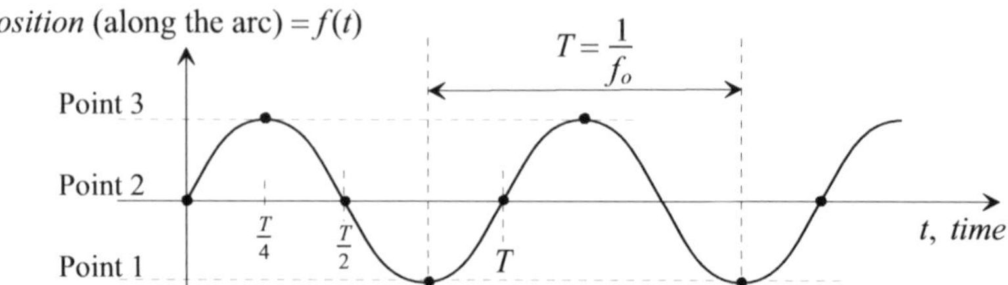

(b) Movement of pendulum along the arc in (a) over time.

Figure 1.1 Physical interpretation of a sinewave.

Next, consider the circle seen in Fig. 1.2. One complete rotation around this circle (360 degrees or 2π) is analogous to one complete movement (swing) of our pendulum. We started plotting the pendulum's position at Point 2 in Fig. 1.1b (Point 2, $t = 0$, in Fig. 1.2). After $T/4$ we reach Point 3 in Fig. 1.1b. This corresponds to a 90 degree, or $\pi/2$, movement in our circle. After another $T/4$ seconds we pass back through Point 2. In the circle we've moved 180 degrees. This continues with each swing of the pendulum corresponding to a complete revolution around the circle. Note that we do have some

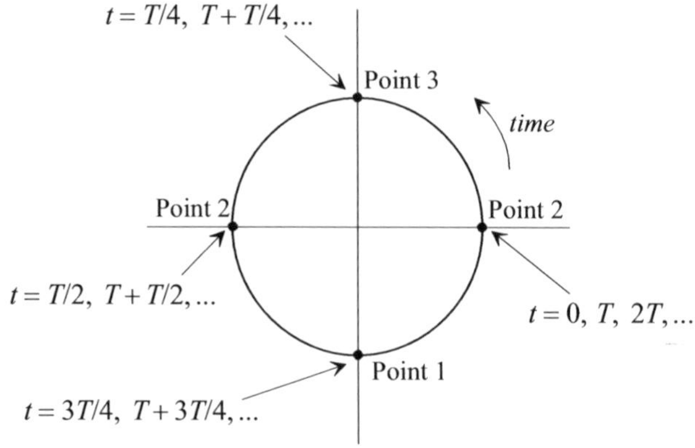

Figure 1.2 Using a circle to describe the movement of the pendulum in Fig. 1.1.

limitations when representing the movement of the pendulum with this circle. For example, what is the amplitude of the sinewave (what is the relative position of the pendulum along the arched path)? We'll address these concerns in a moment. For now let's write, assuming we are using radian angular units,

$$Position = \sin\left(2\pi \cdot \frac{t}{T}\right) = \sin(2\pi f_o \cdot t) \tag{1.3}$$

This function, the sine function, tells us our relative position along the arc (the argument of this function is the angle which relates to the position on the circle in Fig. 1.2). Point 2 corresponds to the function having a value of 0 (and times, $t = 0, T/2, T, 3T/2, 2T, \ldots$), Point 3 to a value of +1, and Point 1 corresponds to −1. Finally, remember that the values of the sine function in Fig. 1.1.b, and Eq. (1.3), are determined empirically from measured data (e.g., plotting the pendulum's position along the arched path against time).

Describing Amplitude in the x-y Plane

Examine the sinewave in Fig. 1.3a. For the moment we won't concern ourselves with the actual distance the pendulum swings. In Fig. 1.3b we represent the sinewave, at Point i (and Point vi), as a zero length vector along the x-axis (the amplitude of the sinewave is 0 at this point in time). As we move towards Point ii in Fig. 1.3a the length of the vector

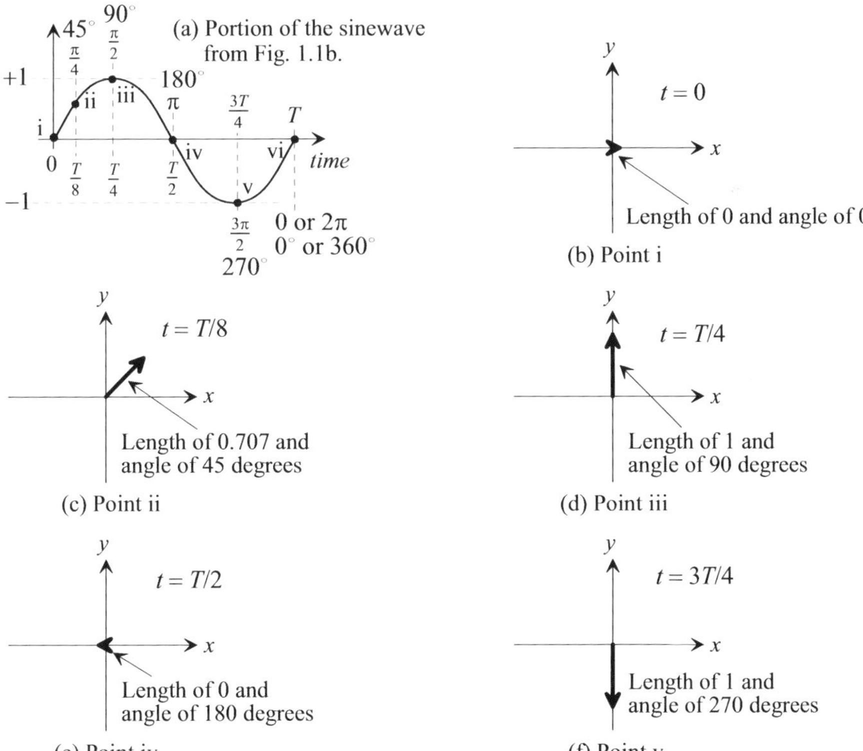

Figure 1.3 A vector swinging around the x-y plane changing both length and angle is used to represent a sinewave.

increases (here indicating an increase in both the x and y directions), Fig. 1.3c. At $T/4$ we are at Point iii in Fig. 1.3a. As seen in Fig. 1.3d the length of the vector is 1 and the angle is 90 degrees. Continuing on towards $T/2$ (Point iv) in Fig. 1.3a the vector is shrinking once again finally reaching a length of 0 and an angle of 180 degrees, Fig. 1.3e. In Fig. 1.3f, Point v, the length is 1 and the angle is 270 degrees. The **key point** here is that a sinewave is represented in the x-y plane by a vector that is changing length and *rotating* around in a circle. Knowing this we can simply represent the sinewave by its peak value and the associated angle between the x-axis and the vector as indicated in Fig. 1.3. For example, as seen in Fig. 1.3d, the peak value of the sinewave occurs when the angle is 90 degrees ($T/4$) so we could write

$$\sin 2\pi f_o \cdot t \rightarrow 1\angle 90^\circ \rightarrow x = 0, y = 1 \tag{1.4}$$

knowing the vector representing the sinewave is actually rotating around in the x-y plane with time (and a frequency f_o). Note that we could plot the sinewave more accurately by adding a third axis, time (the z-axis), and showing the corresponding, corkscrew looking, 3-dimensional plot.

In-Phase and Quadrature Signals

Consider moving the sinusoid in Fig. 1.3a a quarter of a cycle (90 degrees, $\pi/2$, or a delay, t_{delay}, of $T/4$) earlier in time, Fig. 1.4a. We write, for this time-shifted signal,

$$\sin\left(\frac{2\pi}{T}\cdot t+\frac{\pi}{2}\right) = \sin\left(\frac{360}{T}\cdot t+90\right) = \sin\left(\frac{2\pi}{T}\cdot\left(t+\frac{T}{4}\right)\right) = \sin\left(2\pi f_o\cdot t+\frac{\pi}{2}\right) \tag{1.5}$$

As the reader probably already knows, this signal function is called cosinusoidal or simply a cosine signal and is described using

$$\sin\left(2\pi f_o\cdot t+\frac{\pi}{2}\right) = \cos\left(2\pi f_o\cdot t\right) \tag{1.6}$$

We can say that the cosine signal in Fig. 1.4a *leads* the sine signal in Fig. 1.3a by a phase shift, θ, of 90 degrees ($= T/4 = \pi/2$). We could also say that the sine signal *lags* the cosine signal by a phase shift of –90 degrees. Note that when we talk about phase shift it's assumed that the frequencies of the two signals are equal. It doesn't make sense to talk about the phase shift between two sinusoids at different frequencies. Finally note that the phase shift is given by

$$\theta = 2\pi\cdot\frac{t_{delay}}{T} = 2\pi f_o\cdot t_{delay} \tag{1.7}$$

where $t_{delay}/T\times 100\,\%$ is the percentage the delay is of the period.

Figure 1.4b shows the x-y plane plot for a cosine signal. The peak value of the cosine signal occurs when the angle is 0 degrees so we could write

$$\cos 2\pi f_o \cdot t \rightarrow 1\angle 0^\circ \rightarrow x = 1, y = 0 \tag{1.8}$$

to represent the signal and not show the rotating vector with time. Note that we can say the cosine signal is *in-phase* (I) since we are representing it above with a zero degree phase shift. We could also say that the sine signal, since it's shifted in time by a quarter cycle, Eq. (1.4), has a *quadrature* (Q) phase shift. Note what happens, looking at the vector representations of the sinusoids in Figs. 1.3b-f and 1.4b, if we create a signal by *adding a sine signal to a cosine signal* or

$$S_{IQ}(t) = \cos 2\pi f_o \cdot t + \sin 2\pi f_o \cdot t \tag{1.9}$$

The resulting signal, when plotted in the x-y plane (Fig. 1.5), results in a vector (sinewave) with a length (peak amplitude) of $\sqrt{1^2 + 1^2}$ or $\sqrt{2}$ that simply rotates around in a circle with a frequency of $1/T$ ($= f_o$). In more general terms, Fig. 1.6,

$$S_{IQ}(t) = A_I \cdot \cos 2\pi f_o \cdot t + A_Q \cdot \sin 2\pi f_o \cdot t = \sqrt{A_I^2 + A_Q^2} \cdot \cos\left(2\pi f_o \cdot t - \tan^{-1}\frac{A_Q}{A_I}\right) \tag{1.10}$$

where we define the magnitude and phase as

$$|S_{IQ}(t)| \angle S_{IQ}(t) = \sqrt{A_I^2 + A_Q^2} \angle -\tan^{-1}\frac{A_Q}{A_I} \tag{1.11}$$

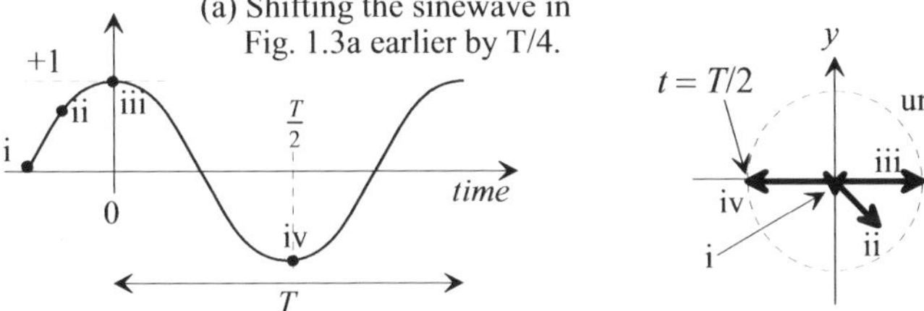

Figure 1.4 Shifting the sinewave in Fig. 1.3a earlier in time by T/4.

Before going too much further, we should ask why we would want to add I and Q signals? The answer is that since the I/Q sinusoids are at the same frequency, shifted in time by a quarter cycle, we can transmit them with, for example, changes in their respective amplitudes and increase the information sent for a given bandwidth. Note that we can't shift the sinewave seen in Fig. 1.3a by 180 degrees, or $T/2$, and add it to an unshifted sinewave since we would get no signal at all!

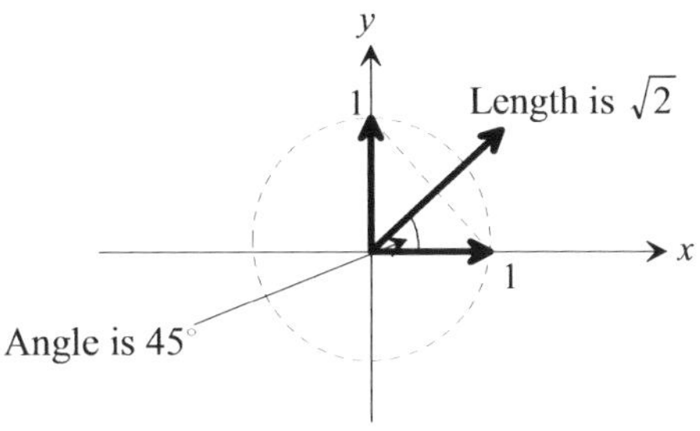

Figure 1.5 Showing how an I/Q signal can be represented in the x-y plane.

Notice that the signals here are very simple. What happens when we start multiplying them together and shifting them in time (changing the phase shift)? We need

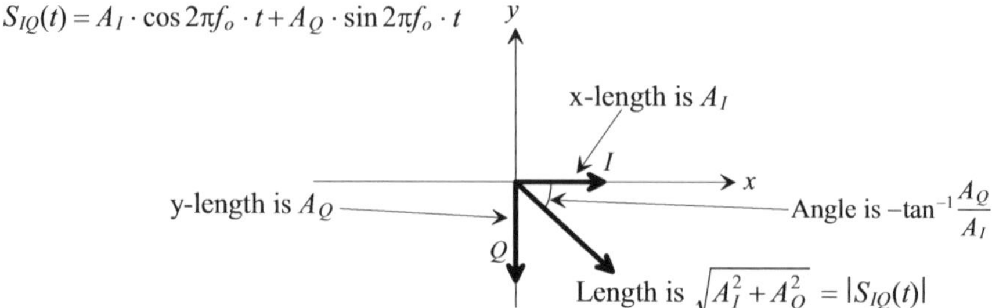

Figure 1.6 Again, showing how an I/Q signal can be represented in the x-y plane.

to simplify the math! To move towards this goal we'll develop the complex, or *z*-, plane and the frequency-domain representation of signals.

1.1.2 The Complex (z-) Plane

Let's attempt (and fail using the *x-y* plane) to simplify our mathematical description of the *IQ* signal given in Eq. (1.10). Recall the following Taylor series expansions

$$e^k = 1 + k + \frac{k^2}{2!} + \frac{k^3}{3!} + \frac{k^4}{4!} + \ldots \tag{1.12}$$

$$\cos k = 1 - \frac{k^2}{2!} + \frac{k^4}{4!} - \frac{k^6}{6!} + \frac{k^8}{8!} - \ldots \tag{1.13}$$

$$\sin k = k - \frac{k^3}{3!} + \frac{k^5}{5!} - \frac{k^7}{7!} + \frac{k^9}{9!} - \ldots \tag{1.14}$$

We can now write

$$\cos k + \sin k = 1 + k - \frac{k^2}{2!} - \frac{k^3}{3!} + \frac{k^4}{4!} + \frac{k^5}{5!} - \frac{k^6}{6!} - \frac{k^7}{7!} + \frac{k^8}{8!} + \frac{k^9}{9!} - \ldots \tag{1.15}$$

Comparing Eq. (1.15) to Eq. (1.12) we see that we are close to writing the Taylor's series for e^k. Why is this important? Perhaps the simplest explanation is that if we can represent sinewaves using exponentiation, then multiplying two sinewaves, or shifting a sinewave in time, can be performed using simple addition (of exponents).

The question now is how do we modify things to ensure that all terms are added so that Eq. (1.15) matches Eq. (1.12)? Let's look at the first discrepancy $(-1) \cdot \frac{k^2}{2!}$. The only way to change the polarity of this term is take the square root of -1 and move it inside with k^2. As the reader may know instead of writing $\sqrt{-1}$ for all of these terms we simplify things and write

$$j = \sqrt{-1} \tag{1.16}$$

Numbers using *j* (or *i*) are called *imaginary or complex numbers* (the reason for using the name imaginary will be explained in Ex. 1.1). Imaginary numbers are invaluable for time-shifting and scaling sinusoidal signals. We now rewrite Eq. (1.12) using *j* as

$$e^{jk} = 1 + jk - \frac{k^2}{2!} - j\frac{k^3}{3!} + \frac{k^4}{4!} + j\frac{k^5}{5!} - \frac{k^6}{6!} - j\frac{k^7}{7!} + \ldots \tag{1.17}$$

or Euler's formula

$$e^{jk} = \cos k + j \cdot \sin k \tag{1.18}$$

where the cosine term is a real number and the sine term is the imaginary component of the complex number. Sometimes the following notation is used

$$Re\{e^{jk}\} = \cos k \text{ and } Im\{e^{jk}\} = \sin k \tag{1.19}$$

Further, with a little algebraic manipulation we can also write

$$\cos k = \frac{e^{jk} + e^{-jk}}{2}, \quad \sin k = \frac{e^{jk} - e^{-jk}}{2j} \tag{1.20}$$

The next thing we have to discuss is plotting complex numbers. Examine Fig. 1.7. Plotting the real component, x, of a complex number, $x + jy$, follows the same methods we've always used. To plot the imaginary component, we now use the y-axis. Note that the factor j in the complex number simply indicates the imaginary component and shouldn't be included when finding the magnitude of the number, $\sqrt{x^2 + y^2}$ or the phase, θ, of the complex number, $\tan^{-1}\frac{y}{x}$.

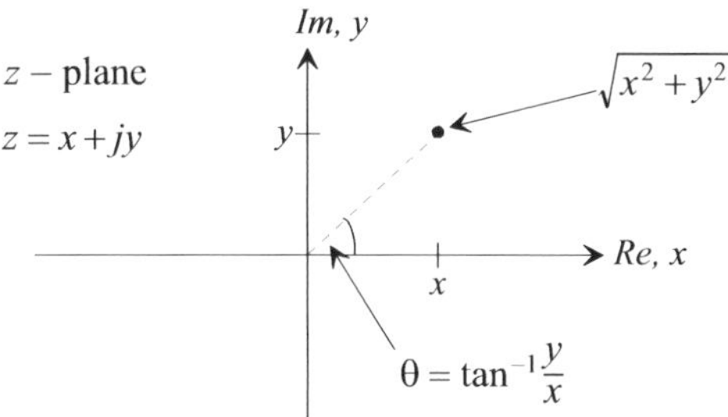

Figure 1.7 The complex plane, plotting the imaginary number

Example 1.1
In the complex plane, plot the signal $e^{j2\pi f_o \cdot t}$. Comment on the resulting plot.

As seen in Fig. 1.8 the magnitude is $|e^{j2\pi f_o \cdot t}| = 1$ and phase shift is $\angle e^{j2\pi f_o \cdot t} = 2\pi f_o \cdot t$. The signal simply swings around and around in the complex

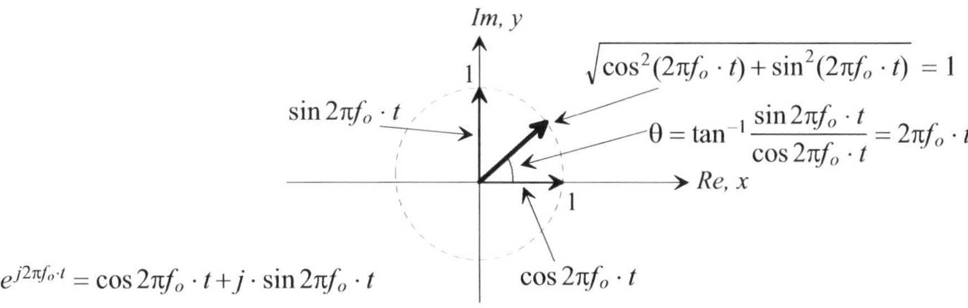

Figure 1.8 Plotting Euler's formula in the z-plane.

plane, on the unit circle, without changing amplitude (both the real and the imaginary components oscillate back and forth between +1 and −1). One complete rotation takes $1/f_o$, or T seconds. We need to pause a moment and ask "If the magnitude of $e^{j2\pi f_o \cdot t}$ is a constant value of 1 isn't it a DC signal?" (answer: no) When a DC signal is represented in the *x-y* plane it doesn't rotate or change amplitude like a sinewave does. (Well, since the frequency of a DC signal, f, is 0, you could say it does rotate around the *x-y* plane just like a sinewave but since $T = \infty$ $(=1/f)$ it never leaves the *x*-axis). What this, $e^{j2\pi f_o \cdot t}$, **imaginary signal** can be used for is to introduce or represent delay (phase shift). ■

To understand this last statement in more detail let's write

$$A \cdot \cos 2\pi f_o \cdot t = Re\{A \cdot e^{j2\pi f_o \cdot t}\} \rightarrow A\angle 0 \qquad (1.21)$$

Suppose we want to delay this signal by t_d. We can represent a delay at a particular frequency, f_o, using

$$e^{j2\pi f_o \cdot (-t_d)} \rightarrow 1\angle 2\pi f_o \cdot (-t_d) \rightarrow 1\angle 2\pi \cdot \frac{-t_d}{T} \qquad (1.22)$$

The delayed cosine signal can be written as

$$A \cdot \cos 2\pi f \cdot (t - t_d) = Re\{A \cdot e^{j2\pi f \cdot t} \cdot e^{j2\pi f \cdot (-t_d)}\} = Re\{A \cdot e^{j2\pi f \cdot (t - t_d)}\} \qquad (1.23)$$

To *simplify the notation* we can drop the real indication, Re{}. We could also describe the delayed cosine signal in terms of angle notation as

$$A\angle 0 \cdot 1\angle 2\pi \cdot \frac{-t_d}{T} = A \cdot 1\angle\left[0 + 2\pi \cdot \frac{-t_d}{T}\right] = A\angle(-2\pi f_o \cdot t_d) \qquad (1.24)$$

Again noting that when we use angle notation, the *x-y* plane, or the complex plane, we need to know the frequency, f_o, of the input signal since this information isn't present in these representations.

1.2 Comb Filters

A delay can be used to construct a simple, but very useful, filter called a comb filter, Fig. 1.9. Before deriving the equations that characterize this filter let's discuss notation. If our input signal is a sinusoid we can represent it using

$$v_{in}(t) = A\cos 2\pi f \cdot t \qquad (1.25)$$

If we were to keep the frequency fixed, $f = f_o$, and vary time, which we've done up to this point, the output signal of a circuit will simply repeat itself (at the same frequency as the input signal). *What's more useful* is varying the input signal's frequency, f, and looking at the output of the circuit or system (how the amplitude and phase shift vary). To move

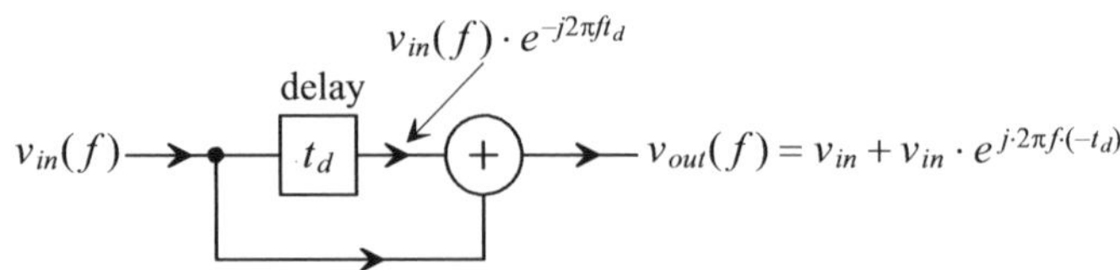

Figure 1.9 A comb filter.

towards this method of characterization let's write the frequency domain description of Eq. (1.25) as $v_{in}(f)$ or simply v_{in}. Representing a signal in the frequency domain will be discussed in the next section.

Returning to our comb filter in Fig. 1.9 we can write

$$v_{out} = v_{in} + v_{in} \cdot e^{j \cdot 2\pi f \cdot (-t_d)} \tag{1.26}$$

or, using Euler's formula,

$$\frac{v_{out}}{v_{in}} = 1 + e^{j \cdot 2\pi f \cdot (-t_d)} = \overbrace{1 + \cos 2\pi f \cdot (-t_d)}^{\text{Real}} + \overbrace{j \cdot \sin 2\pi f \cdot (-t_d)}^{\text{Imaginary}} \tag{1.27}$$

The magnitude response of this filter is

$$\left|\frac{v_{out}}{v_{in}}\right| = \sqrt{(1 + \cos 2\pi f \cdot (-t_d))^2 + (\sin 2\pi f \cdot (-t_d))^2} = \sqrt{2(1 + \cos 2\pi f \cdot t_d)} \tag{1.28}$$

or using

$$1 + \cos x = 2\cos^2 \frac{x}{2} \tag{1.29}$$

simply

$$\left|\frac{v_{out}}{v_{in}}\right| = 2 \cdot |\cos \pi f \cdot t_d| \tag{1.30}$$

The phase response is given by

$$\angle \frac{v_{out}}{v_{in}} = \tan^{-1}\left[\frac{\sin 2\pi f \cdot (-t_d)}{1 + \cos 2\pi f \cdot (-t_d)}\right] \tag{1.31}$$

Notice at $f = 1/(2t_d)$ the phase is $\tan^{-1}(0/0)$, which evaluates to $\pm$ 90 degrees. Using Eq. (1.29) and

$$\sin x = 2\sin \frac{x}{2} \cdot \cos \frac{x}{2} \tag{1.32}$$

the phase response is

$$\angle \frac{v_{out}}{v_{in}} = \pi(-t_d) \cdot f \ \text{ for } f < 1/(2t_d) \tag{1.33}$$

Note that the *phase response is linear* indicating constant delay through the filter, important for distortionless filtering. Figure 1.10 shows the magnitude and frequency response for a comb filter (at this point the reason the filter is called a comb filter should be obvious).

Example 1.2

Figure 1.11 shows one possible implementation of a comb filter. A 50 Ω characteristic impedance co-ax transmission line with an electrical length of 5 ns (delay) is used for the delay element (note how the transmission line is terminated with a 50 ohm resistor and it is assumed that 50 Ω << 5kΩ so that the 5kΩ resistors don't load the output of the transmission line). Determine the comb filter's characteristics (transfer function, v_{out}/v_{in}). Verify your answer with SPICE.

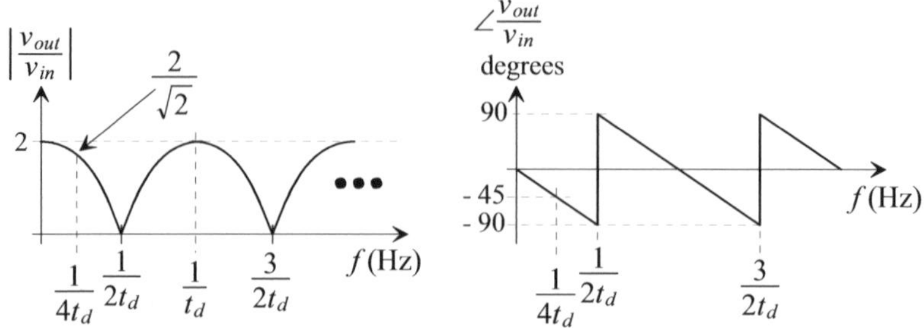

Figure 1.10 Magnitude and phase response of the comb filter in Fig. 1.9.

In order to perform the addition operation seen in Fig. 1.9 we use the two 5k resistors seen in Fig. 1.11. The current through the top resistor must equal the current through the bottom resistor or

$$\frac{v_{in}e^{j2\pi f(-t_d)} - v_{out}}{5k} = \frac{v_{out} - v_{in}}{5k} \tag{1.34}$$

and thus v_{out} is the average of the signals on the top and bottom of the resistors

$$v_{out} = \frac{v_{in} + v_{in}e^{j2\pi f(-t_d)}}{2} \tag{1.35}$$

The result is our derivation of the comb filter's magnitude response in Eq. (1.30) is scaled by 2 or

$$\left|\frac{v_{out}}{v_{in}}\right| = |\cos \pi f \cdot t_d| = \left|\cos \pi \frac{f}{200\text{ MHz}}\right| \tag{1.36}$$

The phase response is still given by Eq. (1.31). SPICE simulation results showing the filter's frequency response, magnitude and phase, are seen in Fig. 1.12. Note how the filter eliminates from its output, input signals at frequencies that are multiples of $1/2t_d$. This filter can be very useful in communication systems where it's used to isolate, and prevent crosstalk between, transmission channels.

In order to ensure that we can sketch time-domain signals from frequency domain plots consider the case when the input has 1 V peak amplitude and a frequency of 50 MHz. We can see from these plots, and the equations, that the

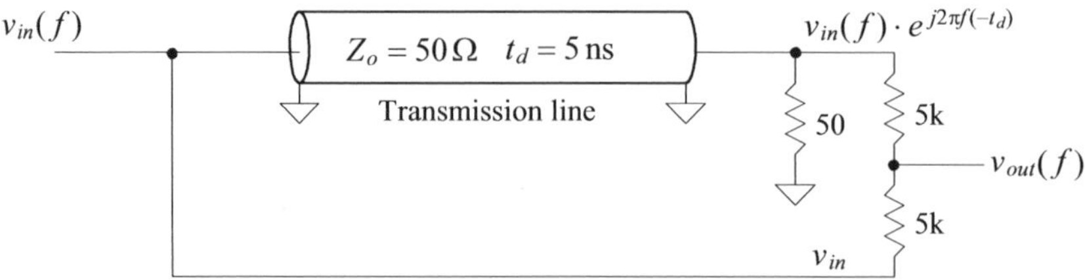

Figure 1.11 Implementation of an analog comb filter.

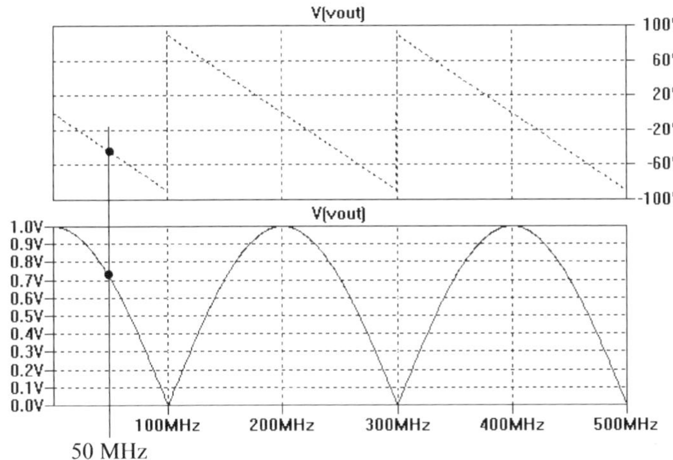

Figure 1.12 Simulating the operation of the comb filter in Fig. 1.11.

output will have an amplitude of 0.707 V and a phase shift of –45 degrees (or the output is lagging the input by 2.5 ns). Time domain simulation results are seen in Fig. 1.13. Note, in this figure, that there is a start-up time, the time it takes the signal to propagate through the transmission line (note the kink at 5 ns), before the filter's behavior follows the equations we derived. ■

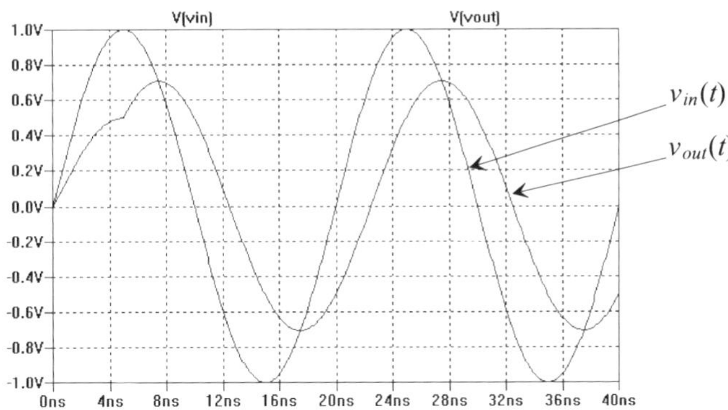

Figure 1.13 Time domain input and output (50 MHz) for the comb filter seen in Fig. 1.11.

1.2.1 The Digital Comb Filter

Notice how, when discussing the comb filter, it was more useful to vary the input signal frequency, f, and look at the output (Fig. 1.12) of the circuit rather than attempting to change the input with time and plot the output. In our comb filter the input signal was delayed by t_d. We represented this time delay in the complex plane using $e^{j2\pi f\cdot(-t_d)}$.

In a digital system we can implement the delay line seen in Figs. 1.9 and 1.11 with a register, Fig. 1.14a. The contents of the register change, in Fig. 1.14a, every T_s seconds

Time domain

In — D Q clk — Out

$f_s = \frac{1}{T_s}$

(a) A register that is clocked at f_s that delays the input signal by T_s.

Frequency domain

$v_{in}(f)$ — z^{-1} — $v_{out}(f) = z^{-1} \cdot v_{in}(f)$

(b) Representing a delay in the complex plane.

$z^{-1} = e^{j2\pi f \cdot (-T_s)} = e^{-j2\pi \cdot \left(\frac{f}{f_s}\right)}$

Figure 1.14 A digital delay.

(so this would correspond to t_d in Fig. 1.11). To simplify the notation, since delays are so common in a digital system that processes signals, we can use

$$z = e^{j2\pi f \cdot T_s} = e^{j2\pi \cdot \frac{f}{f_s}} \text{ or, for a delay, } z^{-1} = e^{j2\pi f \cdot (-T_s)} = e^{-j2\pi \cdot \left(\frac{f}{f_s}\right)} \quad (1.37)$$

This representation for a delay, in the frequency domain, is seen in Fig. 1.14b (**some authors** use Z^{-1} to indicate a delay, $e^{j2\pi f \cdot (-T_s)}$, and to differentiate between the general situation where $z = x + jy$). Figure 1.15 shows how Eq. (1.37) is plotted in the z-plane (the dashed circle). The magnitude of $e^{j2\pi f \cdot T_s}$ is one and its phase shift is $2\pi f \cdot T_s$ or $2\pi \cdot \frac{f}{f_s}$ (plotting $e^{j2\pi f \cdot T_s}$ in the z-plane simply plots the unit circle). Note that plotting a delay, z^{-1} or $e^{j2\pi f \cdot (-T_s)}$, also results in a magnitude of one but the phase shift is now $-2\pi f \cdot T_s$. *As we move through the book we will regularly use Eq. (*1.37*) to evaluate the frequency response of a discrete-time system, see Fig.* 1.21 *and the associated discussion, and to relate the complex number, z to the frequency, f.*

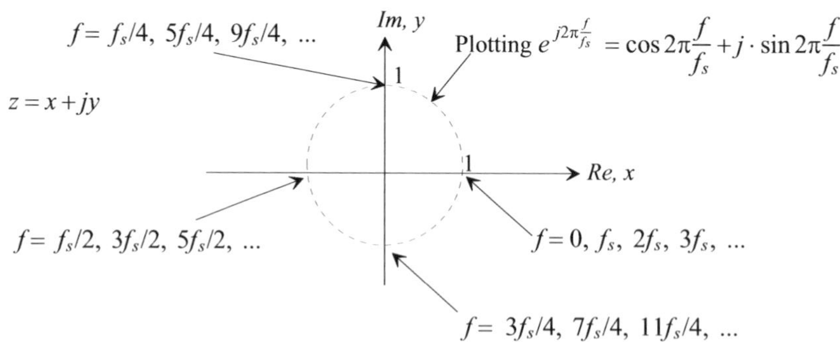

Figure 1.15 Plotting a delay in the complex plane.

Figure 1.16 shows the digital implementation of the comb filter. We can write

$$v_{out} = v_{in}(1 + z^{-1}) \quad (1.38)$$

and thus the transfer function is

$$\frac{v_{out}}{v_{in}} = 1 + z^{-1} = \frac{z+1}{z} \quad (1.39)$$

or, after reviewing Eqs. (1.27) to (1.33), we get

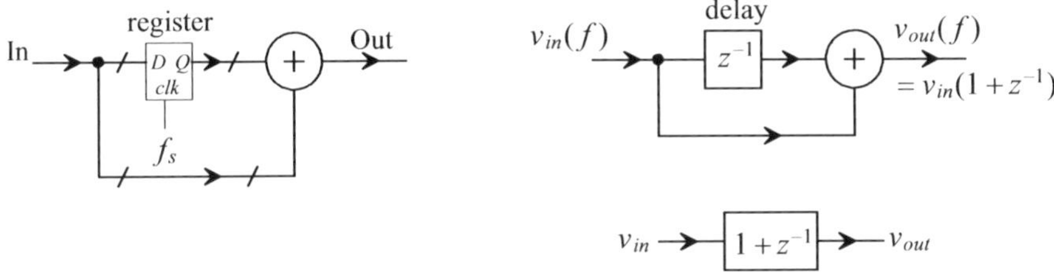

Figure 1.16 A digital comb filter using one delay. Note how the output is the average (sum) of consecutive inputs.

$$\left|\frac{v_{out}}{v_{in}}\right| = 2 \cdot \left|\cos \pi \frac{f}{f_s}\right| \text{ and } \angle \frac{v_{out}}{v_{in}} = -\pi \cdot \frac{f}{f_s} \text{ for } f < \frac{f_s}{2} \tag{1.40}$$

The magnitude and phase response for this digital comb filter is seen in Fig. 1.17. Note that if we apply a constant (DC, $f = 0$) value of 1 to the input of our digital filter, the output goes to 2 verifying that what we get with Eq. (1.40), at least at DC, matches what we see in Fig. 1.17.

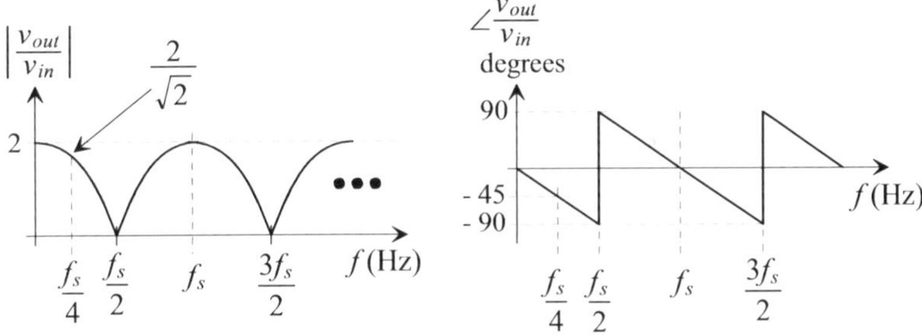

Figure 1.17 Magnitude and phase response of the comb filter in Fig. 1.16.

Note that if we divide the output of the digital comb filter in Fig. 1.16 by two we can think of it as an *averaging filter* (taking two inputs, summing them, and then dividing by two). Let's pause and think about this for a moment. If the input is 8-bits then the output, which is the sum of two 8-bit words, is a 9-bit word. To avoid lowering the resolution of the filter's output (the number of bits in the filter's output word) we can keep all 9-bits and not perform the divide by two (but still call the filter an "averaging filter"...or comb filter). This is **important** because we will regularly perform averaging (which is simply lowpass filtering a signal) to increase the signal's resolution (so the word size should increase!) Note that our definition of averaging here is simply adding a number of input words and outputting the result.

Finally, to show how it's possible for the output of the comb filter in Fig. 1.16 to go to zero, consider averaging two points of an input signal, at a frequency of $f_s/2$, spaced apart by T_s, Fig. 1.18. It may be a good exercise, at this point, to vary the input signal frequency and plot the corresponding outputs to verify the results seen in Fig. 1.17.

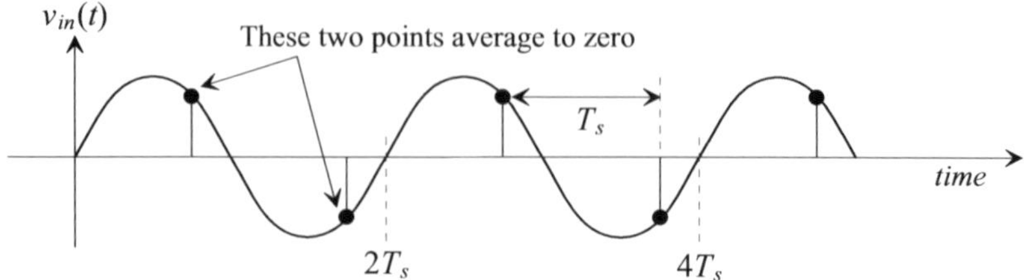

Figure 1.18 Showing how the averaging filter in Fig. 1.16 can have zero output at $f_s/2$.

1.2.2 The Digital Differentiator

The digital comb filter seen in Fig. 1.16 could also be called an averaging filter since its output is the sum of adjacent inputs. The gain of this filter at DC was, as seen in Fig. 1.17, 2. We can also construct a comb filter using a digital *differentiator* or *differencer*, Fig. 1.19. This filter outputs the difference between adjacent input signals.

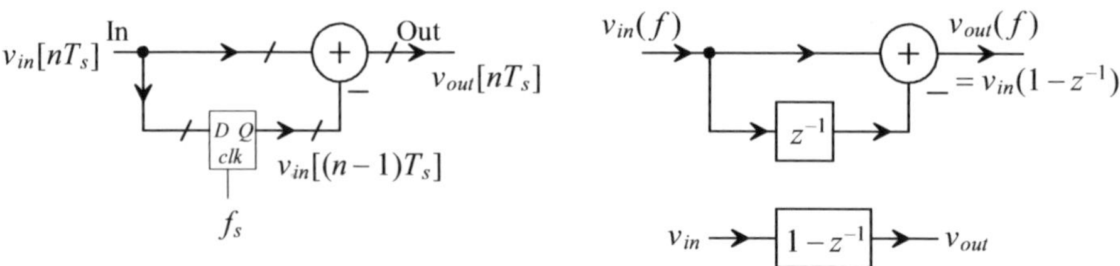

Figure 1.19 A digital comb filter using a digital differentiator.

In the time domain, with an input signal of $v_{in}(t)$ sampled at discrete times nT_s, we can write the output of the digital differentiator filter as

$$v_{out}[nT_s] = v_{in}[nT_s] - v_{in}[(n-1)T_s] \tag{1.41}$$

or, in the frequency domain,

$$v_{out}(f) = v_{in}(f) \cdot (1 - z^{-1}) \tag{1.42}$$

and so the transfer function of the filter is

$$\frac{v_{out}}{v_{in}} = 1 - z^{-1} = \frac{z-1}{z} \tag{1.43}$$

The magnitude response of the digital differentiator is

$$\left|\frac{v_{out}}{v_{in}}\right| = \sqrt{2\left(1 - \cos 2\pi\frac{f}{f_s}\right)} \tag{1.44}$$

or using

$$1 - \cos x = 2\sin^2\frac{x}{2} \tag{1.45}$$

we get

$$\left|\frac{v_{out}}{v_{in}}\right| = 2\left|\sin \pi\frac{f}{f_s}\right| \tag{1.46}$$

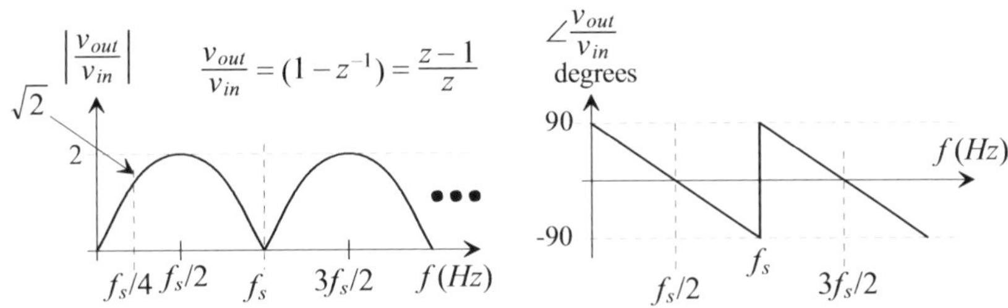

Figure 1.20 Magnitude and phase response of the digital differentiator (also a comb filter).

The phase response is

$$\angle\frac{v_{out}}{v_{in}} = \frac{\pi}{2} - \pi\frac{f}{f_s} \text{ for } 0 < f < f_s \tag{1.47}$$

Figure 1.20 shows the magnitude and phase responses for the digital differentiator in Fig. 1.19. Notice how, for this comb filter, the gain at DC is 0.

1.2.3 An Intuitive Discussion of the *z*-Plane

It will be very helpful in our discussion of mixed-signal circuits and systems to gain an intuitive feel for the frequency response of a discrete-time system, like the digital filters we've discussed in this section (or the switched-capacitor filters in Ch. 3 that are discrete-time, continuous-amplitude, circuits), by looking at the *z*-plane representation of the system. To move toward this goal, consider the transfer function of the simple digital averager seen in Fig. 1.16 with a transfer function of

$$H(f) = \frac{v_{out}(f)}{v_{in}(f)} = H(z) = 1 + z^{-1} = \frac{z+1}{z} \tag{1.48}$$

It is very useful, for an intuitive understanding of the frequency response of a discrete-time system, to plot the poles and zeroes of the transfer function in the *z*-plane. Figure

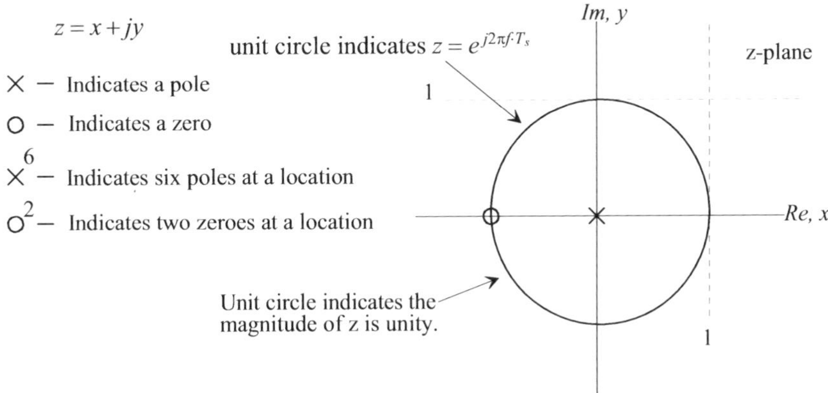

Figure 1.21 The z-plane.

1.21 also shows how Eq. (1.48) can be displayed on the z-plane. A pole is located at $z = 0$ (at the location the denominator goes to zero and the transfer function goes to infinity) and a zero is located at $z = -1$ (at the location where the numerator goes to zero).

The z-plane is usually used to describe the frequency response of a discrete time system, $H(f)$, by assuming the input to the system is a unit magnitude sinusoid with varying frequency, f. This input, $1 \cdot e^{j2\pi \frac{f}{f_s}}$, evaluates the output of the system or

$$H(f) = v_{out}(f) \text{ when } v_{in}(f) = \overbrace{1}^{\text{magnitude}} \cdot \overbrace{e^{j2\pi \frac{f}{f_s}}}^{\text{phase}} \tag{1.49}$$

We should now see that the unit circle, shown in Fig. 1.21, indicates the relationship between z and f when specified by Eq. (1.49) (*this is important*). Therefore, to determine $H(f)$ from a plot of a transfer function's poles and zeroes in the z-plane, we simply evaluate $H(z)$ along the unit circle. To show how this transfer function evaluation is performed, consider Eq. (1.48) and the corresponding plot of its pole and zero shown in Fig. 1.22 along with the magnitude of Eq. (1.48) or Eq. (1.40) plotted against frequency in Fig. 1.17. At DC ($f = 0$ and $z = 1 \cdot e^0 = 1\angle 0$) point A in Fig. 1.22, the gain of the circuit is two and is calculated using

$$|H(f)| = \frac{\text{distance to zero}}{\text{distance to pole}} \tag{1.50}$$

The distance from the zero to point A is 2 while the distance between the pole to point A is 1. Therefore, as shown in Fig. 1.17, the magnitude of $H(f)$ is 2. The phase of the transfer function is calculated along the positive x-axis using

$$\angle H(f) = \angle \text{ of zero} - \angle \text{ of pole} \tag{1.51}$$

which, as seen in Fig. 1.17, results in a phase angle of zero. Next consider evaluating $H(z)$ at $f_s/4$ $\left(f = f_s/4 \text{ and } z = 1 \cdot e^{j\frac{\pi}{2}} = 1\angle 90\right)$, point B in Fig. 1.22. The distance from the pole to point B is 1 while the distance from the zero is $\sqrt{2}$ resulting in a magnitude $\sqrt{2}$. The

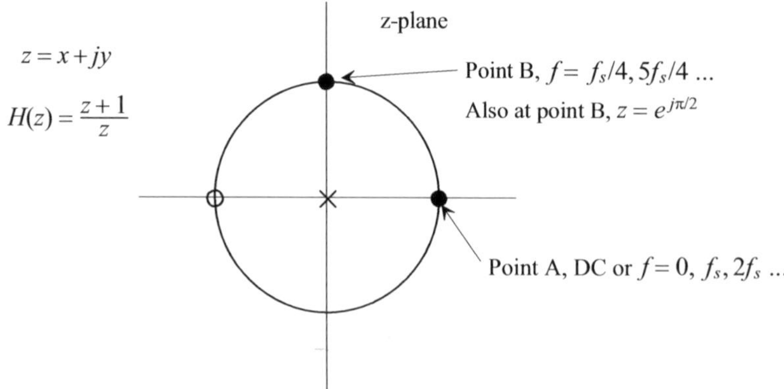

Figure 1.22 The z-plane pole and zero for Eq. (1.48).

angle from the pole along the x-axis to point B is 90°, while the angle from the zero is 45° resulting in an overall phase response of –45° (verify with Fig. 1.17).

Note that (1) any digital filter's or discrete-time system's frequency response is periodic with period f_s (one complete revolution around the unit circle), (2) we normally are only concerned with evaluating $H(z)$ over the top half of the unit circle (from DC to $f_s/2$ [the Nyquist frequency, f_n]), and (3) a pole at the origin has no effect on the magnitude response of $H(z)$ but does affect the phase response (z^{-1}, a pole at the origin, is a delay of one clock cycle). Finally note that the number of poles in $H(z)$ must be greater than or equal to the number of zeroes if the digital filter/system is to be realizable in hardware (the output of the system cannot occur before the system's input).

Example 1.3
Determine, using the graphical approach just discussed, the magnitude and phase of the transfer function indicated by the poles and zeroes plotted in Fig. 1.23 at a frequency of $f_s/4$.

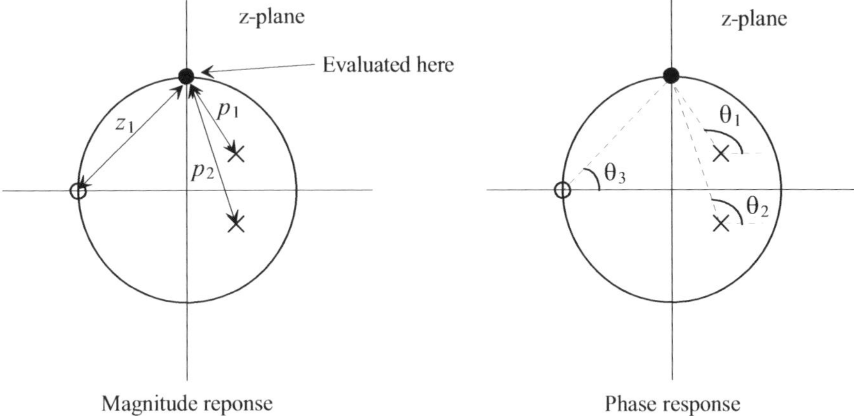

Figure 1.23 The z-plane pole and zero plot for Ex. 1.3.

If we label the length from a pole (zero) to the evaluation point p (z), then the magnitude of the transfer function is given by

$$|H(z)| = \frac{z_1}{p_1 \cdot p_2}$$

Labeling the angles for the poles and zeroes as indicated in the figure, we can write the phase response as

$$\angle H(z) = \theta_3 - \theta_2 - \theta_1 \blacksquare$$

1.2.4 Comb Filters with Multiple Delay Elements

Examine the digital filter seen in Fig. 1.24. The transfer function for this filter is given by

$$H(z) = 1 - z^{-K} = \frac{z^K - 1}{z^K} \tag{1.52}$$

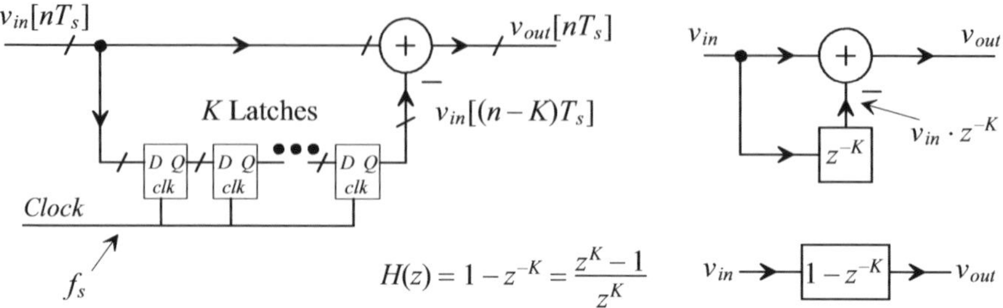

Figure 1.24 A digital comb filter using multiple delay elements.

The magnitude response for this filter is (see Eq. [1.46])

$$\left|\frac{v_{out}}{v_{in}}\right| = 2\left|\sin K\pi \cdot \frac{f}{f_s}\right| \tag{1.53}$$

and the phase response is

$$\angle\frac{v_{out}}{v_{in}} = \frac{\pi}{2} - K\pi \cdot \frac{f}{f_s} \text{ for } 0 < f < f_s/K \tag{1.54}$$

Figure 1.25 shows the magnitude responses and pole-zero plots for comb filters with varying numbers of delay elements, K. For $K = 4$, for example, the zeroes are located at: $1, j, -1$, and $-j$. In other words the transfer function goes to zero at, DC, $f_s/4$, $f_s/2$, $3f_s/4$, f_s,

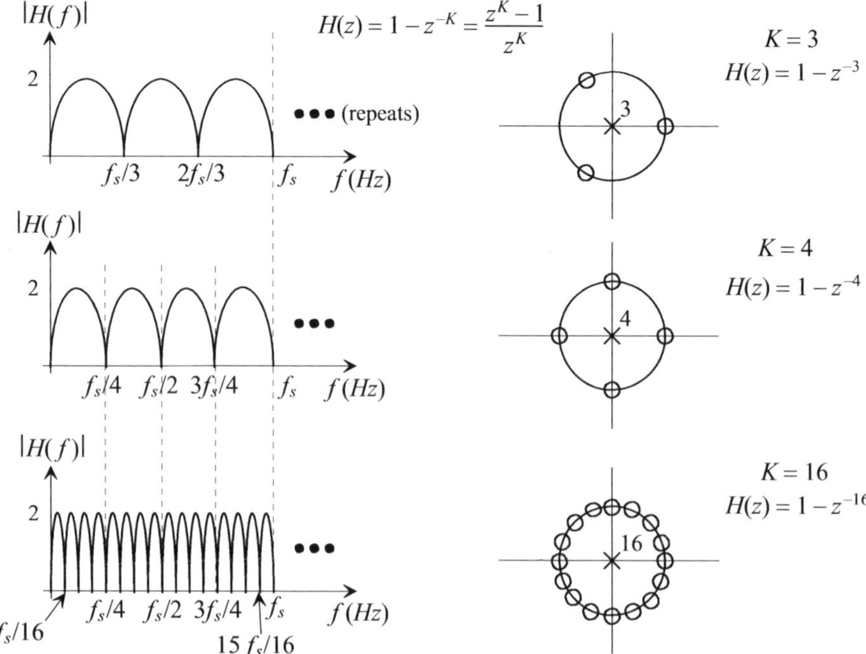

Figure 1.25 Frequency response and z-plane plots for various values of K in a comb filter.

$f_s + f_s/4$, etc.). For $K = 3$ the zeroes are located at: 1, $-\frac{1}{2} + j \cdot \frac{\sqrt{3}}{2}$, and $-\frac{1}{2} - j \cdot \frac{\sqrt{3}}{2}$. Because of their simplicity, comb filters are very useful in mixed-signal design (and used to design lowpass, bandpass, and highpass filters that will be discussed in Ch. 4).

1.2.5 The Digital Integrator

Examine the circuit seen in Fig. 1.26. The output of the circuit, in the time domain, is given by

$$v_{out}[nT_s] = v_{in}[nT_s] + v_{out}[(n-1)T_s] \tag{1.55}$$

or, in the frequency domain,

$$H(z) = H(f) = \frac{v_{out}}{v_{in}} = \frac{1}{1 - z^{-1}} = \frac{z}{z-1} \tag{1.56}$$

This circuit adds the current input to the sum of the past inputs, performing integration. Note that there isn't a delay in series with the input signal so this circuit is often called a *non-delaying integrator* or *accumulator*. We'll see in a moment that a *delaying integrator* does delay the input signal so it has a transfer function of $z^{-1}/(1 - z^{-1})$.

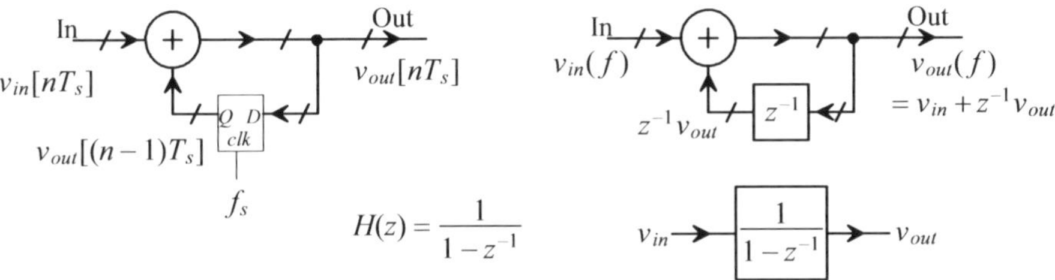

Figure 1.26 The non-delaying digital integrator.

Figure 1.27 shows the z-plane representation of the non-delaying integrator along with its magnitude and phase responses. To determine equations for the magnitude and phase responses we can write

$$H(z) = \frac{z}{z-1} \to \overbrace{e^{j2\pi\frac{f}{fs}}}^{\text{Phase shift}} \cdot \frac{1}{e^{j2\pi\frac{f}{fs}} - 1} = e^{j2\pi\frac{f}{fs}} \cdot \frac{1}{(-1 + \cos 2\pi\frac{f}{f_s}) + j\sin 2\pi\frac{f}{f_s}} \tag{1.57}$$

noting $\left|e^{j2\pi\frac{f}{fs}}\right| = 1$ and $\angle e^{j2\pi\frac{f}{fs}} = 2\pi\frac{f}{fs}$. This certainly isn't in the form, $x + jy$, that we are used to plotting, Fig. 1.7, for magnitude and phase responses. Rather it's in the form

$$\frac{1}{a + jb} \tag{1.58}$$

To get Eq. (1.57) into the form $x + jy$ we can multiply it by its complex conjugate or, using Eq. (1.58) to simplify the notation,

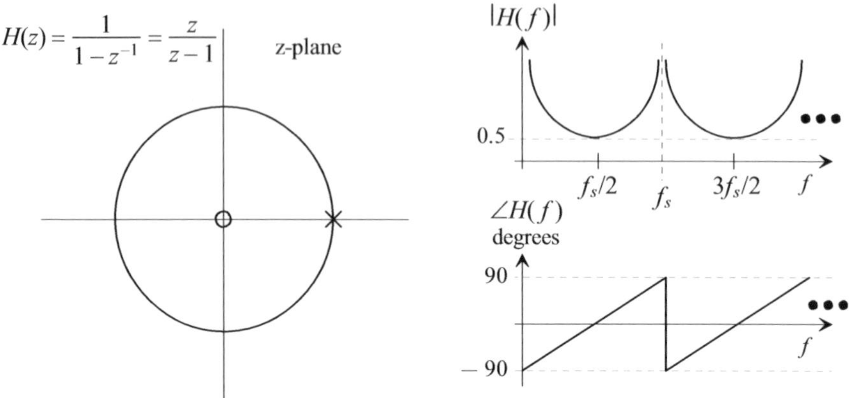

Figure 1.27 The z-plane representation along with magnitude and phase response for a non-delaying digital integrator.

$$\frac{1}{a+jb}\cdot\underbrace{\frac{a-jb}{a-jb}}_{=1} = \underbrace{\frac{a}{a^2+b^2}}_{\text{Real, } x} + j\underbrace{\frac{-b}{a^2+b^2}}_{\text{Imaginary, } y} \tag{1.59}$$

which results, after reviewing Fig. 1.7, in

$$\left|\frac{1}{a+jb}\right| = \frac{1}{\sqrt{a^2+b^2}} \text{ and } \angle\frac{1}{a+jb} = -\tan^{-1}\frac{b}{a} \tag{1.60}$$

We can now write

$$|H(f)| = \frac{1}{\sqrt{\left(-1+\cos 2\pi\frac{f}{f_s}\right)^2 + \left(\sin 2\pi\frac{f}{f_s}\right)^2}} = \frac{1}{\sqrt{2(1-\cos 2\pi\frac{f}{f_s})}} \tag{1.61}$$

or

$$|H(f)| = \frac{1}{2\left|\sin \pi\frac{f}{f_s}\right|} \tag{1.62}$$

Evaluating the phase response directly from the z-plane plot,

$$\angle H(f) = \overbrace{2\pi\frac{f}{f_s}}^{\text{From zero}} - \overbrace{\left(\pi\frac{f}{f_s}+\frac{\pi}{2}\right)}^{\text{From pole}} = 180\frac{f}{f_s} - 90 \text{ (degrees) for } 0 < f < f_s \tag{1.63}$$

At DC the phase contribution from the zero is 0°, while the phase contribution from the pole, at a frequency just above DC, is 90°. The result is an overall phase response of −90°. At $f_s/4$ the phase contribution from the zero is 90°, while the phase contribution from the pole is 135°, resulting in an overall phase response of −45°.

The Delaying Integrator

Figure 1.28 shows the delaying integrator. We get this topology by moving the delay from the feedback path in Fig. 1.26 to the forward path. The output is related to the input using

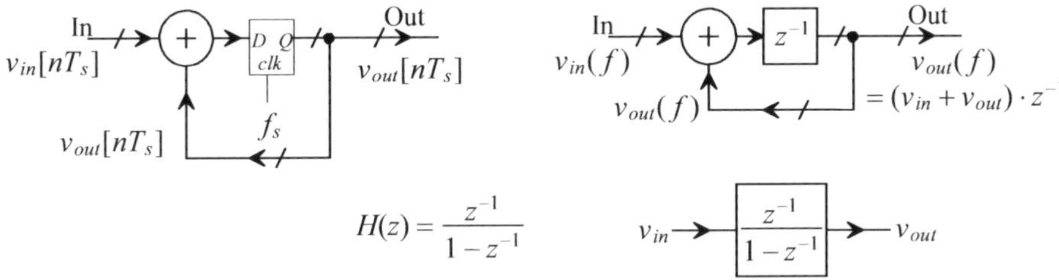

Figure 1.28 The delaying digital integrator.

$$v_{out}[nT_s] = v_{in}[(n-1)T_s] + v_{out}[(n-1)T_s] \tag{1.64}$$

or

$$H(z) = H(f) = \frac{v_{out}}{v_{in}} = \frac{z^{-1}}{1 - z^{-1}} = \frac{1}{z - 1} \tag{1.65}$$

The magnitude response is exactly the same as the non-delaying integrator, Fig. 1.27 and Eq. (1.62). The phase response is given by

$$\angle H(f) = -\pi\frac{f}{f_s} - \frac{\pi}{2} = -180\frac{f}{f_s} - 90 \text{ for } 0 < f < f_s \tag{1.66}$$

While the equations describing the integrators in this section were derived using digital implementations, we'll see in the next chapter that they also apply to discrete analog integrators (DAIs).

An Important Note

When a digital signal is generated from an analog signal using an ADC (analog-to-digital converter) signal aliasing can occur. Signal aliasing comes from the sampling process and is discussed in detail in the next chapter. The importance here in this section is that if we use the output of an ADC, clocked at f_s, to provide the input signal to a digital filter we have to restrict ourselves to input frequencies below the Nyquist frequency, $f_s/2$, in order to avoid aliasing. Simulation examples using the filters in this section are available at CMOSedu.com.

1.3 Representing Signals

In the past two sections we've talked about representing sinusoids using the *x-y* plane with fixed frequency and changing time, Figs. 1.3 and 1.4, and then using the complex, or *z*-plane, Secs. 1.1.2 and 1.2.3 with variable time and fixed frequency, Fig. 1.8, and then with varying frequency, Fig. 1.22. In this section we discuss representing signals, periodic and non-periodic, that are not single tone (single frequency) sinusoids. Any periodic function can be represented by a sum of sinusoids with varying amplitudes and frequencies (but all sinusoid's frequencies are integer multiples, *n*, of the original periodic signal we are modeling with the sum of sinewaves). In the first section we discuss the exponential Fourier series representation of a periodic signal. Next we present the Fourier transform which is used to represent non-periodic signals in the frequency domain. It's assumed this material is a review for the reader. Our presentation is focused on making the material useful for reference later in the book.

1.3.1 Exponential Fourier Series

The exponential Fourier series representation of a periodic function with period T (= $1/f$) can be expressed using

$$g(t) = \sum_{n=-\infty}^{\infty} c_n \cdot e^{j2\pi nft} \tag{1.67}$$

where, as just mentioned, $n \cdot f$ represents the integer multiples of the original signal's frequency f. The weighting of the sinusoids, c_n, is calculated using

$$c_n = \frac{1}{T} \int_{t}^{t+T} g(t) \cdot e^{-j2\pi nft} dt \tag{1.68}$$

The key thing to note is that we use the Fourier Series to represent signals that are periodic. Examples include the outputs of the sample-and-hold and track-and-hold discussed in the next chapter. Using the Fourier Transform to look at the spectrums of periodic signals can be very challenging for all but the simplest waveforms.

As an example of the use of the Fourier Series consider the squarewave seen in Fig. 1.29. The amplitude and period of this waveform are A and T_s respectively. The time the output pulse is high is T_p. When $T_p = T_s/2$ the squarewave has a 50% duty cycle. Using Eq. (1.68) to calculate the coefficients, with $T_s = 1/f_s$, we get

$$c_n = \frac{1}{T_s} \int_{0}^{T_p} A \cdot e^{-j2\pi nf_s t} dt = \frac{A}{-j2\pi n} \cdot (e^{-j2\pi nf_s T_p} - 1) \tag{1.69}$$

When $n = 0$ we can use l'Hospital's rule to get

$$c_0 = \frac{A}{2} \tag{1.70}$$

Using the results from Sec. 1.2.2 we can write

$$c_n = \frac{-Aj}{2\pi n} \cdot (1 - e^{-j2\pi nf_s T_p}) = \frac{-Aj}{2\pi n} \cdot \sqrt{2(1 - \cos 2\pi nf_s T_p)} \tag{1.71}$$

For the case when $T_p = T_s/2$ this simplifies to

$$c_n = \frac{-Aj}{2\pi n} \cdot \sqrt{2(1 - \cos \pi n)} \tag{1.72}$$

so that the coefficients are zero when n is nonzero and even. When n is odd

$$c_n = \frac{-A}{\pi n} \cdot j \text{ or } |c_n| = \frac{A}{\pi n} \text{ and } \angle c_n = -\frac{\pi}{2} \tag{1.73}$$

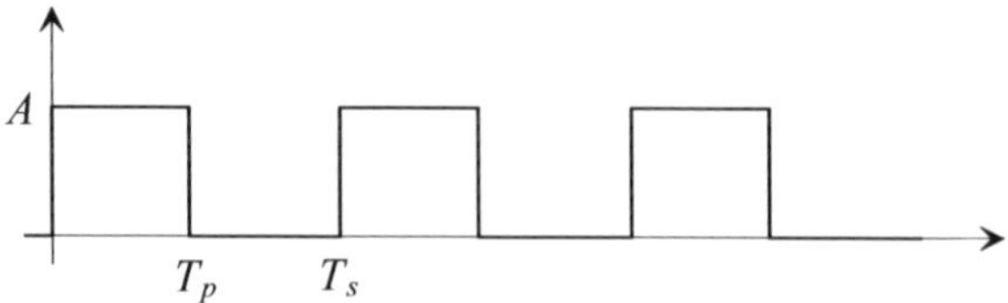

Figure 1.29 Representing a squarewave using exponential Fourier Series.

1.3.2 Fourier Transform

The Fourier Transform of a function $g(t)$ is given by

$$G(f) = \int_{-\infty}^{\infty} g(t) \cdot e^{-j2\pi ft} dt \tag{1.74}$$

while the inverse Fourier Transform is given by

$$g(t) = \int_{-\infty}^{\infty} G(f) \cdot e^{j2\pi ft} df \tag{1.75}$$

Dirac Delta Function (Unit Impulse Response)

The Dirac delta function, sometimes called the unit impulse response, is defined as

$$\delta(t - t_0) = \infty \text{ when } t = t_0 \text{ and } 0 \text{ if } t \neq t_0 \tag{1.76}$$

The discrete version (defined only at discrete time points, that is not continuously) of the Dirac delta function is the Kronecker delta function, or simply delta function, and is given by

$$\delta(t - nT_s) = 1 \text{ when } t = nT_s \text{ and } 0 \text{ if } t \neq nT_s \tag{1.77}$$

Some properties of the Dirac delta function are that

$$\int_{-\infty}^{\infty} \delta(t - t_0) dt = 1 \tag{1.78}$$

As the amplitude tends towards ∞ the width of the function moves to 0 keeping the area equal to 1. Also,

$$\int_{-\infty}^{\infty} f(t) \cdot \delta(t - t_0) dt = f(t_0) \tag{1.79}$$

A few other key properties are that the Fourier Transform of a constant is

$$\int_{-\infty}^{\infty} K \cdot e^{-j2\pi ft} \cdot dt = K \cdot \delta(f) \tag{1.80}$$

The Fourier transform of a delayed signal, $x(t - t_0)$, is

$$Fourier\{x(t - t_0)\} = X(f) \cdot e^{-j2\pi f t_0} \tag{1.81}$$

where the magnitude of the delayed signal is $|X(f)|$ (the delay doesn't affect the magnitude) and the phase shift is $\angle X(f) - 2\pi \cdot \frac{t_0}{T} = \angle X(f) - 2\pi f \cdot t_0$. The Fourier Transform of a sinusoid is determined by first writing

$$V_P \cdot \sin(2\pi f_1 t) = V_P \cdot \frac{e^{j2\pi f_1 t} - e^{-j2\pi f_1 t}}{2j} \tag{1.82}$$

then, using Eq. (1.74)

$$Fourier\{V_P \sin(2\pi f_1 t)\} = V_P \int_{-\infty}^{\infty} \frac{e^{j2\pi f_1 t} - e^{-j2\pi f_1 t}}{2j} \cdot e^{-j2\pi ft} dt = V_P \int_{-\infty}^{\infty} \frac{e^{j2\pi(f_1 - f)t} - e^{-j2\pi(f_1 - f)t}}{2j} dt \tag{1.83}$$

and thus

$$=\frac{V_P}{2j}\cdot[\delta(f-f_1)-\delta(f+f_1)]=0+j\cdot\frac{V_P}{2}\cdot[\delta(f+f_1)-\delta(f-f_1)] \quad (1.84)$$

The magnitude of this equation is

$$\frac{V_P}{2}\cdot[\delta(f+f_1)-\delta(f-f_1)] \quad (1.85)$$

and the phase response is

$$\tan^{-1}\frac{\frac{V_P}{2}\cdot[\delta(f+f_1)-\delta(f-f_1)]}{0}=\tan^{-1}\infty=90^\circ \quad (1.86)$$

Note that using a single-sided spectrum with positive frequencies only we could rewrite the result as

$$=j\cdot V_P\cdot\delta(f-f_1) \quad (1.87)$$

Note also that

$$Fourier\{V_P\cos(2\pi f_1 t)\}=\frac{V_P}{2}\cdot[\delta(f+f_1)+\delta(f-f_1)] \quad (1.88)$$

Lastly note that *negative frequencies* shouldn't confuse us since they simply represent a phase shift. For example,

$$\sin(2\pi(-f)t)=-\sin(2\pi ft)=\sin(2\pi ft+\pi)=\sin(2\pi ft-\pi) \quad (1.89)$$

ADDITIONAL READING

[1] M. Weeks, *Digital Signal Processing Using MATLAB and Wavelets*, Infinity Science Press, 2007. ISBN 978-0977858200

[2] S. Haykin and M. Moher, *An Introduction to Analog and Digital Communications*, Second Edition, John Wiley and Sons, 2006. ISBN 978-0471432227

[3] R. G. Lyons, *Understanding Digital Signal Processing*, Second Edition, Prentice-Hall, 2004. ISBN 978-0131089891

[4] P. A. Lynn and W. Fuerst, *Introductory Digital Signal Processing*, Second Edition, John Wiley and Sons, 1998. ISBN 978-0471976318

[5] L. W. Couch, *Modern Communication Systems: Principles and Applications*, Prentice-Hall, 1995. ISBN 978-0023252860

[6] E. P. Cunningham, *Digital Filtering: An Introduction*, John Wiley and Sons, 1995. ISBN 978-0471124757

QUESTIONS

1.1 Suggest an alternate physical example, to the swinging pendulum in Fig. 1.1, of sinusoidal motion.

1.2 Add the *z*-axis (time) to the representations of the sinewave seen in Figs. 1.3b-f as discussed following Eq. (1.4).

1.3 Suppose an *IQ* signal is generated using an in-phase component having an amplitude of 0.5 V and a quadrature component having an amplitude of 1 V. Sketch the resulting waveform, the *IQ* signal, in the time domain.

1.4 Figure 1.7 shows how the magnitude and phase are calculated for an imaginary number that resides in the first quadrant of the plane (both real and imaginary components are positive). Show how we calculate the magnitude and phase of an imaginary number in the other quadrants.

1.5 If the output of a system occurs after the corresponding input to the system, is the phase shift positive or negative? Why? What does linear phase indicate?

1.6 Using the SPICE files found at CMOSedu.com, verify, in the time-domain, the frequency response information seen in Fig. 1.10 for input frequencies of $f = 0$ (DC), $1/4t_d$, and $1/2t_d$.

1.7 Repeat question 1.6 for the digital comb filter (averager) seen in Fig. 1.17 for input frequencies of DC, $f_s/4$, and $f_s/2$.

1.8 Plot the magnitude and phase frequency responses of a discrete-time system having the transfer function $(1+z^{-1})/z^{-2}$. Next, show the location of this system's poles and zeros in the complex plane and verify, using the intuitive method discussed in Sec. 1.2.3, the gain and phase of the response match the frequency response plots when the input signal frequency is 0.

1.9 For the 3 delay element comb filter seen in Fig. 1.25, repeat question 1.6 for input frequencies of 0, $f_s/6$, and $f_s/3$.

1.10 Show how to plot $1/(4+3j)$ in the complex plane. What is the magnitude and phase shift of this complex number?

1.11 Determine the z-domain representation for the circuit seen in Fig. 1.30. Also, plot the frequency response, both magnitude and phase, and the location of poles and zeroes for this system.

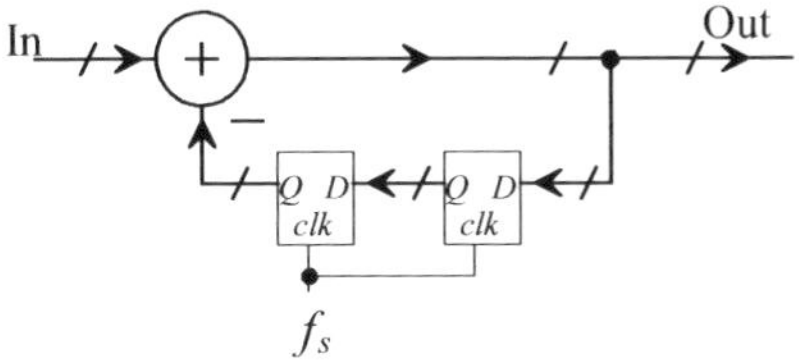

Figure 1.30 Circuit used in question 1.11.

1.12 Repeat question 1.11, and sketch the resulting circuit, if a delay is added to the forward path of the circuit seen in Fig. 1.30.

1.13 Determine the exponential Fourier series representation for the squarewave seen in Fig. 1.29 if it is centered around ground.

1.14 Determine the exponential Fourier series representation for the squarewave seen in Fig. 1.29 for the general case where $T_p \neq T_s/2$.

1.15 What is the Fourier transform of the signal seen in Fig. 1.29?

1.16 What is the area under the Dirac delta function bordered by the x-axis? Why?

1.17 Show how to take the Fourier transform of $\sin(2\pi f_0 t + \theta)$ and $\cos(2\pi f_0 t + \phi)$. Plot the magnitude and phase responses of the transforms.

Chapter 2

Sampling and Aliasing

The block diagram for one example of a mixed-signal circuit design used to process analog signals is seen in Fig. 2.1. The system's input and output are continuous-time (analog) signals. The input signal is applied to an analog filter, called an anti-aliasing filter (AAF) that limits the input spectral content to avoid aliasing (discussed in Section 2.1) when the signal is sampled and held using the sample-and-hold, S/H (also called a zero-order hold, ZOH), circuit. The output of the S/H is connected to the analog-to-digital converter (ADC) where it is converted to a digital word that changes with time. The digital word is then passed through a digital signal processing (DSP) block where it is manipulated (e.g. lowpass filtered using a digital filter like the one seen in Fig. 1.16). The output of the DSP block is connected to a digital-to-analog converter (DAC) that changes the digital words back into an analog voltage. At this point the signal does have a changing amplitude but it is still discrete in time (jagged as seen in Fig. 2.1). The output of the DAC is fed to a reconstruction filter (RCF) to smooth it out. We'll see that this step removes the spectral content above the Nyquist frequency, f_n or $f_s/2$.

Note that analog circuits are an important component of this system. The AAF and RCF must be completely analog in design. Further, the ADC and DAC may contain precise analog circuits.

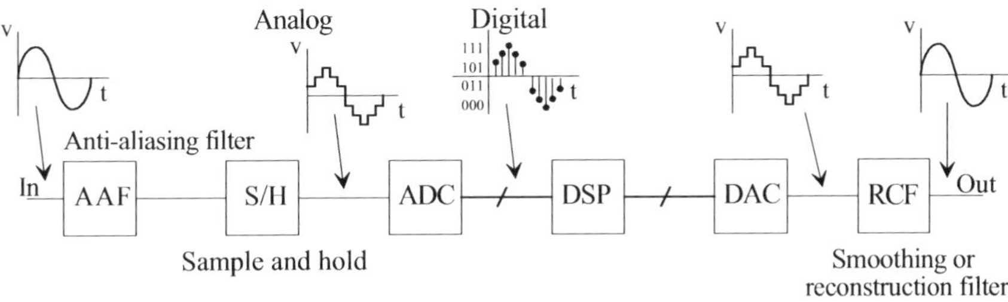

Figure 2.1 Signals resulting from A/D and D/A conversion in a mixed-signal system.

2.1 Sampling

In this section we discuss how sampling a signal changes its spectrum. Impulse sampling, decimation, the sample-and-hold (S/H), the track-and-hold (T/H), interpolation, and K-path sampling methods are discussed.

2.1.1 Impulse Sampling

Consider the sampling block diagram shown in Fig. 2.2. Let's assume that we apply a sinewave input, $x(t)$, to this sampling gate of the form, $V_p \sin(2\pi f_{in} \cdot t)$ (for the moment, a single frequency input). The output of the sampling gate (a.k.a. sampler), $y(t)$, is the product of the input and a sampling unit impulse signal (Dirac delta function), $\delta(t-nT_s)$ or

$$y(t) = \sum_{n=-\infty}^{\infty} V_p \sin(2\pi f_{in} \cdot t) \cdot \delta(t-nT_s) \tag{2.1}$$

Noting that the frequency of the input is f_{in} while the sampling frequency is $f_s(=1/T_s)$, the spectrum of the input signal is seen in Fig. 2.3a. Note that the Fourier series representation of the series of sampling impulses above is

$$\sum_{n=-\infty}^{\infty} \delta(t-nT_s) = \frac{1}{T_s} \sum_{n=-\infty}^{\infty} e^{j2\pi n \cdot t/T_s} = \frac{1}{T_s} \sum_{n=-\infty}^{\infty} e^{j2\pi n f_s \cdot t} \tag{2.2}$$

If we take the Fourier transform of the input signal after sampling (we take the Fourier transform of Eq. [2.1]), that is, we look at the spectrum on the output of the sampler, we get

$$Y(f) = \frac{V_p}{2jT_s} \sum_{k=-\infty}^{\infty} \int_{-\infty}^{\infty} [e^{j2\pi(f_{in}+kf_s)t} - e^{-j2\pi(f_{in}-kf_s)t}] \cdot e^{-j2\pi f \cdot t} \cdot dt \tag{2.3}$$

or, knowing the Fourier transform of $e^{j2\pi f_o \cdot t}$ is $\delta(f-f_o)$

$$Y(f) = \frac{V_p}{2jT_s} \cdot \sum_{k=-\infty}^{\infty} [\delta(f-f_{in}-kf_s) - \delta(f+f_{in}-kf_s)] \tag{2.4}$$

The sampled spectrum is repeated, at intervals of f_s, as seen in Fig. 2.3b (the one-sided spectrum is shown, which is what we will use throughout the book). Note that if an ideal lowpass filter (LPF) is applied to the output spectrum of the sampler (the output of the

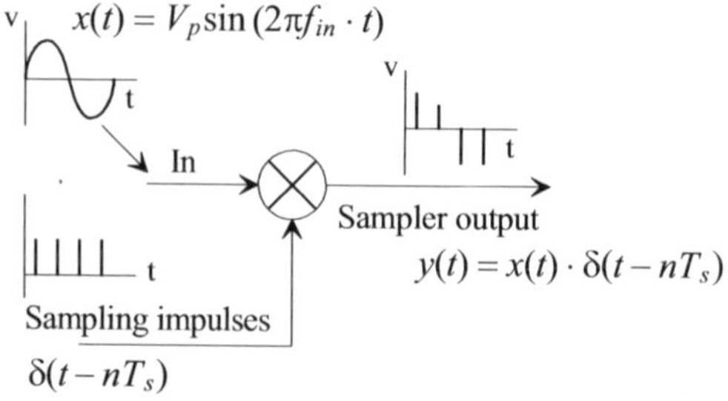

Figure 2.2 Impulse sampling a signal.

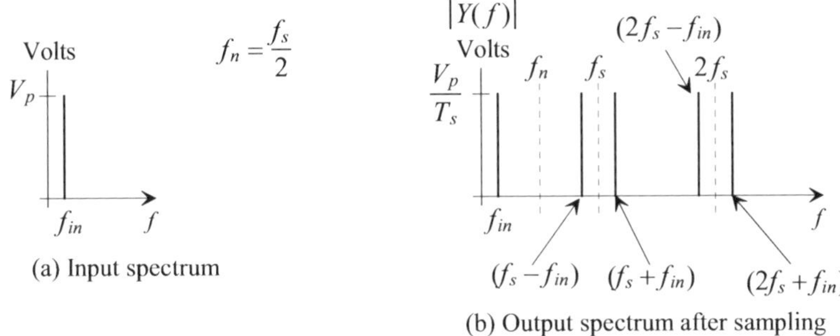

Figure 2.3 One-sided spectrum of a sinewave (a) before and (b) after sampling.

sampler is connected to an ideal LPF) with a bandwidth greater than f_{in} (and lower than f_n [the Nyquist frequency]), then the higher-order frequency components can be removed so that only f_{in} remains (this is our smoothing or reconstruction filter shown in Fig. 2.1).

Example 2.1

A sampling gate is strobed with an impulse train running at a frequency of 100 MHz (f_s = 100 MHz and the time in between the impulses, T_s, is 10 ns). Sketch the resulting output frequency spectrum if a 60 MHz sinewave is applied to the sampler. Also, sketch the time domain input and output of the sampler.

The resulting frequency spectrum is shown in Fig. 2.4. Notice how connecting the output of the sampler through an LPF, with an ideal abrupt cutoff frequency of f_n, results in an output sinewave with a frequency of 40 MHz. In order to avoid this situation, that is, to avoid ending up with the wrong, or alias, signal after sampling and reconstructing, we need to ensure that the signal frequencies applied to the sampler are less than $f_s/2$ (the Nyquist frequency, again, f_n). Reviewing Fig. 2.1, we see that this is the purpose of the antialiasing filter (AAF). Notice how, ideally, both the AAF and RCF (reconstruction filter) in Fig. 2.1 are both ideal LPFs with a cutoff frequency equal to half the sample frequency (the Nyquist frequency). Figure 2.5 shows the time domain sketch of the sampler's output. ■

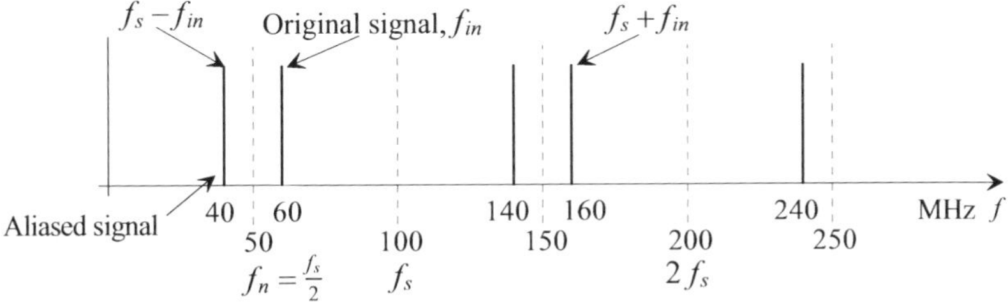

Figure 2.4 Spectrum of a 60 MHz sinewave sampled at 100 MHz.

It should be clear from the preceding discussion that: (1) sampling a signal results in a reproduction of the sampled signal's spectrum at DC, f_s, $2f_s$, $3f_s$, etc., (2) the input signal's spectrum should have no significant spectral content above f_n in order to avoid

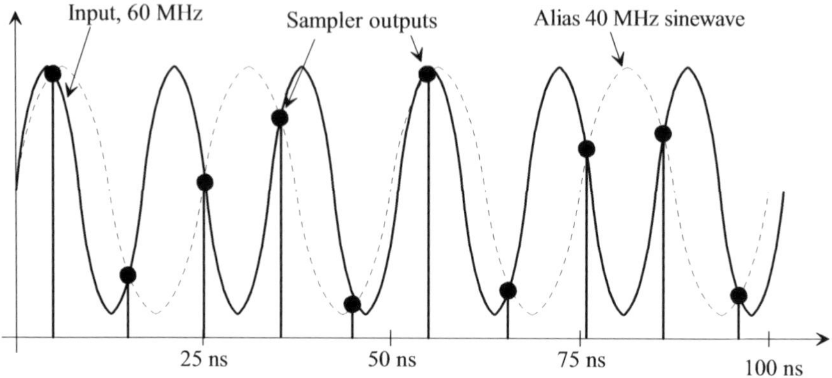

Figure 2.5 Time domain input and output for Ex. 2.1.

aliasing, (3) to avoid aliasing both filtering the input signal using an AAF and increasing the sampling frequency should be used, and (4) to reproduce the sampled signal from the output of the sampler (which is nonzero only during the sampling impulse times) a lowpass RCF should be used.

Note that our discussion illustrates the operation of a sampling gate driven with impulse signals. As shown in Fig. 2.1, a practical system would have other building blocks. We would rarely, if ever, sample a signal and then reconstruct it without processing it first.

A Note Concerning the AAF and the RCF

Before going any further, we should discuss the ideal characteristics of the AAF and the RCF. The ideal characteristics of these filters are shown in Fig. 2.6. Note that both of these filters must be analog by design. The ideal cutoff frequency for the filters can be no greater than f_n (assuming the sampling rate on the input of the system is the same as the sampling rate on the system's output) and the filters should ideally have linear phase. Let's discuss these two ideal characteristics.

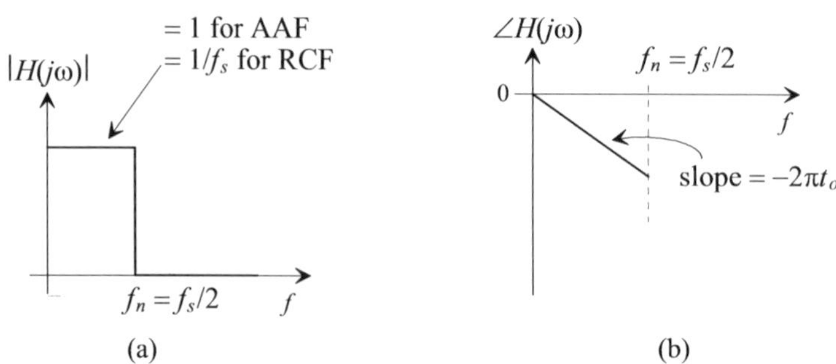

Figure 2.6 (a) Ideal magnitude and (b) phase responses for the AAF and RCF.

The ideal magnitude response, shown in Fig. 2.6a, passes all spectral content below the Nyquist frequency while removing all signals above this frequency. The ideal phase response, shown in Fig. 2.6b, provides a constant delay, t_o, to all signals below f_n. In other words, the filters remove all unwanted signals while not distorting the wanted signals.

Time Domain Description of Reconstruction

In this section we show why the filter shown in Fig. 2.6, an ideal brick wall lowpass filter with linear phase response, is the ideal RCF on the output of our impulse sampler. Shown in Fig. 2.7 is a 20 MHz sinewave sampled at 100 MHz. Suppose we want to reconstruct the original input 20 MHz sinewave from the sampler output (the weighted impulse functions). After reconstruction, the output of the RCF should be a single-tone, 20 MHz sinewave (it should be an exact replica of the sampler input). To determine what happens when the output of our sampler is applied to the ideal RCF, we need to determine the time-domain response of the RCF when its input signal is an impulse.

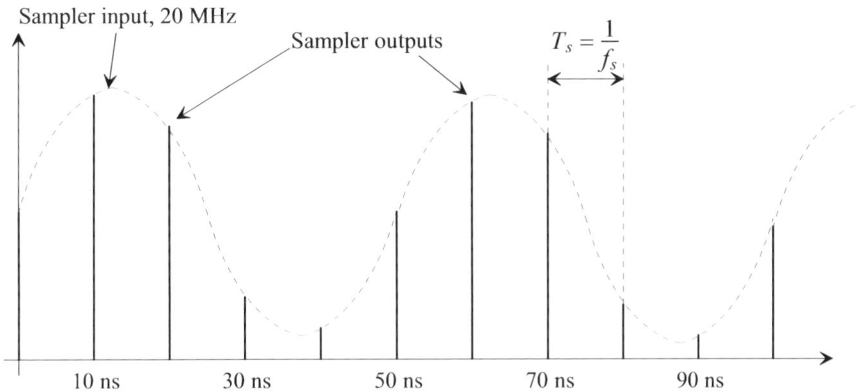

Figure 2.7 Impulse sampling, at 100 MHz, a sinewave at 20 MHz.

The transfer function of a system is the Fourier transform of the system's time domain impulse response (what we are trying to find here). In other words, to determine the transfer function of the system, we apply an impulse to the input of the system (a very large amplitude, very short time duration pulse, Fig. 2.8). We then look at the system's output in the time domain. Taking the Fourier transform of this output gives the system's

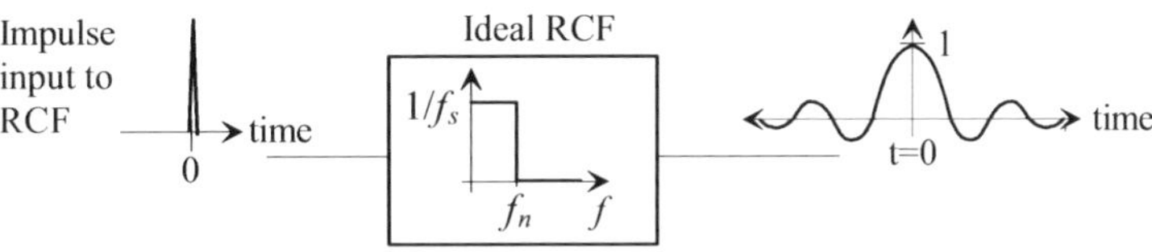

Figure 2.8 Time domain impulse response of the ideal RCF.

transfer function. Therefore (in the reverse order), to determine the time-domain impulse response of the ideal RCF, given the transfer function, we take the inverse Fourier transform of the transfer function. The ideal RCF's transfer function (Fig. 2.6) can be defined by

$$|H(f)| = 1/f_s \text{ for } |f| < f_n \text{ else } |H(f)| = 0 \tag{2.5}$$

The time-domain impulse response is then given, remembering $2f_n = f_s$, by

$$h(t) = \int_{-f_n}^{f_n} \frac{1}{f_s} \cdot e^{j\cdot 2\pi\cdot f\cdot t} \cdot df = \frac{e^{j\cdot 2\pi\cdot f_n\cdot t} - e^{-j\cdot 2\pi\cdot f_n\cdot t}}{j\cdot 2\pi\cdot f_s\cdot t} =$$

$$\frac{\sin 2\pi f_n \cdot t}{\pi f_s \cdot t} = \frac{\sin \pi f_s \cdot t}{\pi f_s \cdot t} = Sinc(\pi f_s \cdot t) \tag{2.6}$$

where

$$\frac{\sin x}{x} \equiv Sinc(x) \tag{2.7}$$

The time-domain impulse response of our ideal RCF is shown in Fig. 2.9. Notice that our impulse is applied to the system's input at $t = 0$ and that the output actually anticipates, or starts, before the application of the input! This indicates that the filter can't be built in a practical analog circuit. Before we discuss the implications of this severe limitation (an ideal reconstruction filter can't actually be built because its impulse response is infinite in time), examine Fig. 2.10. Figure 2.10 shows the individual impulse response outputs of an ideal RCF with the impulse train of Fig. 2.7 as the input. The output of the RCF is the weighted sum of the individual responses. While this figure is "busy," the basic concept of reconstruction can be seen.

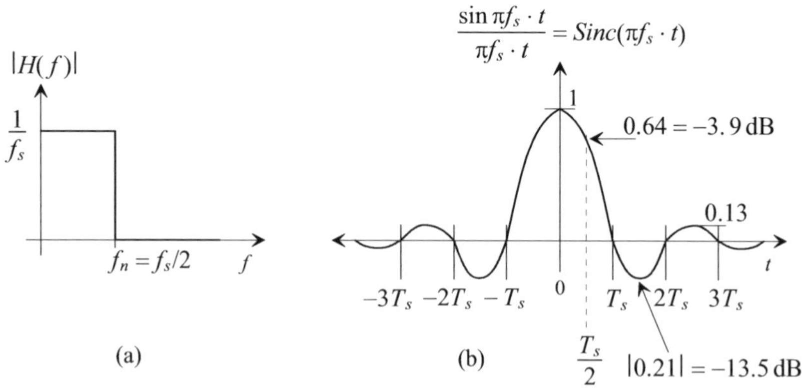

Figure 2.9 (a) Ideal RCF frequency response and (b) impulse response (time).

What can we do to ease the requirements on the RCF? One answer is to increase f_s so it's much larger than the maximum sampled frequency. This solution is the basis for *oversampled data converters* or noise-shaping data converters (studied in detail later in the book). This solution also eases the requirements placed on the AAF. Another idea for easing the design of the RCF is to increase the sample rate of the digital data coming out of the DSP block in Fig. 2.1. This technique is called *interpolation.*

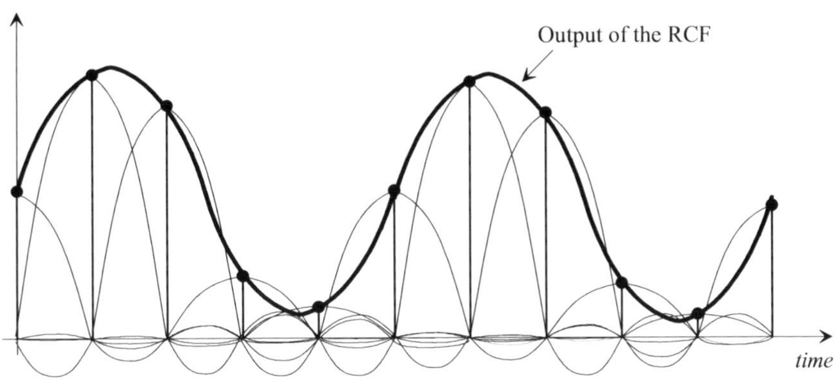

Figure 2.10 Reconstructing the 20 MHz sinewave of Fig. 2.7.

An Important Note

It is important to note that our impulse sampler quantizes[1] the input signal in *time* but not amplitude (unlike an analog-to-digital converter which quantizes the input in both time and amplitude). The amplitude out of the ideal impulse sampler is exactly the same as the amplitude input to the sampler at the sampling impulse times. The z-transform can be used to describe systems using both quantization in time as well as in amplitude. In other words, whether we are discussing digital words, in a binary format, or sampled-analog waveforms with amplitudes of volts, amps, or coulombs, we can use the z-transform to represent the discrete-time systems that process the signals. Laplace-transforms are used for continuous-time systems.

2.1.2 Decimation

In the last section we focused on sampling analog signals. We can apply the same concepts to digital signals. When a digital signal is "down sampled" its sample rate goes from f_s to a lower rate of f_s/K where K is generally, but not necessarily, a power of 2 (e.g., 2, 4, 8, 16, etc.). This reduction in the effective sampling frequency is termed *decimation* and is illustrated with the block diagram shown in Fig. 2.11. The term decimation (or

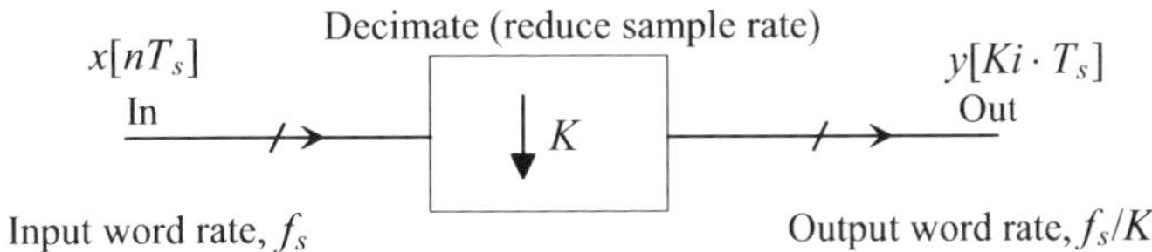

Figure 2.11 Block diagram of a decimation block.

[1] Quantize: to limit the possible values of a quantity to a discrete set of values. Quantizing in time, for example, means that the output amplitude is only defined at certain discrete times (such as the sampling impulse times for the ideal impulse sampler) or that the amplitude is unchanging during certain discrete time intervals (such as seen in the output of the ideal sample-and-hold discussed in the next section).

decimate) can be confusing since, among other uses, the dictionary definition is, "to select by lot and kill one in every ten." The origin of the word comes from a method of punishing military troops by selecting one in every ten for execution. Our much more kind-hearted definition will mean that we are passing the input word through a lowpass digital filter and then down-sampling the result (discarding samples). This procedure is effectively passing the digital data through an antialiasing filter and then resampling the result at a lower rate, Fig. 2.12. Note that the sampling gate is simply a register so it is trivial to implement decimation.

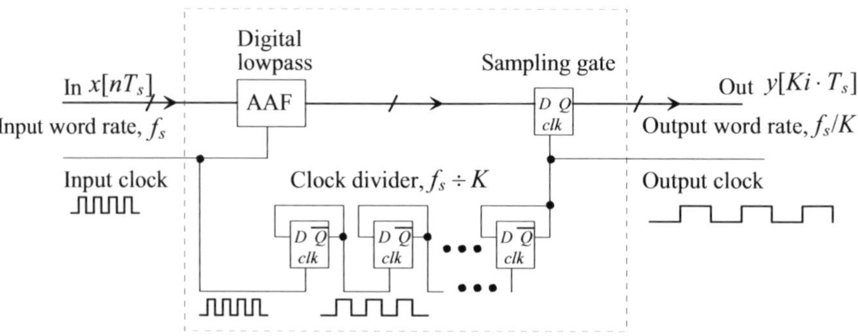

Figure 2.12 Components of a decimation block.

To illustrate the aliasing concerns when using decimation, and $K = 8$, examine Fig. 2.13. The input spectrum of the digital data, in (a), repeats every f_s. The digital input data in (a) is first passed through a digital lowpass filter that is used as an anti-aliasing filter, Fig. 2.13b. The input and output of the digital filter is clocked at f_s in both (a) and (b). In (c) we re-time the input signal at the slower rate. Decimation is used to lower power (because of the reduced clock frequency), simplify circuitry (e.g., serial multipliers can be used), and to lower the amount of data storage required (fewer words to store). Note that to use decimation the *wanted* input spectral content can't extend beyond $f_s/2K$.

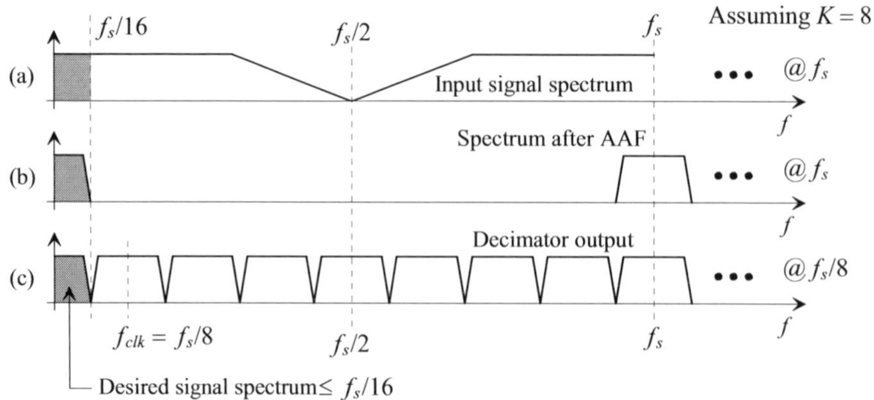

Figure 2.13 Example spectrums when decimation is employed.

2.1.3 The Sample-and-Hold (S/H)

Understanding the operation of the impulse sampler in Sec. 2.1.1 is important in understanding the concepts of aliasing and reconstruction. However, as seen in Fig. 2.1, most mixed-signal systems employ a sample-and-hold (S/H) rather than an impulse sampler so that the sampled waveform is available at times other than the sampling impulse times. Having the samples "held" in between the sampling impulse times is important for proper ADC operation. The disadvantage of using the S/H, as we shall shortly see, is that it will introduce distortion into our signal.

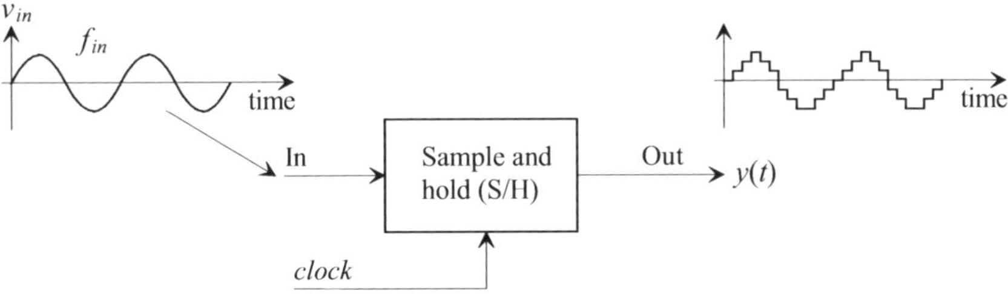

Figure 2.14 Sampling and holding an input sinewave.

S/H Spectral Response

Consider the application of a sinewave, at a frequency f_{in}, to the ideal S/H shown in Fig. 2.14. To make the discussion as general as possible, assume that the output of the S/H can return-to-zero (RZ) as shown in Fig. 2.15 (which shows coarse time quantization for a simpler figure and illustration of the concept of RZ). Note that as T approaches T_s we get the operation of the S/H in Fig. 2.14. The output of the ideal S/H is given by

$$y(t) = \sum_{n=-\infty}^{\infty} [V_p \sin(2\pi f_{in} \cdot nT_s) \cdot [u(t-nT_s) - u(t-nT_s-T)]] \tag{2.8}$$

As depicted in this equation the input is sampled at the instants nT_s and then held at the sampled value $v_{in}(nT_s)$ for a pulse length of T. We can represent the resulting signal using convolution, see Fig. 2.16a, as

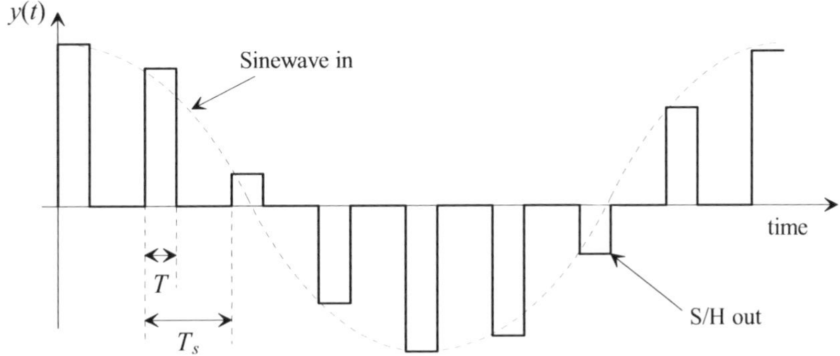

Figure 2.15 Sample-and-hold output with return to zero format.

$$y(t) = \sum_{n=-\infty}^{\infty} \overbrace{[V_p \sin(2\pi f_{in} \cdot t) \cdot \delta(t - nT_s)]}^{v_{in}(nT_s)} \otimes [u(t) - u(t-T)]$$

$$= \left[\overbrace{V_p \sin(2\pi f_{in} \cdot t)}^{v_{in}(t)} \cdot \overbrace{\sum_{n=-\infty}^{\infty} (\delta(t - nT_s))}^{p(t)} \right] \otimes \overbrace{[u(t) - u(t-T)]}^{h(t)} \qquad (2.9)$$

or

$$Y(f) = (v_{in}(f) \otimes P(f)) \cdot H(f) \qquad (2.10)$$

where, see Eq. (2.4),

$$v_{in}(f) \otimes P(f) = \frac{V_p}{2jT_s} \cdot \sum_{k=-\infty}^{\infty} [\delta(f - f_{in} - kf_s) - \delta(f + f_{in} - kf_s)] \qquad (2.11)$$

We can determine the spectrum of the sampling pulse $h(t)$ seen in Fig. 2.16a, $|H(f)|$, by taking its Fourier transform

$$H(f) = \int_0^{T_s} [u(t) - u(t-T)] e^{-j \cdot 2\pi \cdot f \cdot t} \cdot dt \qquad (2.12)$$

which is evaluated as

$$H(f) = \frac{e^{-j \cdot 2\pi \cdot f \cdot T} - 1}{-j \cdot 2\pi \cdot f} = e^{-j \cdot \pi \cdot f \cdot T} \cdot \frac{e^{j \cdot \pi \cdot f \cdot T} - e^{-j \cdot \pi \cdot f \cdot T}}{j \cdot 2\pi \cdot f} = \overbrace{e^{-j \cdot \pi \cdot f \cdot T}}^{\text{phase shift}} \cdot \overbrace{T \cdot Sinc(\pi \cdot f \cdot T)}^{\text{magnitude}}$$

$$(2.13)$$

The magnitude of Eq. (2.13), $|H(f)|$, is plotted in Fig. 2.16b. The phase response corresponds to a shift in time of $T/2$. Since

$$Y(f) = (v_{in}(f) \otimes P(f)) \cdot H(f) \qquad (2.14)$$

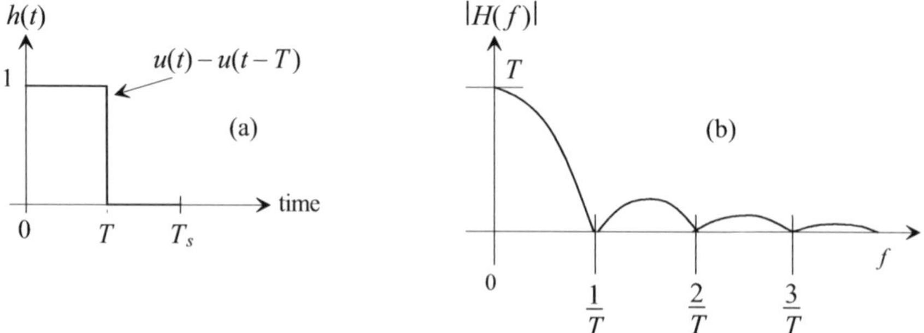

Figure 2.16 (a) Sampling pulse and (b) its spectrum.

$$Y(f) = \left(\frac{1}{T_s}\sum_{k=-\infty}^{\infty} v_{in}(f-kf_s)\right)\cdot T\cdot Sinc(\pi\cdot f\cdot T)\cdot e^{-j\cdot\pi\cdot f\cdot T}$$
$$= \frac{V_p}{2jT_s}\cdot\sum_{k=-\infty}^{\infty}[\delta(f-f_{in}-kf_s)-\delta(f+f_{in}-kf_s)]\cdot T\cdot Sinc(\pi\cdot f\cdot T)\cdot e^{-j\cdot 2\pi\cdot f\cdot\frac{T}{2}}$$
(2.15)

or

$$|Y(f)| = \overbrace{T\cdot|Sinc(\pi\cdot T\cdot f)|}^{\text{Weighting from S/H, }|H(f)|}\cdot\overbrace{\left[\frac{V_p}{2T_s}\cdot\sum_{k=-\infty}^{\infty}[\delta(f-f_{in}-kf_s)-\delta(f+f_{in}-kf_s)]\right]}^{\text{Ideal impulse sampler response}} \quad (2.16)$$

As $T\to 0$ the output spectrum of the sample-and-hold approaches the ideal impulse sampler's spectrum seen in Sec. 2.1.1. Note that using an RZ format (making $T < T_s$) can reduce the amount of attenuation (and thus distortion) introduced by the S/H. $|H(f)|$, as seen in Fig. 2.16, doesn't roll off as fast. However, the cost for this is a reduction in the sampled signal power (ultimately we get no signal power out of the S/H as $T\to 0$). Generally, reducing signal-to-noise ratio, SNR, by returning the S/H's output to ground (RZ), to improve distortion performance, is not a good idea. Reducing SNR to improve distortion can be useful in some situations, e.g., digital data transmission, where good SNR isn't as much of a problem as distortion.

For most circuit designs, $T = T_s$ so that, as Eq. (2.16) and Fig. 2.17 show, the sample-and-hold operation weights the amplitude of the ideal impulse sampler's frequency response by $Sinc\,\frac{\pi f}{f_s}$ or $Sinc\,\frac{\pi f}{2f_n}$. Note that at the sampling frequency ($f_s = 1/T_s$) the output of the ideal S/H goes to zero. Also note that at the Nyquist frequency, f_n, the input signal is attenuated by 0.64 or $-$ 3.9 dB. This "droop" in the S/H's response, as mentioned above, adds distortion to the input signal. Let's illustrate the effects of the S/H's frequency response using an example.

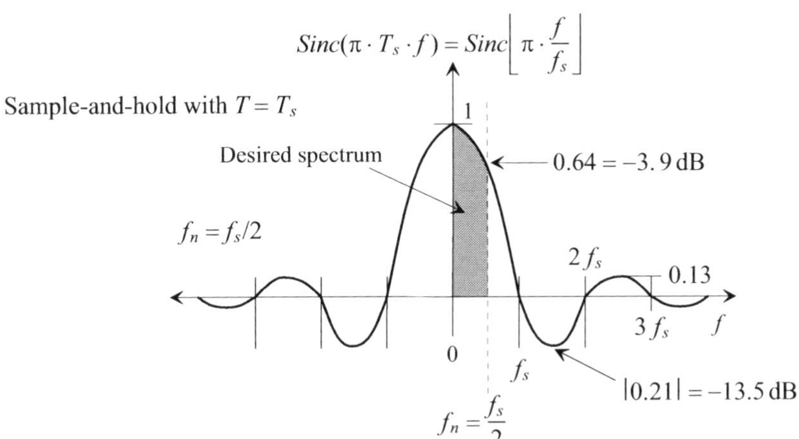

Figure 2.17 The frequency response of a S/H.

Example 2.2

Using an ideal SPICE model for the S/H show, and discuss, the spectrum resulting from sampling a 3 MHz sinewave at 100 Msamples/s.

The results of passing a 0.5 V (peak) sinewave centered at 0.5 V (–9 dB using RMS voltages) through the ideal S/H are shown in the SPICE simulation results seen in Fig. 2.18. Note that SPICE uses a one-sided spectrum so, for example, we must multiply Eq. (2.16) by two to ensure proper signal levels. The attenuation the 97 MHz image sees is

$$\text{Attenuation} = Sinc\left(\frac{\pi \cdot 97}{100}\right) = 0.031 = -30.2\ dB$$

The amplitude of the 97 MHz image is –9 dB below the attenuation resulting from using a S/H or –39.2 dB. At the Nyquist frequency of 50 MHz, an input signal is attenuated by –3.9 dB as seen in Fig. 2.17.

Note that the S/H cannot be used as an AAF since any aliasing that occurred using the impulse sampler still occurs using the S/H. For example, sampling a 60 MHz sinewave at 100 MHz still results in a 40 MHz alias signal in the base spectrum (the spectrum from DC to f_n), as shown in Fig. 2.3. Now, however, the signal is attenuated by the S/H (the attenuation is –2.4 dB at 40 MHz when sampling at 100 Msamples/s.) In other words, the S/H can be thought of as an ideal impulse sampler followed by a *Sinc* response filter.

An important thing to note is that repetitively sampling and holding a signal results in only one S/H attenuation hit (assuming the timing is such that a sampling operation is not occurring when the previous S/H stage's output is changing). This means that topologies that use several S/H operations on an input signal, such as a pipeline ADC, only attenuate the signal by $Sinc(\pi f/f_s)$ once. This is important to understand. ■

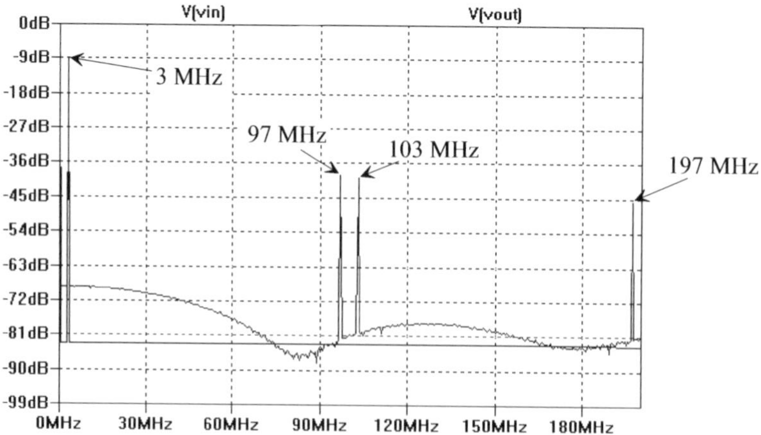

Figure 2.18 Output of a S/H after sampling a 3 MHz sinewave at 100 MHz.

The Reconstruction Filter (RCF)

Ultimately the (processed) output of the S/H (assuming $T = T_s$) should be passed through an RCF to recover the input signal. The spectral shape of the ideal RCF is seen in Fig. 2.19. The response peaks at the Nyquist frequency to compensate for the attenuation introduced by the S/H Sinc response, Fig. 2.17. Note how using the RZ format modifies the requirements placed on the reconstruction filter to the point, when using impulse sampling, of having the brick wall RCF seen in Fig. 2.6. Therefore there is no need for the peaking type response seen below (but using an RZ format reduces the S/H output signal power). Note that in many situations a digital filter may be used to compensate for the droop introduced by the S/H process.

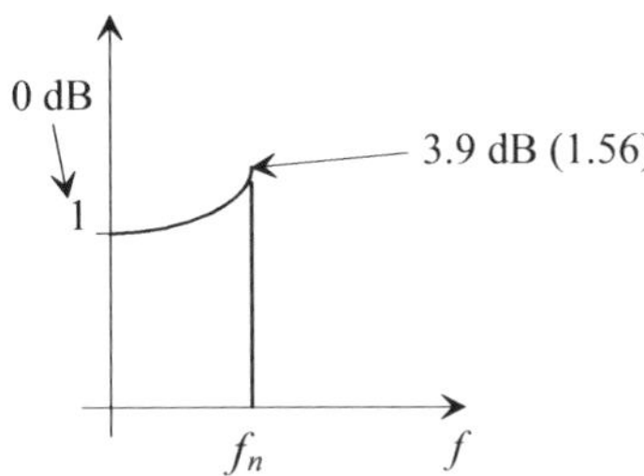

Figure 2.19 Ideal reconstruction filter frequency response for a S/H.

Circuit Concerns for Implementing the S/H

Figure 2.20 shows a single-ended S/H implementation. At the time t_0, the ϕ_1 and ϕ_2 switches are closed while the ϕ_3 switches are open. At this time the input signal charges the bottom plate (left side or, here, the plate closest to the substrate) of the hold capacitor, C_H, while the right side of the capacitor is held at ground by the feedback around the op-amp through the ϕ_1 switch. At t_1 the ϕ_1 switch opens and for a very short time (set by $t_3 - t_1$) the op-amp operates open loop (no feedback). It's assumed that this time is so short that the capacitor doesn't charge or discharge. As the top plate (right side of the capacitor) is always at ground at t_1, the charge injection and capacitive feedthrough

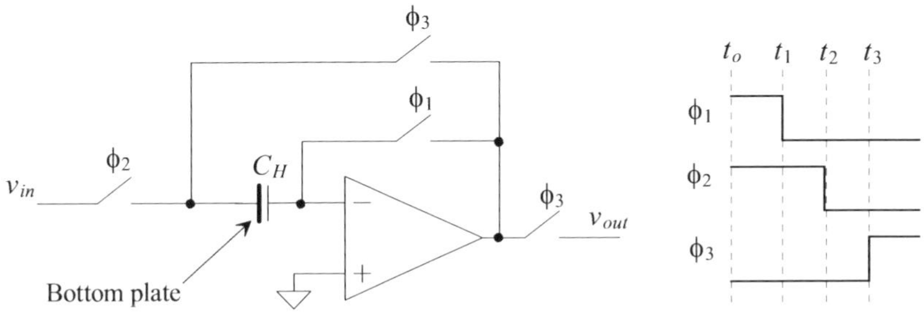

Figure 2.20 Single-ended S/H operation.

resulting from the ϕ_1 switch turning off is independent of the input signal. When the ϕ_2 switch turns off its charge will, ideally, be injected into the low-impedance input, v_{in}, since the impedance looking into the right of the ϕ_2 switch is large. This leaves the voltage across the hold capacitor unaffected. The sequence of turning off the switch to the right of C_H (the top plate) followed by turning off the switch connecting v_{in} to C_H (the bottom plate) is often called *bottom plate sampling*.

Bottom plate sampling is illustrated in its simplest form in Fig. 2.21. In this figure the switch connected to the bottom plate of the capacitor, the ϕ_1 switch, is turned off first. When this happens the charge is injected into the circuit independent of the input signal (each side of the switch is at ground). When the ϕ_2 switch turns off, its charge can be injected into the low-impedance node, the input v_{in}, or into the series combination of C_H and the off ϕ_1 switch. Again, the charge takes the lowest impedance path to ground and thus most of the charge injection resulting from the ϕ_2 switch turning off flows through v_{in}, leaving the voltage across the hold capacitor unaffected. The name "bottom plate sampling" can be confusing. Reviewing Fig. 2.20, we see that the (physical) top plate of the hold capacitor is connected to the ϕ_1 switch while, in Fig. 2.21, the (schematic representation) bottom plate of the hold capacitor is connected to the ϕ_1 switch.

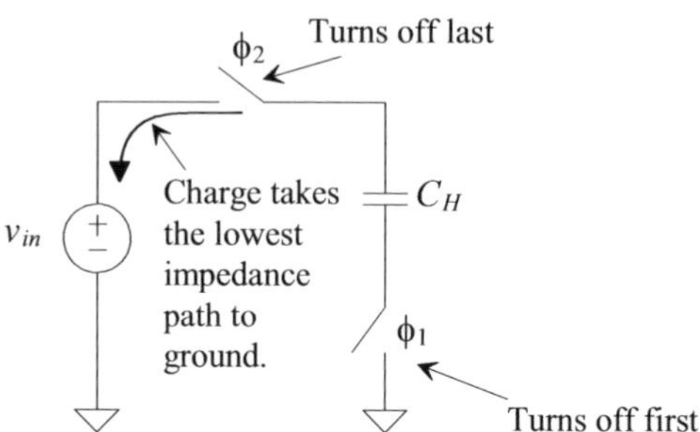

Figure 2.21 Bottom plate sampling.

An Example

Before leaving this section let's give an example of the spectrums associated with (ideal) sampling and reconstruction using a S/H, Figs. 2.1 and 2.22. In Fig. 2.22a we represent an input signal as a continuous spectrum that is not bandlimited (that is, the spectrum is completely occupied). In a real spectrum the spectral components don't have a constant amplitude (e.g. noise at high frequencies may be considerably smaller than desired content at low frequencies) but here, to simplify things, we assume a constant amplitude.

The first step in our example is to pass the input signal through an ideal AAF to limit the spectral content to the Nyquist frequency, $f_s/2$, Fig. 2.22b. At this point the spectrum hasn't been sampled so it doesn't repeat at multiples of f_s. The output of the AAF is then passed through the S/H. In Fig. 2.22c we show the Sinc weighting from the sample-and-hold process but we don't show the effects of sampling. Note the droop in the response at frequencies approaching $f_s/2$. In (d) the output of the S/H is seen. Note how the entire spectrum is occupied. Finally (e) shows the output after passing the S/H's output through an ideal RCF, Fig. 2.19. Note that the spectrum is no longer periodic.

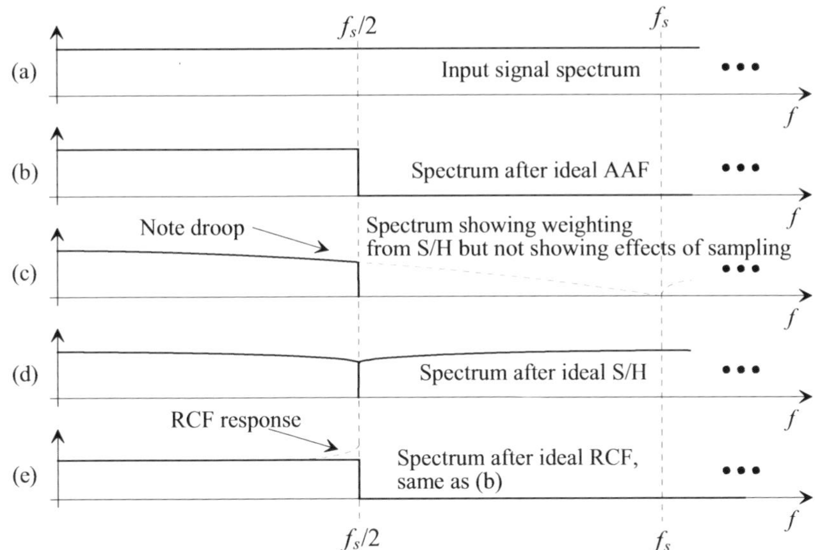

Figure 2.22 Example spectrums when ideal, AAF, S/H, and RCF are used.

2.1.4 The Track-and-Hold (T/H)

Another sampling circuit that is useful in mixed-signal circuits, especially those employing both analog- and digital-signal processing, is the track-and-hold, Fig. 2.23. The T/H is implemented using a sampling gate, here a MOSFET, and a storage capacitor. When the gate of the MOSFET is driven high it turns on and allows the input signal to directly drive the capacitor (the T/H's output). In the following discussion, we are assuming that the product of the MOSFET's on resistance and the hold capacitor, $R_{ch} \cdot C_H$, is much smaller than the period of the input signal.

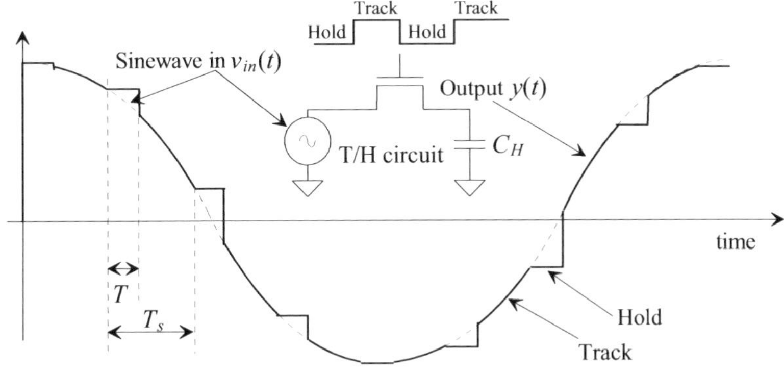

Figure 2.23 Track-and-hold output.

To determine how the T/H affects a sampled signal let's first notice that the *hold* portion of the output is exactly the same as the S/H with RZ format seen in Fig. 2.15, Eq. (2.15). Knowing this we can focus on the *track* portion of the output and then sum the responses to get the overall T/H response. For the track portion of the ideal T/H we can write, see Eq. (2.8),

$$y_t(t) = \sum_{n=-\infty}^{\infty} \{V_p \sin(2\pi f_{in} \cdot t) \cdot [u(t - nT_s - T) - u(t - nT_s - T_s)]\}$$

$$= V_p \sin(2\pi f_{in} \cdot t) \cdot \left\{ h_t(t) \otimes \sum_{n=-\infty}^{\infty} \delta(t - nT_s) \right\} \tag{2.17}$$

Knowing

$$\sum_{n=-\infty}^{\infty} \delta(t - nT_s) \text{ has a Fourier transform of } f_s \sum_{k=-\infty}^{\infty} \delta(f - kf_s) \tag{2.18}$$

we can write

$$Y_t(f) = v_{in}(f) \otimes \left\{ H_t(f) \cdot f_s \sum_{k=-\infty}^{\infty} \delta(f - kf_s) \right\} \tag{2.19}$$

Reviewing Eqs. (2.12) and (2.13) we can write

$$H_t(f) = e^{-j2\pi f \cdot \frac{(T_s+T)}{2}} \cdot (T_s - T) \cdot Sinc(\pi \cdot f \cdot (T_s - T)) \tag{2.20}$$

and so, since $f_s = 1/T_s$,

$$H_t(f) \cdot f_s \sum_{k=-\infty}^{\infty} \delta(f - kf_s) = \frac{T_s - T}{T_s} \cdot \sum_{k=-\infty}^{\infty} e^{-j2\pi \cdot kf_s \cdot \frac{(T_s+T)}{2}} \cdot Sinc(\pi \cdot kf_s \cdot (T_s - T)) \cdot \delta(f - kf_s) \tag{2.21}$$

The track portion of the T/H's output spectrum for an input signal, $v_{in}(f)$, is

$$Y_t(f) = v_{in}(f) \otimes \left[\frac{T_s - T}{T_s} \cdot \sum_{k=-\infty}^{\infty} e^{-j2\pi \cdot kf_s \cdot \frac{(T_s+T)}{2}} \cdot Sinc(\pi \cdot kf_s \cdot (T_s - T)) \cdot \delta(f - kf_s) \right]$$

$$= \frac{T_s - T}{T_s} \cdot \sum_{k=-\infty}^{\infty} \left(e^{-j2\pi \cdot kf_s \cdot \frac{(T_s+T)}{2}} \cdot Sinc(\pi \cdot kf_s \cdot (T_s - T)) \cdot V_{in}(f - kf_s) \right) \tag{2.22}$$

For the single-tone input sinusoid used in Eq. (2.17)

$$Y_t(f) =$$

$$\frac{V_p(T_s - T)}{2jT_s} \cdot \sum_{k=-\infty}^{\infty} \left(Sinc(\pi \cdot kf_s \cdot (T_s - T)) \cdot [\delta(f - f_{in} - kf_s) - \delta(f + f_{in} - kf_s)] \cdot e^{-j2\pi \cdot kf_s \cdot \frac{(T_s+T)}{2}} \right) \tag{2.23}$$

or

$$|Y_t(f)| = \sum_{k=-\infty}^{\infty} \left[\overbrace{(T_s - T) \cdot Sinc(\pi \cdot kf_s \cdot (T_s - T))}^{\text{Weighting from } h_t(t)} \cdot \overbrace{\frac{V_p}{2T_s} \cdot [\delta(f - f_{in} - kf_s) - \delta(f + f_{in} - kf_s)]}^{\text{Ideal impulse sampler response}} \right] \tag{2.24}$$

The total T/H output spectrum, $Y_{T/H}(f)$, for a general input signal $V_{in}(f)$, is the sum of Eqs. (2.15) and (2.22) or

$$Y_{T/H}(f) = \frac{T}{T_s} \cdot Sinc(\pi \cdot f \cdot T) \cdot e^{-j \cdot 2\pi \cdot f \cdot \frac{T}{2}} \cdot \sum_{k=-\infty}^{\infty} v_{in}(f - kf_s) +$$

$$\frac{T_s - T}{T_s} \cdot \sum_{k=-\infty}^{\infty} \left(e^{-j2\pi \cdot kf_s \cdot \frac{(T_s+T)}{2}} \cdot Sinc(\pi \cdot kf_s \cdot (T_s - T)) \cdot v_{in}(f - kf_s) \right) \quad (2.25)$$

If $T = T_s/2 = 1/2f_s$ then we can write this equation as

$$Y_{T/H}(f) = \sum_{k=-\infty}^{\infty} \left(\left\{ \frac{1}{2} \cdot Sinc\left(\frac{\pi}{2} \cdot \frac{f}{f_s}\right) \cdot e^{-j \cdot \frac{\pi}{2} \cdot \frac{f}{f_s}} + \frac{1}{2} \cdot Sinc\left(\frac{\pi}{2} \cdot k\right) \cdot e^{-j\frac{3\pi}{2} \cdot k} \right\} \cdot v_{in}(f - kf_s) \right) \quad (2.26)$$

We know from Fig. 2.17 that a sinewave at nearly f_n will see an attenuation of 0.64 (–3.9 dB) when using a S/H with $T = T_s$. Using the T/H with $T = T_s/2$ the attenuation the sinewave sees at nearly f_n ($f \to f_s/2$) where $k = 0$ (the frequencies from DC to $f_s/2$) is

$$|Y_{T/H}(f)| = \frac{1}{2} \left| Sinc\left(\frac{\pi}{4}\right) \cdot e^{-j \cdot \frac{\pi}{4}} + 1 \cdot e^{0} \right| \quad (2.27)$$

or

$$|Y_{T/H}(f)| = \frac{1}{2} \left| 0.9 \angle \left(-\frac{\pi}{4}\right) + 1 \angle 0 \right| \quad (2.28)$$

We know we can't directly add the polar representation of these numbers so let's rewrite this equation, noting $0.9 \cos\left(-\frac{\pi}{4}\right) = 0.636$ and $0.9 \sin\left(-\frac{\pi}{4}\right) = -0.636$, using the Cartesian representation as

$$|Y_{T/H}(f)| = \frac{1}{2} |0.636 + j(-0.636) + 1 + j0| = 0.877 \to -1.1\text{dB} \quad (2.29)$$

Note that if we are sampling an analog waveform for Nyquist-rate, analog-to-digital conversion then we have to use a S/H (or the hold portion of the T/H). In this situation we don't want the input to our quantizer (ADC) to vary, as it does in the T/H during the track portion, since this will cause an error called aperture uncertainty (discussed in Sec. 5.2.1).

2.1.5 Interpolation

One of the main assumptions when reconstructing the output signals, in the preceding discussions, is that an RCF is available with a brickwall like shape (Figs. 2.9 and 2.19) and a cutoff frequency of $f_s/2$ (the Nyquist frequency). This, however, is a very challenging analog filter design problem. To make the design of the RCF less challenging we can increase f_s while keeping the desired spectrum limited in bandwidth. In other words, by increasing f_s the effective Nyquist frequency becomes larger than the maximum wanted frequency of interest. This is called *oversampling*.

Decreasing the clock frequency (decimation) was covered in Sec. 2.1.2. Here we cover increasing the clock frequency (called interpolation). Both decimation and interpolation can be used in discrete-time analog signal processing. Here, in this introductory section, we focus on digital signals. When a digital signal is "up sampled" its rate goes from f_s to a higher rate of $K \cdot f_s$. As with decimation, K is generally, but not necessarily, a power of 2. Interpolation is represented as seen in Fig. 2.24.

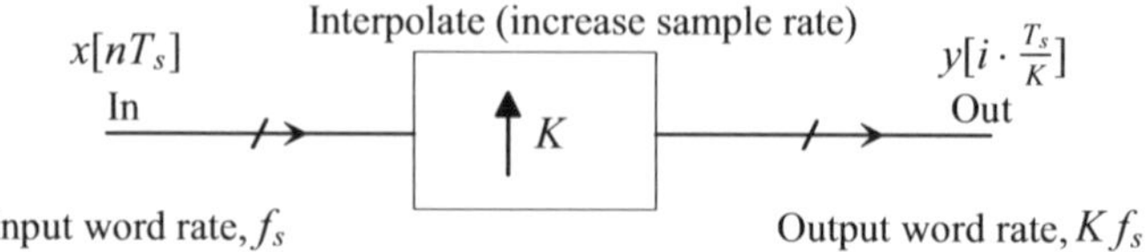

Figure 2.24 Block diagram of an interpolation block.

There are three basic interpolation schemes (ways of increasing the sampling, or clocking, frequency): zero padding, using a hold register, and linear interpolation, Fig. 2.25. Figure 2.25a shows an example input to an interpolator. In (b)-(d) we show interpolation with $K = 4$.

Zero Padding

The benefit of using zero padding, Figs. 2.25b and 2.26, is that the desired spectrum remains unchanged. We'll discuss this more in a moment. The drawback of zero padding is that the amplitude of the output signal drops by K. If our input is a constant value of 1 and we insert 3 zeroes ($K = 4$) then our output has an average value of 0.25 (average of 1, 0, 0, 0). This is normally not a problem for a digital signal since we can simply increase

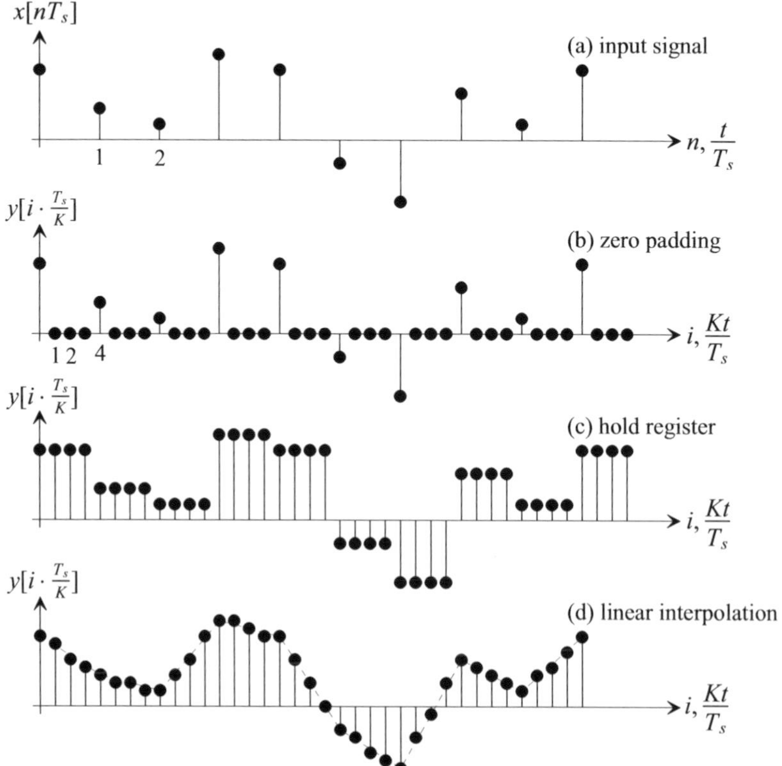

Figure 2.25 Types of interpolation.

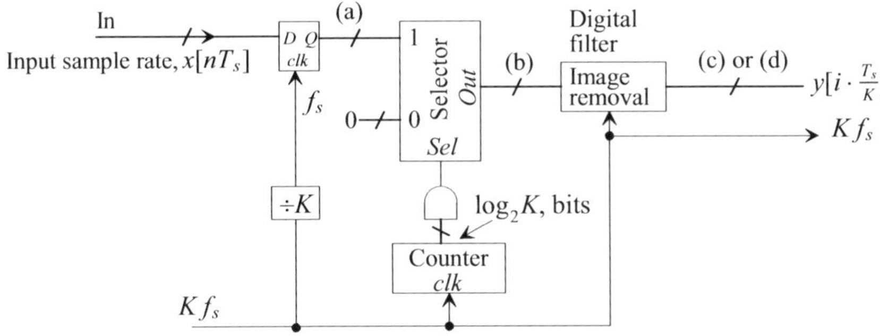

Figure 2.26 A zero-padding interpolation block, see spectrums in Fig. 2.27.

the word size and shift each nonzero sample towards the MSB by $\log_2 K$ (assuming K is a power of 2). For the present example we would shift the constant value of 1 to the left two places resulting in 4. When we average the 4 samples (4, 0, 0, 0) we get 1 (our original input). In the following, and in Fig. 2.26, we won't explicitly show the shifting in the input or output words to compensate for the signal reduction when using zero padding.

Returning to Fig. 2.26, we see that our input signal changes every T_s seconds and has the assumed spectrum seen in Fig. 2.27a (marked (a) in Fig. 2.26). A $\log_2 K$ size counter and AND gate are used to select the input signal one of K times so we get the zero padded waveform seen in Fig. 2.25b (spectrum shown in Fig. 2.27b). Note that the Nyquist frequency has moved from $f_s/2$ to $Kf_s/2$. The problem is that the images are still

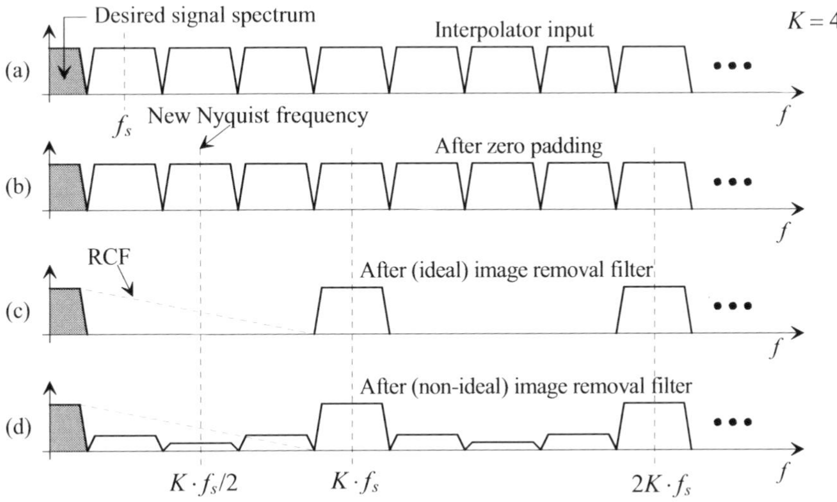

Figure 2.27 Example spectrums when zero padding interpolation is employed.

present in the waveform and thus need to be removed. This is the point of using the image removal filter seen in Fig. 2.26. Figure 2.27c shows the spectrum after increasing the sample rate and passing the signal through an ideal image removal filter. Also seen in this figure is the response of the RCF. It now starts rolling off at $f_s/2$ and can extend up to $Kf_s - f_s/2$. Note that the RCF's response seen in Fig. 2.27a must cut off abruptly at $f_s/2$. Finally, Fig. 2.27d shows the spectrums when a non-ideal digital image removal filter is used. The figure shows the nonzero spectral content remaining in the spectrum after filtering.

Hold Register

Figure 2.28 shows the block diagram of interpolation using an input hold register (which may simply be the input of a digital filter). The benefit of this topology is simplicity. We simply clock the input word K times faster into a register to increase the clocking frequency as seen in Fig. 2.25c. Note how the waveform seen in Fig. 2.25c looks similar to the output of the S/H discussed earlier in the chapter (so we should be expecting some sort of Sinc response effect on our input signal as seen in Eq. [2.16]). In order to quantify this last comment we can write the output of the hold register in terms of its input using

$$y_u\left[i \cdot \frac{T_s}{K}\right] = \sum_{i=K\cdot n}^{K\cdot(n+1)-1} \frac{1}{K} \cdot x\left[i \cdot \frac{T_s}{K}\right] = x[nT_s] \tag{2.30}$$

noting, $K \cdot n \leq i \leq K \cdot (n+1) - 1$, or

$$K \cdot y_u[nT_s] = x\left[Kn \cdot \frac{T_s}{K}\right] + x\left[(Kn+1)\frac{T_s}{K}\right] + x\left[(Kn+2)\frac{T_s}{K}\right] + \ldots + x\left[[K\cdot(n+1)-1]\frac{T_s}{K}\right] \tag{2.31}$$

Writing the z-domain representation (using the input clock as the reference for the delays, or, $z = e^{j2\pi f \cdot T_s}$) of this equation gives

$$Y_u(z) \cdot z^n \cdot K = [1 + z^{1/K} + z^{2/K} + \ldots + z^{(K-1)/K}] \cdot z^n \cdot X(z) \tag{2.32}$$

or, where the z^n represents a shift in time of nT_s,

$$\frac{Y_u(z)}{X(z)} = \frac{1}{K} \cdot [1 + z^{1/K} + z^{2/K} + \ldots + z^{(K-1)/K}] \tag{2.33}$$

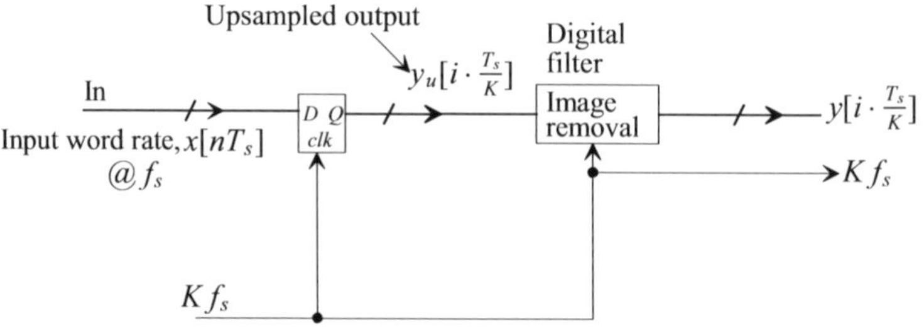

Figure 2.28 An interpolation block using a hold register, see spectrums in Fig. 2.32.

If we multiply the top and bottom of this equation by $1-z^{1/K}$ we get

$$H_u(z)=\frac{Y_u(z)}{X(z)}=\frac{1}{K}\cdot\frac{1-z}{1-z^{1/K}} \tag{2.34}$$

or

$$H_u(f)=\frac{1}{K}\cdot\frac{1-e^{j\cdot 2\pi\frac{f}{f_s}}}{1-e^{j\cdot 2\pi\frac{f}{Kf_s}}} \tag{2.35}$$

Knowing $|1-e^{jx}|=|(1-\cos x)-j\sin x|=\sqrt{(1-\cos x)^2+(-\sin x)^2}=\sqrt{2(1-\cos x)}$ and $1-\cos x=2\sin^2\frac{x}{2}$ we get

$$|H_u(f)|=\left|\frac{1}{K}\cdot\frac{1-e^{j\cdot 2\pi\frac{f}{f_s}}}{1-e^{j\cdot 2\pi\cdot\frac{f}{Kf_s}}}\right|=\frac{1}{K}\cdot\frac{\sqrt{2(1-\cos 2\pi\cdot\frac{f}{f_s})}}{\sqrt{2(1-\cos 2\pi\cdot\frac{f}{Kf_s})}}=\frac{1}{K}\cdot\left|\frac{\sin\left(\pi\cdot\frac{f}{f_s}\right)}{\sin\left(\pi\cdot\frac{f}{Kf_s}\right)}\right| \tag{2.36}$$

If we label the interpolator's output frequency $f_{s,new}=K\cdot f_s$ then

$$|H_u(f)|=\frac{1}{K}\cdot\left|\frac{\sin\left(\pi\cdot\frac{K\cdot f}{f_{s,new}}\right)}{\sin\left(\pi\cdot\frac{f}{f_{s,new}}\right)}\right|=\left|\frac{Sinc\left(\pi\frac{K\cdot f}{f_{s,new}}\right)}{Sinc\left(\pi\frac{f}{f_{s,new}}\right)}\right| \tag{2.37}$$

This equation is sketched in Fig. 2.29. This isn't exactly a Sinc response, Fig. 2.17, but a similar-shaped response. We'll find that this equation also comes up when discussing Sinc response digital filters so let's spend a moment characterizing it. The ratio of the main lobe to the first sidelobe can be determined by evaluating the response at $1.5f_{s,new}/K$ or

$$\left|\frac{\text{Main lobe}}{\text{First sidelobe}}\right|=K\cdot\sin\left(\frac{1.5\pi}{K}\right) \tag{2.38}$$

This equation is plotted in Fig. 2.30 for varying K. Note how the maximum amount of attenuation approaches 13.5 dB (the same as the Sinc response seen in Fig. 2.17).

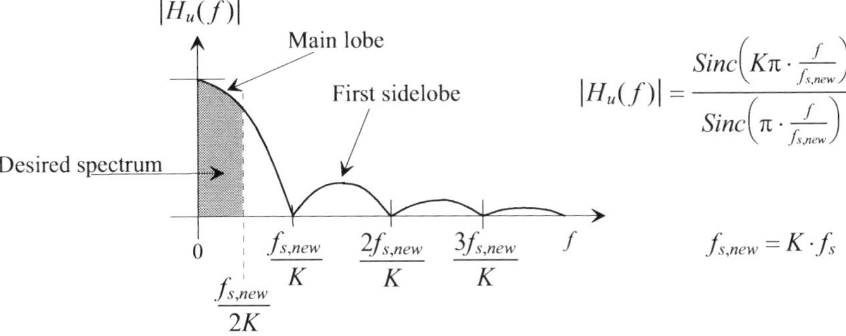

Figure 2.29 Frequency response interpolated data sees using a hold register.

It's also of interest to determine how much droop the filter will introduce into the signal frequencies of interest at the Nyquist frequency. Figure 2.31 shows the droop (attenuation) at the maximum desirable input frequency of $f_s/2$ or $f_{s,new}/(2K)$. We can calculate the amount of droop, again using Eq. (2.37) when $f = f_{s,new}/(2K)$, as

$$\text{Droop} = \frac{1}{K \cdot \sin\left(\frac{\pi}{2K}\right)} \tag{2.39}$$

This equation is plotted in Fig. 2.31. Note that as K gets large the amount of droop approaches 3.9 dB (= 0.64) or the same as the Sinc response seen in Fig. 2.17.

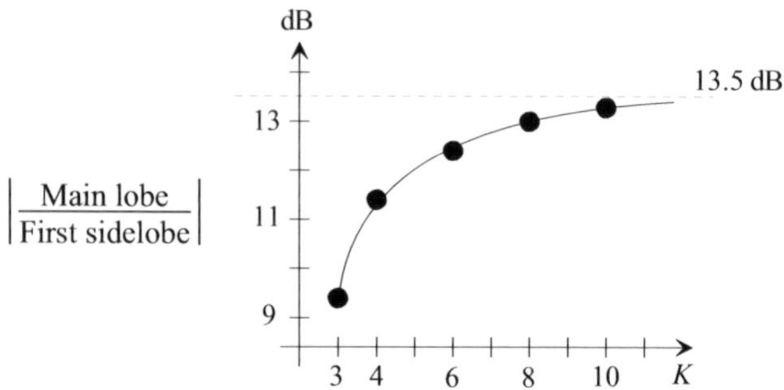

Figure 2.30 Attenuation versus K.

Figure 2.32 shows some example spectrums when $K = 4$ and using an input hold register for interpolation. In (a) we see a representative input spectrum that repeats every f_s. In (b) the input is upsampled to a clocking rate of $K \cdot f_s$. The effects of the input holding register's Sinc response are seen. At this point we have distorted our desired information with the drooping response seen in Fig. 2.31. In (c) the signal's spectrum, after passing through an ideal image removal filter to remove the undesired frequency components, is seen (see also Fig. 2.28). The RCF can have a slow roll-off ultimately passing negligible content at frequencies above $K \cdot f_s - f_s/2$ where the image around the new clocking frequency, $K \cdot f_s$, exists. Finally, in (d) we show what the spectrums may look like with a non-ideal image removal filter. The design of the RCF depends on the allowed amount of unwanted spectral content that can be tolerated in the final output spectrum.

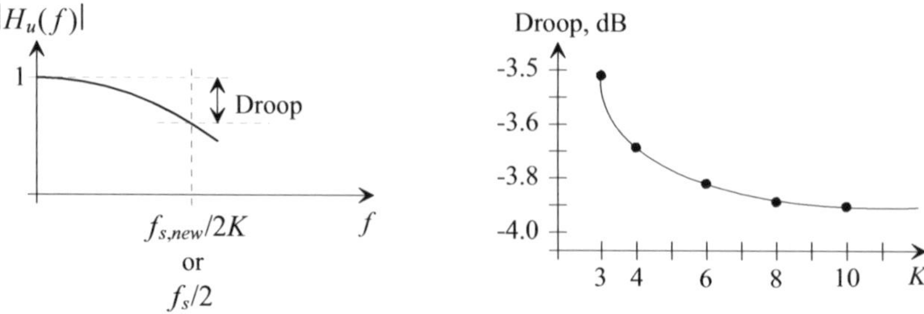

Figure 2.31 Droop at edge of signal bandwidth.

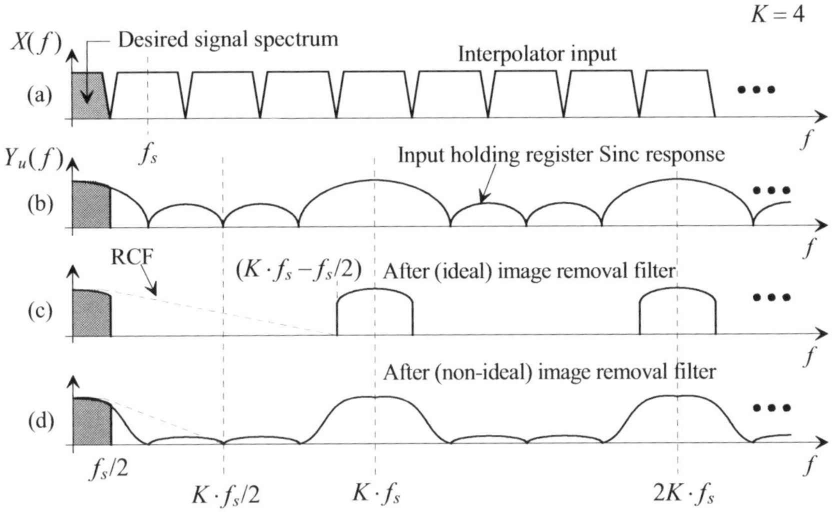

Figure 2.32 Example spectrums when interpolation using a hold register is employed.

Linear Interpolation

The final interpolation scheme we'll look at is linear interpolation, adding samples in between the interpolator's inputs that linearly change with time, Figs. 2.25 and 2.33. We can describe linear interpolation mathematically using

$$y_u\left[(i+1)\cdot\frac{T_s}{K}\right]=y_u\left[i\cdot\frac{T_s}{K}\right]+\frac{x[(n+1)T_s]-x[n\cdot T_s]}{K} \tag{2.40}$$

where $K\cdot n\le i<K\cdot(n+1)$. Note that when $i=K\cdot n$, $y_u\left[i\cdot\frac{T_s}{K}\right]=x[nT_s]=y_u[nT_s]$. Writing this equation in the z-domain we get

$$Y_u(z)\cdot z^{n+1/K}=z^n\cdot\left[Y_u(z)+\frac{X(z)\cdot z-X(z)}{K}\right] \tag{2.41}$$

$$\frac{Y_u(z)}{X(z)}=\frac{1}{K}\cdot\frac{1-z}{1-z^{1/K}} \tag{2.42}$$

which is the same as Eq. (2.34). After reviewing Fig. 2.25 we might have expected linear interpolation to introduce less distortion into the input signal than an interpolator made

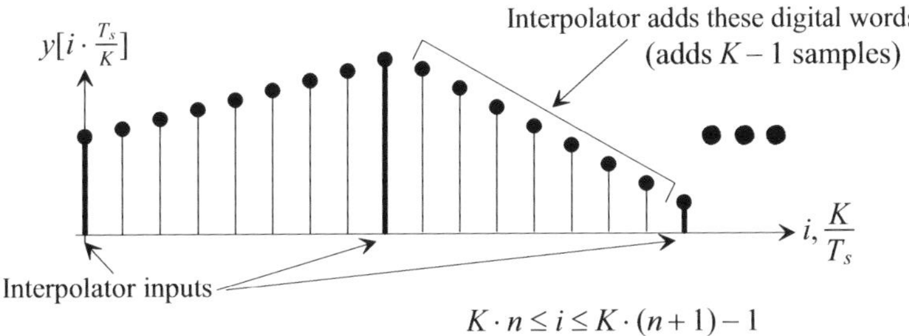

Figure 2.33 Using linear interpolation.

using a hold register. However, the fact that the output of an interpolator made using a hold register sees abrupt changes or steps over time (indicating higher spectral content) results in the equivalence in both interpolator's spectral responses. Note that this type of interpolator is more commonly found in analog signal processing using discrete-time analog circuits.

2.1.6 *K*-Path Sampling

In the last section discussing interpolation (upsampling our signal or increasing the clock rate) the input clock rate was f_s and the output clock rate was Kf_s. If we define z in terms of the *input clock*, $z \equiv e^{j2\pi \cdot \frac{f}{f_s}}$, then a delay on the output of the interpolator is written as $z^{-1/K} = e^{-j2\pi f \cdot \frac{T_s}{K}}$ and a delay on the input of the interpolator is written as z^{-1}. If, on the other hand, we define z in terms of the *output clock*, $z \equiv e^{j2\pi \cdot \frac{f}{Kf_s}}$ then a delay on the output is written z^{-1} while a delay on the input of the interpolator is written as z^{-K}. In simpler terms, a delay on the input lasts T_s seconds while a delay on the output lasts T_s/K seconds. In analog signal processing we can get the same type of behavior, an increase in the output clock frequency (or upsampling the input signal) by using more than one path.

In order to understand this last statement, consider the parallel paths of S/Hs seen in Fig. 2.34. Each S/H is clocked on an opposite phase of an input clock. The resistors are used to sum the outputs of the S/Hs. If the input clock is f_s, then the output signal will change every $T_s/2$ seconds or at a rate of $2f_s$ Hz. By using two-paths we effectively realize an interpolation rate, K, of 2. In other words, we can think of the two-path S/H topology as a single path topology clocked at $2f_s$ Hz. For K paths we can write, again,

$$f_{s,new} = K \cdot f_s \tag{2.43}$$

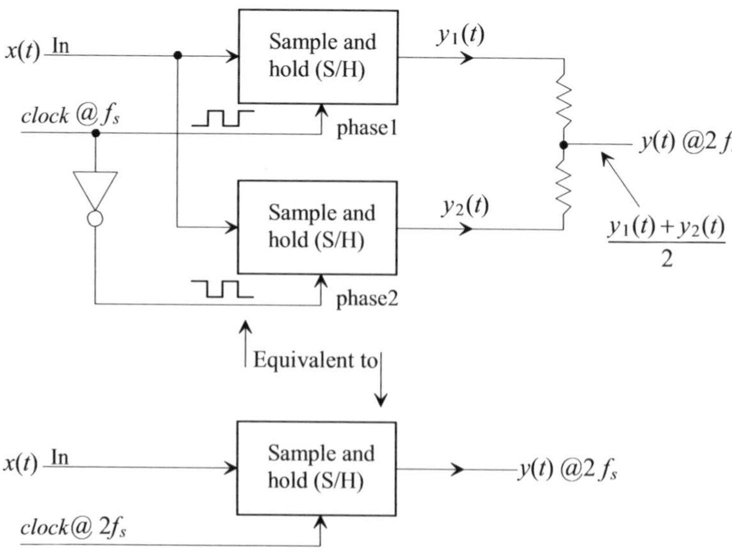

Figure 2.34 Using two S/H paths.

Switched-Capacitor Circuits

Let's show how to apply K-path design to switched-capacitor circuits (commonly used for analog signal processing). Examine the switched-capacitor circuit seen in Fig. 2.35. A switch is closed when a non-overlapping clock (meaning that the clock signals are never high at the same time) signal is high. When the ϕ_1 switch opens at $(n-1/2)T_s$ the charge on the capacitor is

$$Q_1 = v_{in}[(n-1/2)T_s] \cdot C_I \tag{2.44}$$

and when the ϕ_2 switch opens at nT_s the charge on the capacitor is

$$Q_2 = v_{out}[nT_s] \cdot (C_I + C_F) = v_{in}[(n-1/2)T_s] \cdot C_I + v_{out}[(n-1)T_s] \cdot C_F \tag{2.45}$$

Writing this equation in the z-domain we get

$$v_{out}(z) \cdot (C_I + C_F) = v_{in}(z) \cdot z^{-1/2} \cdot C_I + v_{out}(z) \cdot z^{-1} \cdot C_F \tag{2.46}$$

or

$$\frac{v_{out}(z)}{v_{in}(z)} = \frac{C_I \cdot z^{-1/2}}{C_I + C_F - C_F \cdot z^{-1}} \tag{2.47}$$

If we input a DC voltage, $f = 0$ or $z = 1$, then the output is equal to the input (the magnitude of Eq. (2.47) is one). This circuit behaves, for input frequencies $<< f_s$ like a simple RC circuit (as seen in Fig. 2.35). To prove this let's write

$$z = e^{j2\pi f/f_s} \approx 1 + j2\pi\frac{f}{f_s} = 1 + \frac{s}{f_s} \text{ for } f << f_s \tag{2.48}$$

and, knowing the $z^{-1/2}$ term in the numerator is simply a phase shift of one-half clock cycle (which is negligible for input frequencies $<< f_s$),

$$\left|\frac{v_{out}(f)}{v_{in}(f)}\right| = \left|\frac{\frac{C_I}{C_F}}{\frac{C_I}{C_F} + 1 - \left(1 - j2\pi\frac{f}{f_s}\right)}\right| \tag{2.49}$$

or

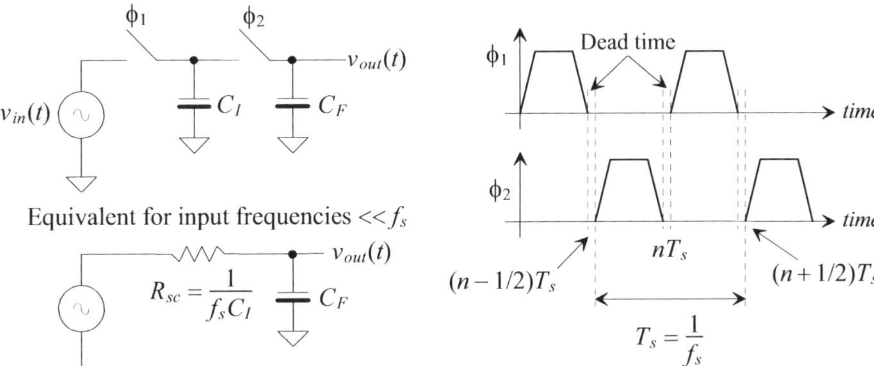

Figure 2.35 Switched-capacitor lowpass filter.

$$\left|\frac{v_{out}(f)}{v_{in}(f)}\right| = \left|\frac{1}{1 + j2\pi C_F \frac{f}{C_I \cdot f_s}}\right| = \frac{1}{\sqrt{1 + (2\pi f \cdot R_{sc} C_F)^2}} \tag{2.50}$$

where we've defined

$$R_{sc} = \frac{1}{f_s C_I} = \frac{T_s}{C_I} \tag{2.51}$$

We can also write the exact frequency response, Eq. (2.47), of this switched-capacitor circuit

$$\frac{v_{out}(f)}{v_{in}(f)} = \frac{\frac{C_I}{C_F} \cdot e^{-j2\pi f/2f_s}}{\frac{C_I}{C_F} + 1 - e^{-j2\pi f/f_s}} = \frac{\frac{C_I}{C_F} \cdot e^{-j2\pi f/2f_s}}{\left(\frac{C_I}{C_F} + 1 - \cos 2\pi f/f_s\right) + j(-\sin 2\pi f/f_s)} \tag{2.52}$$

or

$$\left|\frac{v_{out}(f)}{v_{in}(f)}\right| = \frac{\frac{C_I}{C_F}}{\sqrt{\left(\frac{C_I}{C_F} + 1 - \cos 2\pi f/f_s\right)^2 + \left(\sin^2 2\pi f/f_s\right)}} \tag{2.53}$$

Note that for $f \ll f_s$ the cosine terms is approximately one and the sine term is approximately $2\pi f/f_s$ so this equation simplifies to Eq. (2.50).

Next let's examine the two-path version of Fig. 2.35 shown in Fig. 2.36. The effect of using two paths is to double the output sampling rate, $f_{s,new} = 2f_s$. Using Eq. (2.47) we can write

$$H_{1\text{-}path}(z) = \frac{C_I \cdot z^{-1/2}}{C_I + C_F - C_F \cdot z^{-1}} \tag{2.54}$$

At the new sampling rate we can write

$$H_{2\text{-}path}(z) = \frac{C_I \cdot z^{-1}}{C_I + C_F - C_F \cdot z^{-2}} \tag{2.55}$$

or, in generic terms of K, replace z in the transfer function of the single-path topology with z^K to get the transfer function in the K-path topology,

$$z \to z^K \tag{2.56}$$

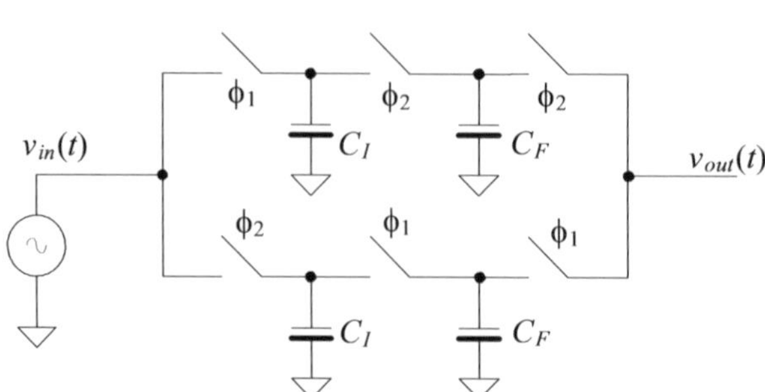

Figure 2.36 Switched-capacitor 2-path lowpass filter.

Figure 2.37 shows K-paths and the equivalent single path topology, a time interleaved topology. Note that at this point there are several important topics we can discuss including path matching and the effects of clock jitter (more later).

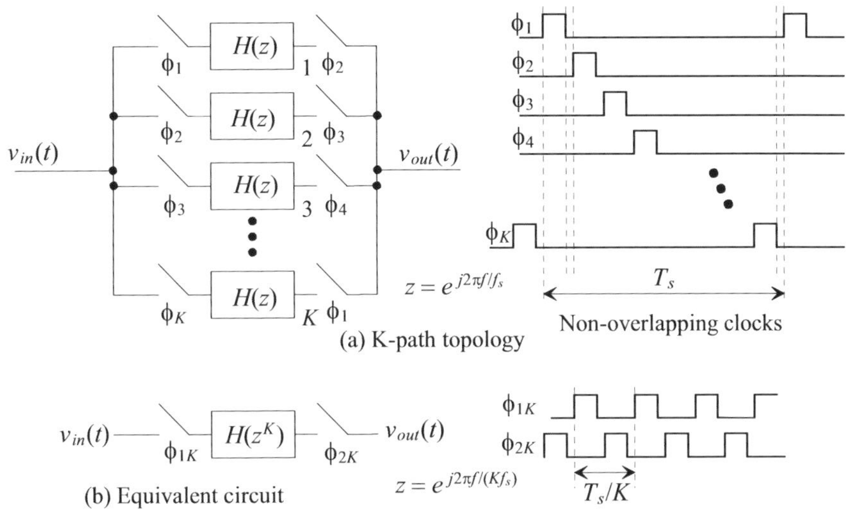

Figure 2.37 A K-path topology and its equivalent circuit.

Non-Overlapping Clock Generation

Figure 2.38 shows a circuit useful for generating four-phase, non-overlapping, clock signals. A shift register is preset so that only one bit is high. The logic block seen in the figure is used to detect if more than one output is high. Note the clear and set inputs of the flip-flops. The amount of non-overlap, or dead time, between pulses is set by the delay through the two inverters connected to the outputs of the NAND gates.

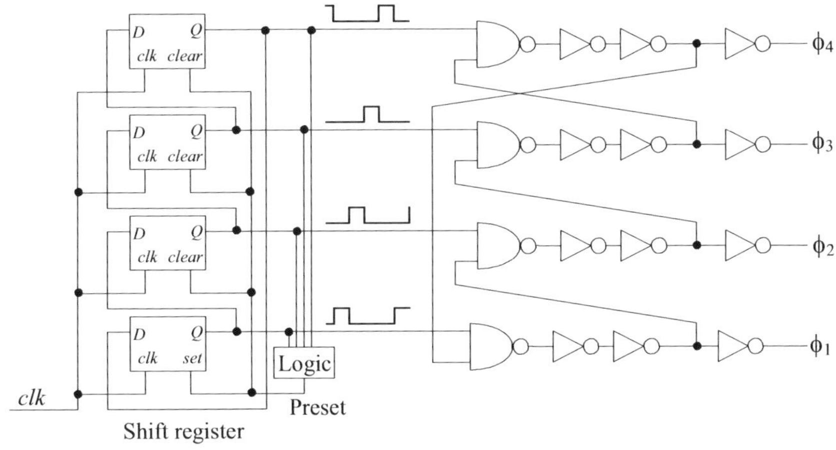

Figure 2.38 Generating a four-phase non-overlapping clock signal.

2.2 Circuits

In this section we discuss the implementation of the S/H and discrete analog integrators (DAIs). The focus is on developing equations and block diagrams for the circuits that will be useful as building blocks in the coming chapters. It's assumed, in this section, that the reader is familiar with the bottom-plate sampling technique seen in Fig. 2.20 and the associated discussion.

2.2.1 Implementing the S/H

A fully-differential mixed-signal S/H based on the topology seen in Fig. 2.20 is seen below in Fig. 2.39. The sample portion of the S/H occurs when the ϕ_1 and ϕ_2 switches are closed and the ϕ_3 switches are open. When the ϕ_3 switches are closed the ϕ_1 and ϕ_2 switches open and the value of the input signal at this instance is "held" until the next time the ϕ_3 switches are closed.

We can determine the relationship between the input of the S/H and its output by writing the charge stored on C_F when the ϕ_1 and ϕ_2 switches are closed (the ϕ_3 switches are open) as

$$Q_F^{\phi_1} = C_F \cdot (v_{in} - V_{CM} \pm V_{OS}) \tag{2.57}$$

where V_{OS} is the offset voltage of the op-amp. When the op-amp is in the follower configuration, the ϕ_1 switches are closed, and the input/output voltages of the op-amp go to $V_{CM} \pm V_{OS}$ (assuming infinite op-amp gain). When the ϕ_3 switches close we can write

$$Q_F^{\phi_3} = C_F \cdot (v_{out} - V_{CM} \pm V_{OS}) \tag{2.58}$$

Since charge must be conserved, $v_{out} = v_{in}$. The input is sampled on the falling edge of ϕ_1.

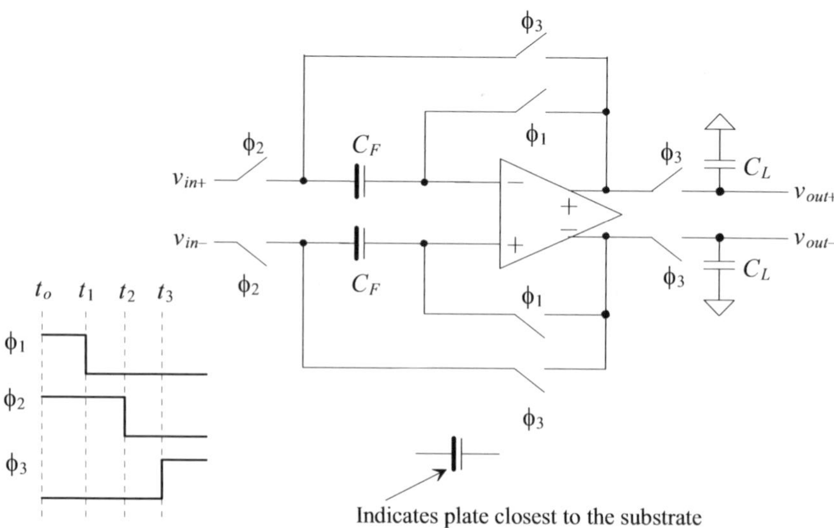

Figure 2.39 Fully-differential S/H differential topology.

Example 2.3
Simulate the operation of the S/H building block seen in Fig. 2.39 assuming C_F = 1 pF and f_s = 100 MHz. Show that the sampled signal isn't affected by the matching of the two capacitors in the S/H or by an op-amp offset.

The simulation results are shown in Fig. 2.40. In part (a) the clock signals are shown. Note the non-overlap time. Unlike the clock signals shown in Fig. 2.39 where the falling edge of ϕ_2 is delayed from ϕ_1, the simulation sets the signals so they go low at the same time. This was to avoid the outputs of the op-amp changing to very large values for the small amount of time the ideal op-amp operates open-loop with an input signal applied.

In part (b) we show the op-amp outputs. Note how, when ϕ_1 goes high, both outputs are set to the common-mode voltage by forcing the op-amp into a follower configuration (which may lead us to use switches to short the terminals of the op-amp to V_{CM} when ϕ_1 is high if offset isn't important, more on this in a moment). When ϕ_3 goes high, the circuit behaves as an S/H. Part (c) of the figure shows the outputs connected through ϕ_3 switches, as seen in Fig. 2.39, driving 10 pF load capacitances.

Introducing a 100% mismatch in the two capacitors by changing one of the values of C_F from 1 pF to 2 pF doesn't affect the simulation results seen in Fig. 2.40. Also, an unrealistically large op-amp offset of 100 mV doesn't affect the S/H's operation. ■

Finite Op-Amp Gain-Bandwidth Product

The previous derivations assumed the op-amp had infinite gain so the one op-amp input was driven to precisely $V_{CM} + V_{OS}$ while the other op-amp input was held at $V_{CM} - V_{OS}$. Op-amp open-loop gain is an important parameter when designing a S/H. Let's write the open-loop gain of an op-amp, assuming a dominant pole, as

$$A_{OL}(f) = \frac{A_{OLDC}}{1 + j \cdot \frac{f}{f_{3dB}}} = \frac{v_{out+} - v_{out-}}{v_{in+} - v_{in-}} \tag{2.59}$$

In nanometer CMOS the DC gain, A_{OLDC}, may be 500 while the $f_{3\mathrm{dB}}$ may be 100 kHz giving a gain-bandwidth product, or unity-gain frequency, of 50 MHz.

The S/H is an example of a feedback system that can be characterized using the classic feedback equation

$$A_{CL} = \frac{A_{OL}}{1 + \beta A_{OL}} \tag{2.60}$$

Since, during the hold operation, all of output signal is fed back to the input $\beta = 1$ and we can write, for DC,

$$A_{CL} = \frac{A_{OLDC}}{1 + A_{OLDC}} \tag{2.61}$$

As the op-amp's open-loop gain becomes very large the S/H's gain moves towards 1. Note that the closed-loop gain will always be less than the desired value. If, in a mixed-signal circuit, VDD is 1-V and a S/H with a resolution better than 1 mV is required, then the open-loop gain of the op-amp can be estimated using

$$\frac{VDD - \text{Resolution}}{VDD} = \frac{1-0.001}{1} < \frac{A_{OLDC}}{1+A_{OLDC}} \tag{2.62}$$

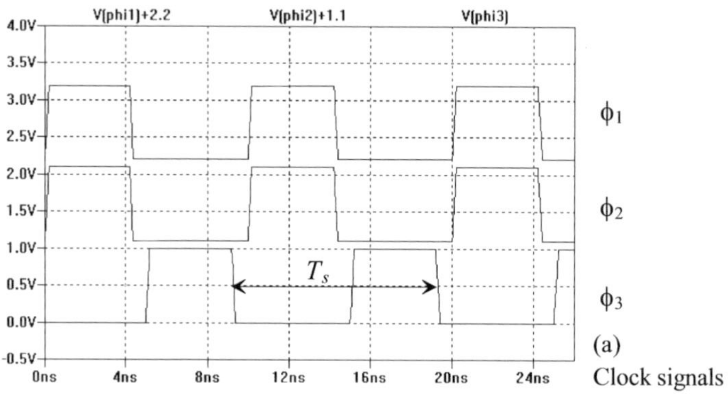

(a)
Clock signals

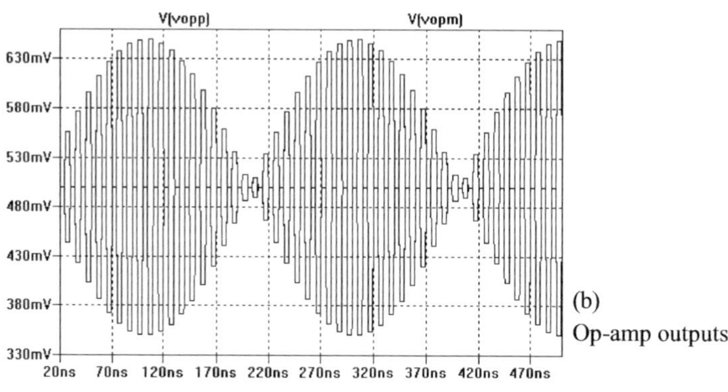

(b)
Op-amp outputs

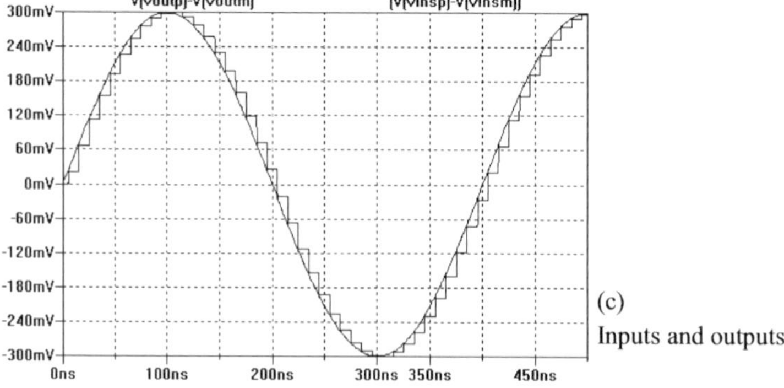

(c)
Inputs and outputs

Figure 2.40 SPICE simulations of the operation of the S/H in Fig. 2.39

or $A_{OLDC} > 1{,}000$. We can write this equation in a more useful form as

$$A_{OLDC} > \frac{VDD}{\text{Resolution}} \tag{2.63}$$

We can estimate the required op-amp gain-bandwidth product (unity-gain frequency f_{un}) $A_{OLDC} f_{3\text{dB}}$, for a S/H by substituting Eq. (2.59) into Eq. (2.60) with $\beta = 1$ (noting that this assumes the op-amp doesn't experience slew-rate limitations and its response is first-order)

$$A_{CL} = \frac{A_{OLDC}}{1 + j \cdot \frac{f}{f_{3dB}} + A_{OLDC}} \tag{2.64}$$

or

$$A_{CL} \approx \frac{1}{1 + j \cdot \frac{f}{A_{OLDC} \cdot f_{3dB}}} = \frac{1}{1 + j \cdot \frac{f}{f_{un}}} \tag{2.65}$$

This first-order system has a time-domain response given by

$$v_{out} = v_{in}(1 - e^{-t \cdot 2\pi \cdot f_{un}}) \tag{2.66}$$

For a given resolution we can write

$$\text{Resolution} = 1 - \frac{v_{out}}{v_{in}} = e^{-t \cdot 2\pi \cdot f_{un}} \tag{2.67}$$

If the settling time must be faster than half of the sampling clock period T_s $(=1/f_s)$, as seen in Fig. 2.40 during ϕ_3, then we can write

$$t_{settling} < \frac{1}{2f_s} = \frac{T_s}{2} \tag{2.68}$$

The minimum gain-bandwidth product of an op-amp used in a S/H is determined using

$$f_{un} = A_{OLDC} \cdot f_{3dB} > \frac{-f_s \cdot \ln(\text{Resolution})}{\pi} \tag{2.69}$$

If the VDD in a mixed-signal system is 1-V, the desired resolution of the S/H is 1 mV, and the sampling frequency is 100 MHz then the unity-gain frequency, f_{un}, must be greater than 210 MHz. If the required DC gain, from Eq. (2.63), is 1,000, then the f_{3dB} of the op-amp is 210 kHz. Of course, the actual DC gain and f_{un} should be much higher than these minimums.

Autozeroing

A single-ended version of the S/H in Fig. 2.39 is seen in Fig. 2.41 including the input referred noise, $V^2_{inoise}(f)$, power spectral density (PSD with units of V^2/Hz) and op-amp offset voltage. We'll represent the input-referred noise in the time-domain using $v_{inoise}(t)$. We've already shown in Ex. 2.3, Eqs. (2.57), and (2.58) that this topology "autozeroes" or removes the offset voltage. When the ϕ_1 switches are closed the op-amp is in the unity-follower configuration and its inputs move to $V_{CM} + V_{OS} + v_{inoise}(t)$. When the ϕ_3 switches close the voltage on the inputs of the op-amp is $V_{CM} + V_{OS} + v_{inoise}(t_3)$. Using Eq. (2.57) we can write, assuming $t_1 \approx t_3$,

$$Q_F^{\phi_1} = C_F \cdot (v_{in}(t_3) - V_{CM} - V_{OS} - v_{inoise}(t_3)) \tag{2.70}$$

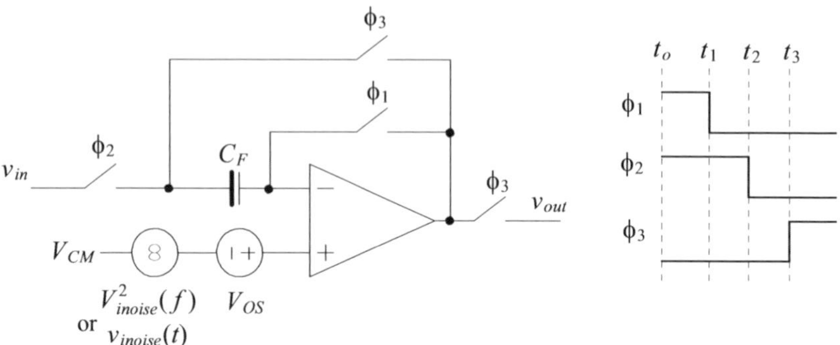

Figure 2.41 S/H with input-referred offset and noise shown.

At a time $T_s/2$ later, the ϕ_3 switches open and the ϕ_1 switches close again to sample the input signal and the noise. Writing the charge on C_F between t_3 and $t_3 + T_s/2$

$$Q_F^{\phi_3} = C_F \cdot (v_{out}(t) - V_{CM} - V_{OS} - v_{inoise}(t)) \tag{2.71}$$

Qualitatively, we can see that if the noise is moving slowly (e.g., Flicker noise) it is removed from the output signal. However, fast moving noise isn't subtracted out during the autozero process. Ultimately the bandwidth of the circuit (say switch resistances and capacitors) and op-amp finite bandwidth limit the frequency content of the noise.

To get a quantitative idea for how the autozero process affects noise in the S/H's output signal we can write

$$v_{out}(t) = v_{in}(t_3) + v_{inoise}(t) - v_{inoise}(t_3) \quad \text{for } t_3 \le t \le t_3 + T_s/2 \tag{2.72}$$

Focusing on the noise and taking the Fourier Transform of each side of this equation gives

$$V_{onoise}(f) \cdot e^{-j2\pi f \cdot t} = V_{inoise}(f) \cdot e^{-j2\pi f \cdot t} - V_{inoise}(f) \cdot e^{-j2\pi f \cdot t_3} \tag{2.73}$$

Note that when t is close to t_3 the output has little noise. The worst case situation is right before the ϕ_3 switches open at a time $t_3 + T_s/2$ (the ϕ_3 switches are on for $T_s/2$ seconds). If we look at this worst-case situation only, then

$$\left| \frac{V_{onoise}}{V_{inoise}} \right| = \left| 1 - e^{-j\pi f \cdot T_s} \right| \tag{2.74}$$

which is essentially the transfer function of a differentiator, Sec. 1.2.2. Note how it would be straightforward to extend this derivation to any arbitrary time that the ϕ_3 switches are on. Borrowing the results seen in Eq. (1.46) we get a noise transfer function, *NTF*, of

$$NTF = \left| \frac{V_{onoise}}{V_{inoise}} \right| = 2 \cdot \left| \sin \frac{\pi}{2} \cdot f \cdot T_s \right| \tag{2.75}$$

This equation is plotted in Fig. 2.42 along with the response of the S/H. Note that at DC (where the op-amp's offset voltage is located) the output of the S/H is noise free. As alluded to earlier, autozeroing works well for reducing the effects of Flicker noise (a low frequency noise that is common in CMOS integrated circuits).

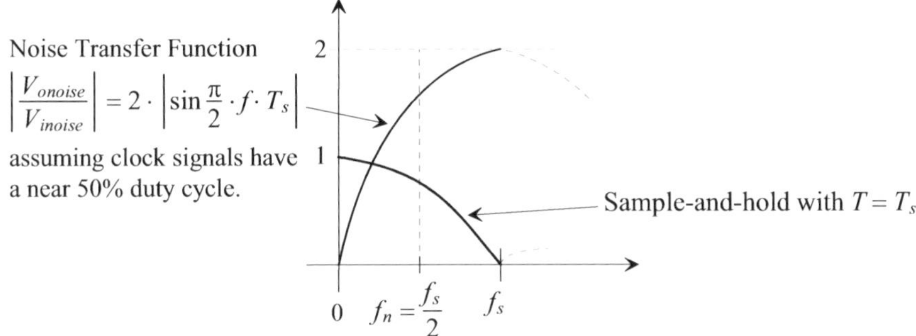

Figure 2.42 The noise transfer function of a S/H.

In the preceding discussion we didn't include the effects of sampling on the output noise spectrum. However, at this point, replicating the noise spectrum at multiples of f_s and weighting the output noise by the S/H response should be straightforward, Fig. 2.43. Having said this, however, note that if our op-amp's input-referred noise PSD isn't bandlimited to the Nyquist frequency, $f_s/2$, as it is in Fig. 2.43 then noise will alias into DC to $f_s/2$ range (the desired range).

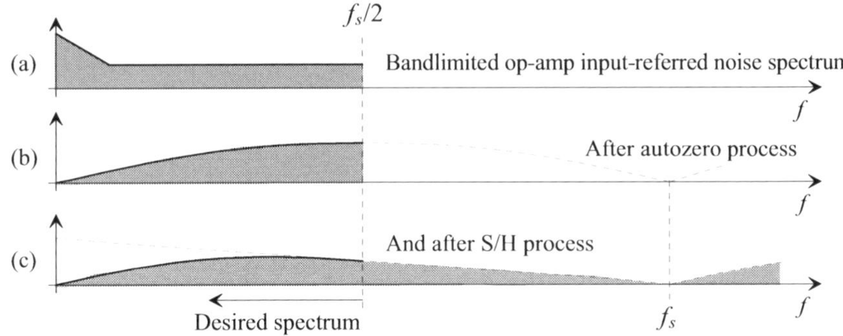

Figure 2.43 Example spectrums when S/H in Fig. 2.39 is used.

Correlated Double Sampling (CDS)

Correlated Double Sampling (CDS) is a name used for describing the autozero process followed by a S/H. The S/H in Fig. 2.39 thus employs CDS. Both the input signal and the noise/offset are sampled (double sample). Then the offset/noise is subtracted from the output signal (the correlation) to give an output signal with less noise and ideally no offset.

As an example of CDS, Fig. 2.44 shows an input-referred noise signal and offset along with a S/H input and output. The op-amp's input referred offset is about –220 mV and the input-referred noise has a peak-to-peak variation of about 80 mV. (Remember offset and noise are always measured on the output of a circuit and referred back to its input.) While we can see noise in the S/H's output it is clear that it has been reduced.

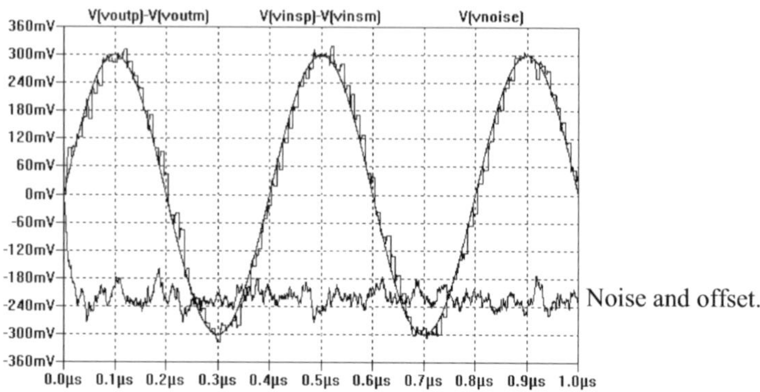

Figure 2.44 Showing how CDS reduces noise and offset in the S/H in Fig. 2.39.

Figure 2.45 shows a S/H that doesn't employ CDS. This topology is used to help ensure the CMFB circuits and biasing in the op-amp are more tolerant to offsets. When the ϕ_1 switches are on, the inputs of the op-amp are shorted to the common-mode voltage, V_{CM}, and the outputs are shorted together. Note that the op-amp settling time isn't a factor in the design during this hold portion of the S/H process. At this point in time we can write

$$Q_F^{\phi_1} = C_F \cdot (v_{in} - V_{CM}) \tag{2.76}$$

When the ϕ_3 switches turn on, the op-amp moves into the follower configuration and the op-amp inputs move to $V_{CM} \pm V_{OS}$. During this time we can write, see Eq. (2.58),

$$Q_F^{\phi_3} = C_F \cdot (v_{out} - V_{CM} \pm V_{OS}) \tag{2.77}$$

Since charge must be conserved

$$v_{out} = v_{in} \pm V_{OS} \tag{2.78}$$

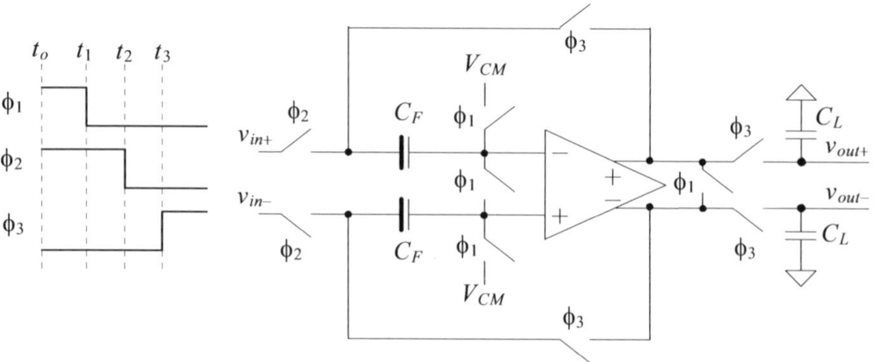

Figure 2.45 Fully-differential S/H differential topology without using CDS.

So why would we use this configuration? Neither the offset or noise would be reduced which is a significant disadvantage over the topology in Fig. 2.39. The answer is that since the inputs of the op-amp are driven to a known voltage (V_{CM}) and the outputs are driven to a known voltage via the op-amp's common-mode feedback circuit (again V_{CM}) ensuring good biasing and stable CMFB loops in the op-amp are easier to attain. Note that in a two-stage op-amp design we would also short the outputs of the first-stage diff-amp together so the inputs of output buffer are forced to a known value (the ideal output voltage of the diff-amp). Design of fully-differential op-amps is discussed in detail, as is our next topic, in the book *CMOS Circuit Design, Layout, and Simulation.*

Selecting Capacitor Sizes

The selection of the capacitor sizes in the S/H is based on thermal noise considerations, *kT/C* (kay tee over cee), and settling time. Small capacitors result in lower power circuits and faster settling times but, at the same time, increase the thermal noise floor. For a detailed discussion of *kT/C* noise, as well as other circuit noise topics, see *CMOS Circuit Design Layout, and Simulation.* Table 2.1 shows the relationship between various capacitor sizes and corresponding thermal noise for quick reference.

Table 2.1 Capacitor size and corresponding *kT/C* noise at 300 °K.

Capacitor size, pF	$\sqrt{kT/C}$, μV RMS	$\sqrt{kT/C}$, mV peak-to-peak
0.01	640	3.84
0.1	200	1.2
1	64	0.384
10	20	0.12
100	6.4	0.038

2.2.2 The S/H with Gain

Figure 2.46 shows a S/H with gain. Following the derivations from the last section we can write

$$Q_{I,F}^{\phi_1} = C_I \cdot (v_{in} - V_{CM} \pm V_{OS}) + C_F \cdot (v_{in} - V_{CM} \pm V_{OS}) \tag{2.79}$$

and when the ϕ_3 switches turn on

$$Q_I^{\phi_3} = C_I \cdot (V_{CM} - V_{CM} \pm V_{OS}) \tag{2.80}$$

and

$$Q_F^{\phi_3} = C_F \cdot (v_{out} - V_{CM} \pm V_{OS}) \tag{2.81}$$

Knowing charge must be conserved

$$Q_F^{\phi_3} = C_F \cdot (v_{out} - V_{CM} \pm V_{OS}) =$$

$$= \overbrace{C_F \cdot (v_{in} - V_{CM} \pm V_{OS})}^{Q_F^{\phi_1}} + \overbrace{C_I \cdot (v_{in} - V_{CM} \pm V_{OS})}^{Q_I^{\phi_1}} - \overbrace{C_I \cdot (V_{CM} - V_{CM} \pm V_{OS})}^{Q_I^{\phi_3}} \tag{2.82}$$

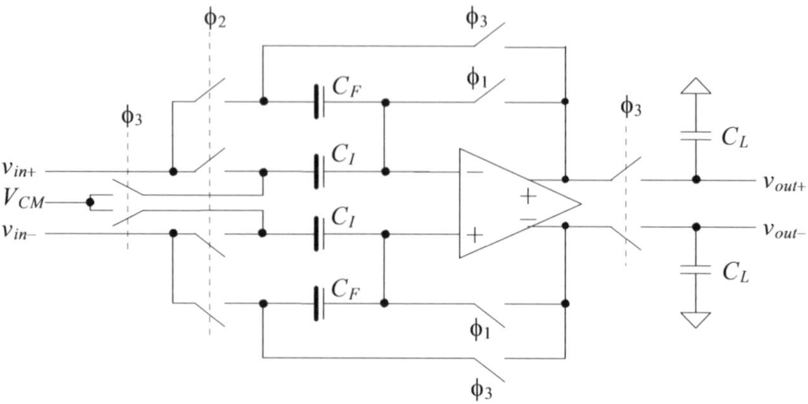

Figure 2.46 A S/H with gain.

or

$$v_{out} = \left(1 + \frac{C_I}{C_F}\right) \cdot v_{in} - \frac{C_I}{C_F} \cdot V_{CM} \tag{2.83}$$

For a fully-differential topology the last term is common to both the inverting and non-inverting inputs of the S/H so we can write

$$v_{out+} - v_{out-} = \left(1 + \frac{C_I}{C_F}\right) \cdot (v_{in+} - v_{in-}) \tag{2.84}$$

A block diagram for the S/H in Fig. 2.46 is shown in Fig. 2.47. Note that this op-amp topology employs CDS. Also note, though we didn't derive it in the last section (because it ends up being negligible in most circuits), the residual offset after autozeroing is V_{OS}/A_{OLDC}. An op-amp DC gain of 1,000 and an op-amp offset of 50 mV results in a 50 μV residual offset when employing CDS.

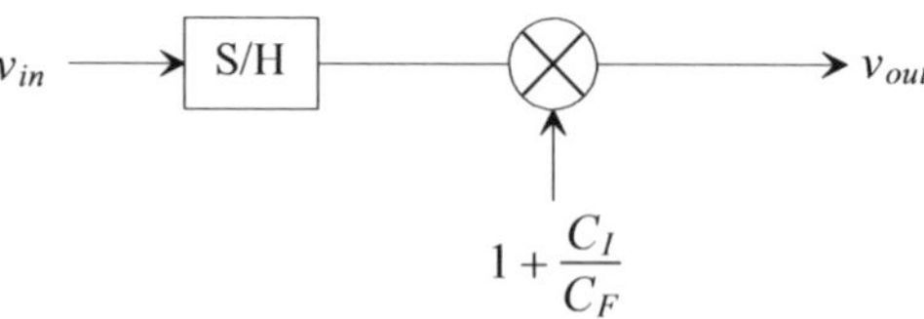

Figure 2.47 Block diagram for the S/H of Fig. 2.46.

Example 2.4

Simulate the operation of the data converter S/H building block shown in Fig. 2.46. Assume $C_I = C_F = 1$ pF and $f_s = 100$ MHz.

The simulation results are shown in Fig. 2.48. The gain, as we would expect, is 2. It may be useful at this point to simulate this circuit with an offset or noise like we did in Ex. 2.3. ■

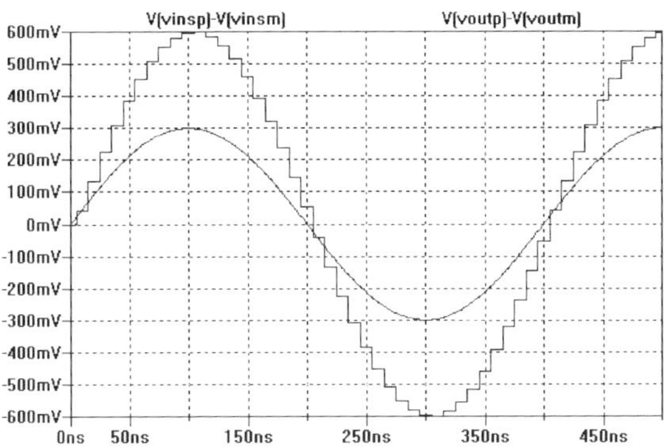

Figure 2.48 Simulating the S/H seen in Fig. 2.46 with a gain of 2.

Implementing Subtraction in the S/H

We'll see later, when covering Nyquist-rate data converters, that it is useful to implement subtraction in the S/H. Consider what would happen if instead of connecting the bottom plates of the C_I capacitors in Fig. 2.46 to V_{CM} we connect them to V_{CI+} and V_{CI-} (see Fig. 2.49). Doing this results in

$$Q_I^{\phi_3} = C_I \cdot (V_{CI+} - V_{CM} \pm V_{OS}) \tag{2.85}$$

or, after reviewing Eqs. (2.82) - (2.83)

$$v_{out+} = \left(1 + \frac{C_I}{C_F}\right) \cdot v_{in+} - \frac{C_I}{C_F} \cdot V_{CI+} \tag{2.86}$$

The differential output voltage is then given by

$$v_{out} = v_{out+} - v_{out-} = \left(1 + \frac{C_I}{C_F}\right) \cdot (v_{in+} - v_{in-}) - \frac{C_I}{C_F} \cdot (V_{CI+} - V_{CI-}) \tag{2.87}$$

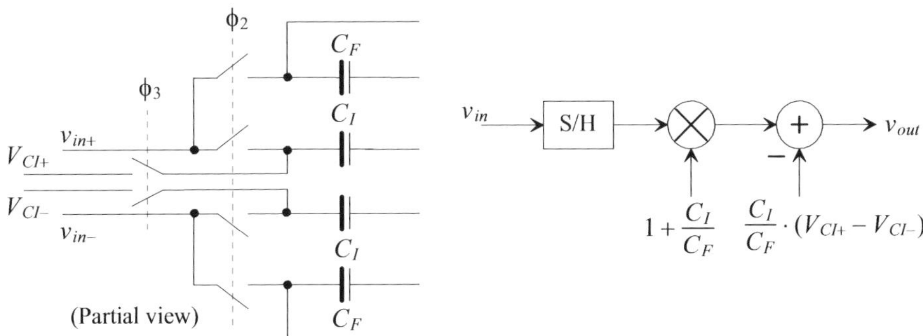

Figure 2.49 Implementing subtraction in the S/H.

Rearranging the block diagram seen in Fig. 2.49 results in the topology seen in Fig. 2.50.

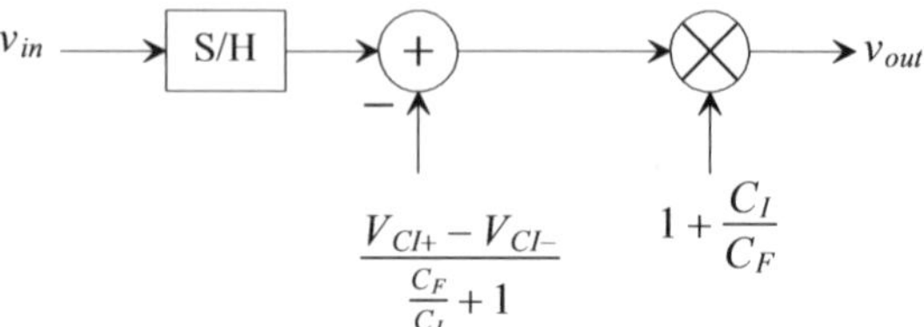

Figure 2.50 Block diagram of Fig. 2.49 with bottom plates of C_I tied to V_{CI}.

Example 2.5

Simulate the operation of the S/H shown in Fig. 2.51 if f_s = 100 MHz, $C_F = C_I = 1$ pF, $V_{CI+} = 1.5V_{CM}$, and V_{CI-} is $0.5V_{CM}$ (V_{CM} = 500 mV). Comment on the resulting output.

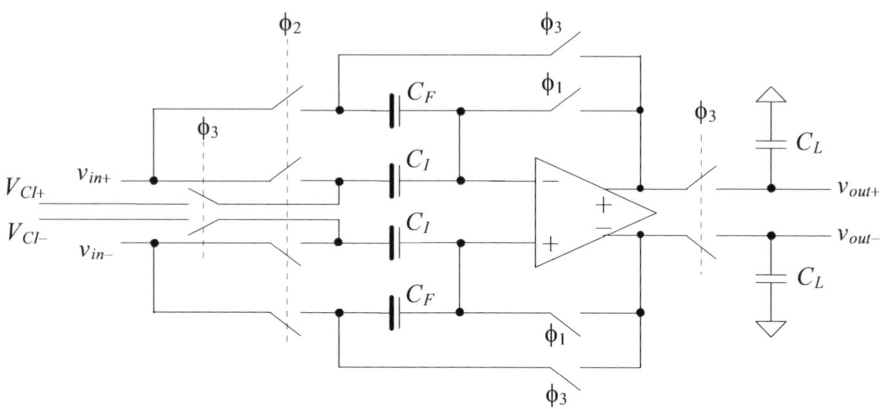

Figure 2.51 S/H used in Ex. 2.5.

The simulation results are shown in Fig. 2.52. We only show the situation when we would want to subtract $V_{CM}/2$ from the differential input signal. The inputs are fully-differential, swinging around the common-mode voltage of 500 mV with an amplitude of 100 mV (so they swing between 600 mV and 400 mV). The largest differential voltage is 600 mV – 400 mV or +200 mV while the smallest differential signal is 400 mV – 600 mV or –200 mV (so the differential signal swings around ground with an amplitude of 200 mV). Reviewing the block diagram in Fig. 2.50 with the values in this problem shows that the circuit takes this input signal, sample-and-holds it, subtract $V_{CM}/2$ (250 mV) then multiplies it by 2. We've scaled the output in Fig. 2.52 to show that this sequence of events is indeed what is happening. Note that if we were to switch V_{CI+} and V_{CI-} we would add $V_{CM}/2$ to the input signal. ■

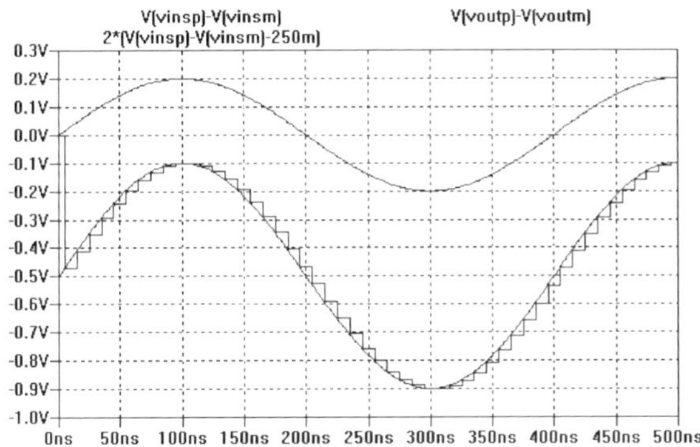

Figure 2.52 Simulation results for Ex. 2.5.

A Single-Ended to Differential Output S/H

Many input signals are single-ended, meaning the (one) input signal swings around V_{CM}. It is desirable in the first stage of the mixed-signal circuit to S/H the signal and then to change it into a fully-differential signal for further processing. A good single-to-differential converter will hold the op-amp's input common-mode voltage at V_{CM} for low distortion (important). In order to meet this goal consider the modified, from Fig. 2.46, S/H seen in Fig. 2.53. Again, we can write, (noting the C_F capacitors are discharged when the ϕ_1 switches are on)

$$Q_{I,F,total}^{\phi_1} = C_{I+} \cdot (v_{in} - V_{CM} \pm V_{OS}) + C_{I-} \cdot (V_{CM} - V_{CM} \pm V_{OS}) \tag{2.88}$$

When the ϕ_3 switches turn on the charge on the C_I capacitors is

$$Q_I^{\phi_3} = C_{I+} \cdot \left(\frac{v_{in}}{2} + \frac{V_{CM}}{2} - V_{CM} \pm V_{OS}\right) + C_{I-} \cdot \left(\frac{v_{in}}{2} + \frac{V_{CM}}{2} - V_{CM} \pm V_{OS}\right) \tag{2.89}$$

noting that the change in charge on the C_I capacitors redistributes through the C_F capacitors. The charge on the feedback capacitors is then

$$Q_F^{\phi_3} = 2C_F \cdot (v_{out} - V_{CM} \pm V_{OS}) \tag{2.90}$$

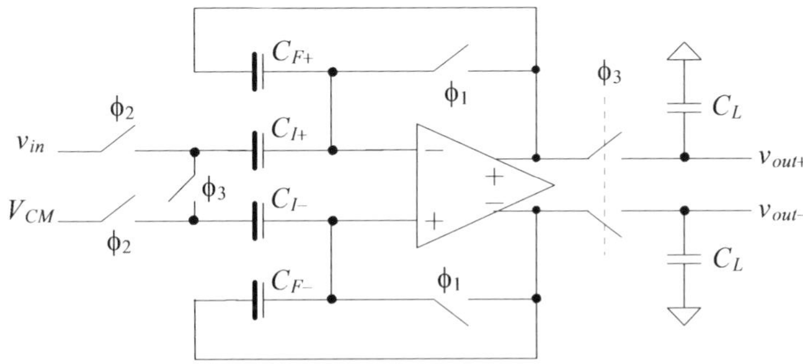

Figure 2.53 S/H for single-ended to differential conversion.

Equating the redistributed charge through C_{F+} and C_{I+}

$$C_{I+}\cdot(v_{in}-V_{CM}\pm V_{OS})-C_{I+}\cdot\left(\frac{v_{in}}{2}+\frac{V_{CM}}{2}-V_{CM}\pm V_{OS}\right)=C_{F+}\cdot(v_{out+}-V_{CM}\pm V_{OS})$$

or, re-writing for both paths (2.91)

$$C_{F+}\cdot(v_{out+}-V_{CM}\pm V_{OS})=C_{I+}\cdot\left(\frac{v_{in}}{2}-\frac{V_{CM}}{2}\right)$$

$$C_{F-}\cdot(v_{out-}-V_{CM}\pm V_{OS})=C_{I-}\cdot\left(-\frac{v_{in}}{2}+\frac{V_{CM}}{2}\right) \quad (2.92)$$

Assuming $C_{I+}=C_{I-}$, $C_{F+}=C_{F-}$, and taking the difference in the S/H outputs gives

$$\frac{(v_{out+}-v_{out-})\pm 2V_{OS}}{v_{in}-V_{CM}}=\frac{C_I}{C_F} \quad (2.93)$$

The output is shifted up or down by the offset. Note that this topology doesn't employ correlated double sampling (CDS). Also note that if we model the op-amp's offset with a single voltage source in series with the non-inverting input of the op-amp then one of the inputs will go to $V_{CM}+V_{OS}/2$ while the other input will go to $V_{CM}-V_{OS}/2$ (we've just indicated that the inputs of the op-amp are at a potential of $V_{CM}\pm V_{OS}$). Hence the factor of two in Eq. (2.93). In other words, if we re-write all of the equations in this chapter by replacing V_{OS} with $V_{OS}/2$ then the factor of two in Eq. (2.93) will go away. Simulations at CMOSedu.com are invaluable to understanding the operation of the circuits in this chapter. For example, see the simulation for Fig. 2.53.

2.2.3 The Discrete Analog Integrator (DAI)

The final sampling circuit we'll discuss in this chapter is an analog building block that we will find useful in implementing our data converters using feedback. The discrete analog integrator, DAI, is shown in Fig. 2.54. The two clocks signals, ϕ_1 and ϕ_2, form nonoverlapping clock signals. The common mode voltage, V_{CM}, falls halfway between the mixed-signal system's high- and low-reference voltages (generally *VDD* and ground). Note that the parasitic capacitance to ground associated with the bottom-plate of C_I is charged back and forth between v_1 and v_2 but doesn't affect the amount of charge transferred to the feedback capacitor, C_F. For this reason this DAI is often called a *parasitic-insensitive integrator*.

Table 2.2 shows the various relationships between the possible inputs and outputs for the DAI of Fig. 2.54. Let's derive the input/output relationships for the most general situations where both v_1 and v_2 are the inputs.

Table 2.2 Discrete analog integrator (DAI) input/output relationships.

Input	Output connected to ϕ_1	Output connected to ϕ_2
v_1 = input and $v_2=V_{CM}$	$\frac{z^{-1}}{1-z^{-1}}\cdot\frac{C_I}{C_F}$	$\frac{z^{-1/2}}{1-z^{-1}}\cdot\frac{C_I}{C_F}$
v_2 = input and $v_1=V_{CM}$	$\frac{-z^{-1/2}}{1-z^{-1}}\cdot\frac{C_I}{C_F}$	$\frac{-1}{1-z^{-1}}\cdot\frac{C_I}{C_F}$
v_1 and v_2 are both inputs	$\frac{V_1(z)\cdot z^{-1}-V_2(z)\cdot z^{-1/2}}{1-z^{-1}}\cdot\frac{C_I}{C_F}$	$\frac{V_1(z)\cdot z^{-1/2}-V_2(z)}{1-z^{-1}}\cdot\frac{C_I}{C_F}$

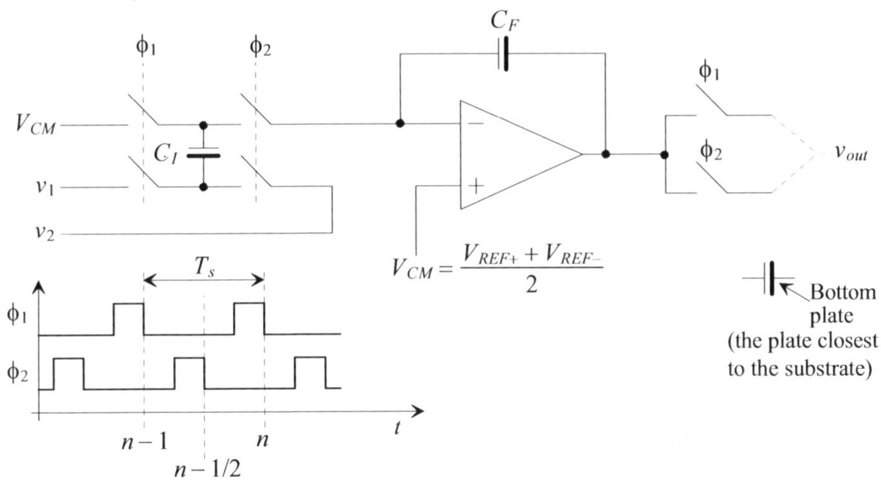

Figure 2.54 Schematic diagram of a discrete analog integrator (DAI).

To begin, let's assume the output of the DAI is connected to the op-amp through the ϕ_1 switch. When the ϕ_1 switches are closed (ϕ_1 is high) at $n-1$ (the instance when the switches shut off), the charge stored on C_I is

$$Q_1 = C_I(V_{CM} - v_1[(n-1)T_s]) \tag{2.94}$$

and the output of the integrator is $v_{out}[(n-1)T_s]$. When the ϕ_2 switches turn on the charge stored on C_I becomes

$$Q_2 = C_I(V_{CM} - v_2[(n-1/2)T_s]) \tag{2.95}$$

remembering that the op-amp holds its noninverting input terminal at V_{CM}. The difference in these charges, $Q_2 - Q_1$, is transferred to the op-amp's feedback capacitor resulting in an output voltage change. This change can be written as

$$(v_{out}[nT_s] - v_{out}[(n-1)T_s])C_F = C_I(v_1[(n-1)T_s] - v_2[(n-1/2)]T_s) \tag{2.96}$$

or writing this equation in the z-domain results in

$$v_{out}(z)(1-z^{-1}) = \frac{C_I}{C_F}(v_1(z)\cdot z^{-1} - v_2(z)\cdot z^{-1/2}) \tag{2.97}$$

The transfer function of the DAI with the output connected to the ϕ_1 switches is then

$$v_{out}(z) = \frac{C_I}{C_F}\cdot\frac{v_1(z)\cdot z^{-1} - v_2(z)\cdot z^{-1/2}}{1-z^{-1}} \tag{2.98}$$

Similarly, if we connect the output through the ϕ_2 switches (the edges we label n in Fig. 2.54 shift in time by $T_s/2$) we can write

$$Q_1 = C_I(V_{CM} - v_1[(n-1/2)T_s]) \tag{2.99}$$

$$Q_2 = C_I(V_{CM} - v_2[nT_s]) \tag{2.100}$$

and

$$(v_{out}[nT_s] - v_{out}[(n-1)T_s])C_F = C_I(v_1[(n-1/2)T_s] - v_2[nT_s]) \tag{2.101}$$

The transfer function of the DAI with the output connected to the ϕ_2 switches is then

$$v_{out}(z) = \frac{C_I}{C_F} \cdot \frac{v_1(z) \cdot z^{-1/2} - v_2(z)}{1 - z^{-1}} \tag{2.102}$$

Note that if $v_2(z) = V_{CM}$, this equation can be written as

$$H(z) = \frac{v_{out}(z)}{v_1(z)} = \frac{C_I}{C_F} \cdot \frac{z^{-1/2}}{1 - z^{-1}} \tag{2.103}$$

which has a frequency response, $|H(f)|$, shown in Fig. 1.27. Note that the factor C_I/C_F simply scales the amplitude response. If this factor is unity then the magnitude response, as shown in Fig. 1.27, is 0.5 at $f_s/2$. The $z^{-1/2}$ term in the numerator simply modifies the phase response of the DAI (delaying the output by $T_s/2$ or -180 degrees) and has no effect on the magnitude response. Note that at this point we could discuss the frequency responses of the transfer functions given in Table 2.2. However, we would see that the discussions and results given in Ch.1 for the digital integrator would apply to the DAI with little, or no, modifications.

Example 2.6

Determine the transfer function of the DAI of Fig. 2.54 *without* the switches on the output of the op-amp.

Reviewing Fig. 2.54 we see that charge is transferred to the feedback capacitor only when the ϕ_2 switches are closed. Therefore, the output only changes states during the time interval when the ϕ_2 switches are closed. The transfer function of the DAI, when no switches are used on the output of the op-amp, is given by Eq. (2.102). Using the ϕ_1 switches simply adds a half clock cycle delay, $z^{-1/2}$, to the integrator's transfer function (instead of the output changing with the rising edge of ϕ_2, the output changes one-half cycle later on the rising edge of ϕ_1). ■

A Note Concerning Block Diagrams

As we draw block diagrams describing our modulator topologies in this chapter and the next we often show a circuit like the one shown in Fig. 2.55. The summation, gain, and integrating blocks are implemented with a single switched-capacitor DAI having the transfer function given by Eq. (2.102). The gain, G, of the DAI is set by the ratio of

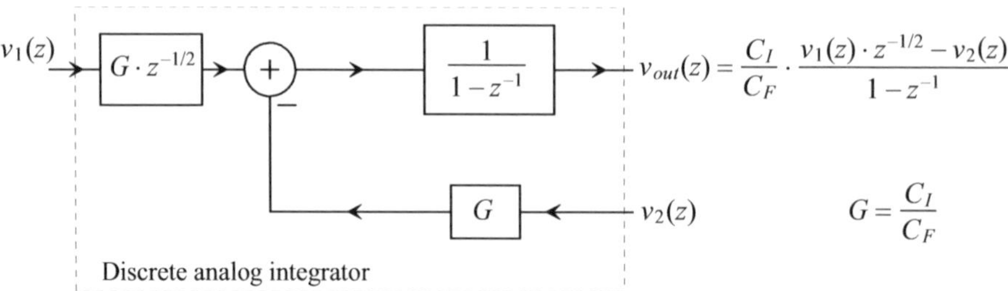

Figure 2.55 Block diagram of a DAI.

capacitors as indicated in the figure. It's important to realize that this circuit is entirely analog and is interfaced to, in general, both ADCs ($v_{out}[z]$ is connected to the input of an ADC) and DACs ($v_2[z]$ is connected to the output of a DAC).

It should be clear from both Fig. 2.54 and Table 2.2 that many different combinations of discrete analog building blocks are possible. Figure 2.56 shows two other possibilities. In part (a) the capacitors used have the same parasitic capacitance on each plate so there is no benefit to using a parasitic insensitive topology. The transfer function of this DAI is

$$v_{out}(z) = \frac{C_I}{C_F} \cdot \frac{z^{-1}}{1-z^{-1}} \cdot (v_2(z) - v_1(z)) \tag{2.104}$$

noting each input signal sees the same delay, i.e., z^{-1} when the outputs are connected through ϕ_1 controlled switches and $z^{-1/2}$ delay when no switches or ϕ_2 controlled switches are used. If the integrator inputs must see the same delay and the capacitors available have asymmetric parasitic capacitance, the topology of Fig. 2.56b can be used. Its transfer function is

$$v_{out}(z) = \frac{z^{-1}}{1-z^{-1}} \cdot \left(\frac{C_{I1}}{C_F} \cdot v_1(z) + \frac{C_{I2}}{C_F} \cdot v_2(z) \right) \tag{2.105}$$

noting the input signals can be scaled independently (a useful feature in filter design and discussed further in the next chapter).

Fully-Differential DAI

While we've derived the equations governing the operation of the DAI using a single-ended topology, in most practical circuits we'll use fully-differential implementations. The same equations apply to both configurations. Figure 2.57 shows the schematic for the fully-differential DAI. Note how we keep the bottom plates of the capacitors away from the op-amp's inputs.

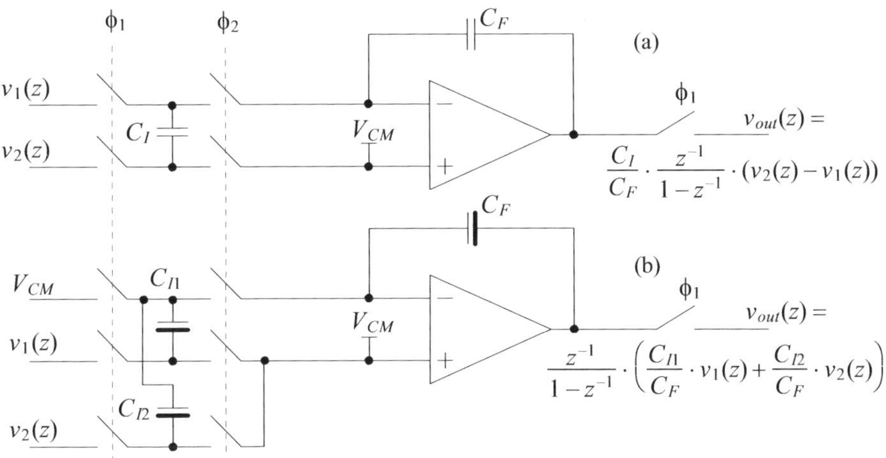

Figure 2.56 Other forms of DAIs.

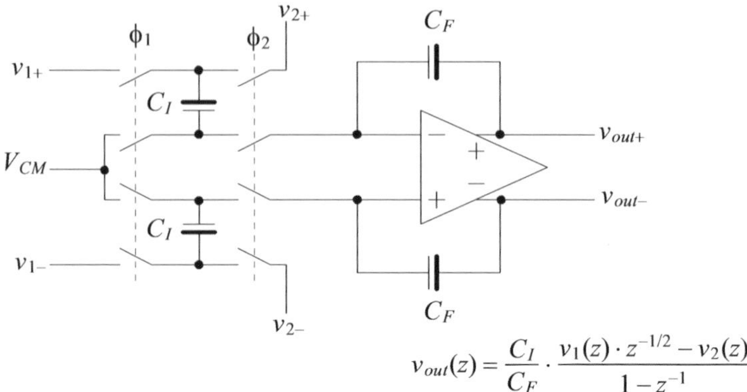

Figure 2.57 Fully-differential discrete-analog integrator (DAI) implementation.

DAI Noise Performance

Figure 2.58 shows the DAI with kT/C noise sources shown (see Table 2.1). The mean squared input- referred noise is given by

$$V_{in,RMS}^2 = \frac{kT}{C_I} \text{ (units of } V^2\text{)} \tag{2.106}$$

in series with both v_1 and v_2. A total of $2kT/C_I$ is sampled onto C_I during each clock cycle. If the input signal can swing from VDD to ground (peak value of the input is $VDD/2$ while the RMS value of this input is $VDD/\left(2\sqrt{2}\right)$), then we can estimate the SNR using

$$\text{SNR} = 20 \log \frac{VDD/\left(2\sqrt{2}\right)}{\sqrt{2kT/C_I}} \tag{2.107}$$

This equation is useful for determining the capacitor values used in a DAI.

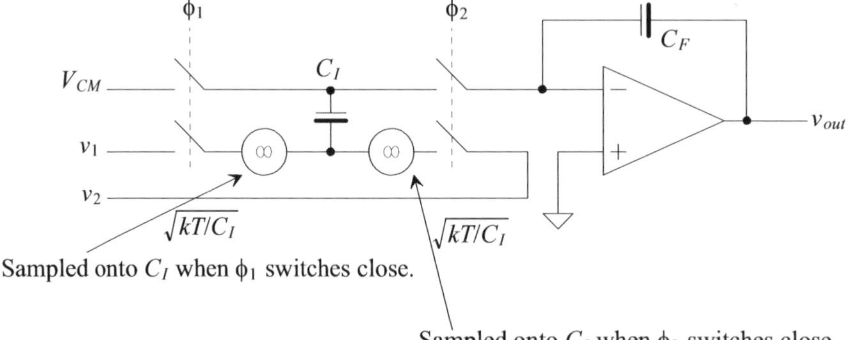

Figure 2.58 Noise performance of the DAI.

ADDITIONAL READING

[1] R. J. Baker, *CMOS: Circuit Design, Layout, and Simulation, Revised Second Edition*, Wiley-IEEE, 2008. ISBN 978-0470229415

[2] C. C. Enz and G. C. Temes, "Circuit Techniques for Reducing the Effects of Op-Amp Imperfections: Autozeroing, Correlated Double Sampling, and Chopper Stabilization," *Proceedings of the IEEE*, Vol. 84, No. 11, pp. 1584-1614, November 1996.

[3] L. W. Couch, *Modern Communication Systems: Principles and Applications*, Prentice-Hall, 1995. ISBN 978-0023252860

QUESTIONS

2.1 Qualitatively, using figures, show how impulse sampling a sinewave can result in an alias of the sampled sinewave at a different frequency.

2.2 Re-sketch Figs. 2.12 and 2.13 when decimating by 5. hint: use a counter and some logic to implement the divide by 5 clock divider.

2.3 Explain why returning the output of the S/H to zero reduces the distortion introduced into a signal. What is the cost for the reduced distortion in a practical circuit?

2.4 Sketch the input and output spectrum for the following block diagram. Assume the DC component of the input is 0.5 V while the AC component is a sinewave at 4 MHz with a peak amplitude of 100 mV. Assume the clock frequency is 100 MHz.

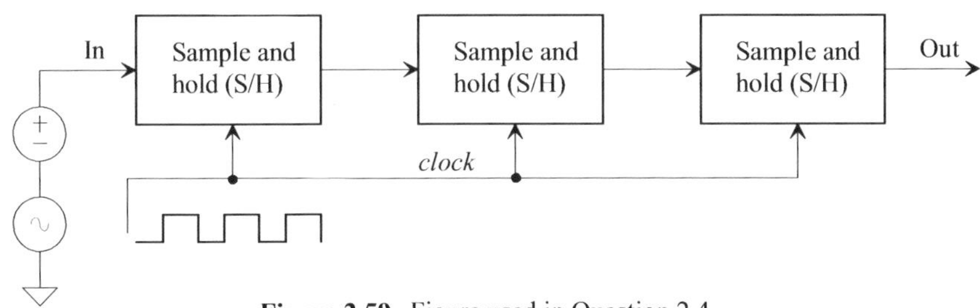

Figure 2.59 Figure used in Question 2.4.

2.5 Repeat Ex. 2.2 with an input sinewave at 30 MHz.

2.6 Re-sketch Fig. 2.22 if the input signal is a sinewave at 10 MHz (no other spectral content).

2.7 Suppose we are interpolating, with $K = 8$, digital data with $f_s = 100$ MHz. Prior to interpolation what is the frequency range of the desired spectrum? After interpolation what is the frequency range of the desired spectrum? What is the interpolator's output clock rate?

2.8 Verify, with simulations, that the topologies seen in Fig. 2.34 are equivalent.

2.9 Determine the transfer function, and verify with simulations, the behavior of 4 paths of the switched-capacitor topology seen in Fig. 2.36.

2.10 In your own words discuss why the ϕ_2 switches are shut off after the ϕ_1 switches in the S/H seen in Fig. 2.39.

2.11 Sketch the op-amp's open loop response, both magnitude and phase, specified by Eq. (2.59).

2.12 What is the voltage across C_F in Fig. 2.41 in terms of the input-referred offset and noise? Verify your answer with simulations commenting on the deviation of the frequency behavior of the input-referred noise to the frequency response of the voltage across the capacitor.

2.13 Provide a quantitative description of how capacitor mismatch will affect the operation of the S/H seen in Fig. 2.46. Verify your descriptions with simulations.

2.14 Is it possible to design a S/H with a gain of 0.5? How can this be done or why can't it be done? Use simulations to verify your answer.

2.15 For the first entry (v_1 = input, $v_2 = V_{CM}$) in Table 2.2 derive the frequency response, magnitude and phase, of the DAI. Use simulations at a few frequencies to verify your derivations.

2.16 Repeat Question 2.15 for the second entry.

2.17 Repeat Question 2.15 for the third entry.

2.18 Does the DAI use CDS? Why or why not? Use simulations to support your answers.

Chapter

3

Analog Filters

In the last chapter we discussed that analog anti-aliasing filters (AAFs) and reconstruction filters (RCFs) are an important component of a mixed-signal system. While we can perform signal processing and filtering in the digital domain, as seen in Fig. 2.1, AAFs and RCFs are still required in our system. Analog continuous-time filters can be faster (have wider bandwidths) and take up less area than their analog discrete-time (e.g., switched-capacitor) counterparts. However, unlike discrete-time filters, continuous-time filters cannot be fabricated with precise transfer functions and must be tuned. This is especially true if passive resistors and capacitors are used. Each one can have a variation of $\pm 20\%$. By using active CMOS integrators in the filter implementations instead of passive elements, we can electrically tune the filters. Also, we can more easily implement higher-order filters while minimizing the effects of loading.

In this chapter we discuss analog filters made using continuous-time analog integrators (CAIs or active-RC integrators), MOSFET-C integrators, transconductor-capacitor (g_m-C) integrators, and discrete-time analog integrators (DAIs). Our focus is on practical analog filters for mixed-signal AAFs and RCFs. These filters may have fully-differential inputs and outputs so the common-mode voltage of the op-amp used in the active filter remains constant (important for noise and distortion). Further, inverting a signal using fully-differential topologies is trivial since we simply swap the filter's outputs. Single-ended topologies where the op-amp's common mode voltage can vary, such as Sallen-Key, and topologies that require separate amplifiers to generate an inversion such as Tow-Thomas biquad, are covered in the excellent books by Franco [1] and Schaumann [2].

3.1 Integrator Building Blocks

3.1.1 Lowpass Filters

In order to methodically develop our understanding of CMOS filters, consider the lowpass filter shown in Fig. 3.1. The transfer function of this filter is

$$\frac{v_{out}(f)}{v_{in}(f)} = \frac{1}{1 + j\omega RC} \tag{3.1}$$

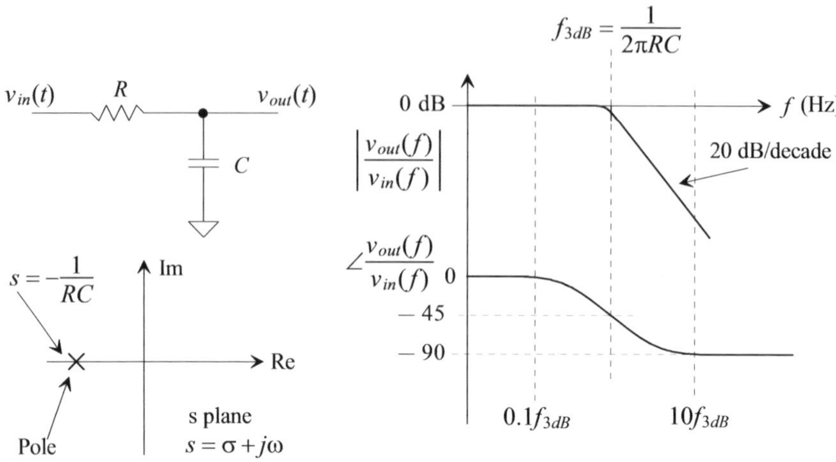

Figure 3.1 First-order lowpass filter.

where $\omega = 2\pi \cdot f$ and f is the frequency of the input (and thus the output). Next, consider the block diagram in Fig. 3.2. This figure shows an integrator and a summing block. The output of the block diagram can be determined by solving

$$v_{out}(f) = \frac{G}{s} \cdot (v_{in}(f) - v_{out}(f)) \tag{3.2}$$

or

$$\frac{v_{out}(f)}{v_{in}(f)} = \frac{1}{1 + s/G} \tag{3.3}$$

where for a sinewave input $s = j\omega$. Comparing this equation to Eq. (3.1), we see that if we set the integrator's gain, G, using

$$G = \frac{1}{RC} \text{ where } f_{3dB} = \frac{G}{2\pi} \tag{3.4}$$

we can use an integrator to implement a lowpass first-order filter (the filter has a single pole).

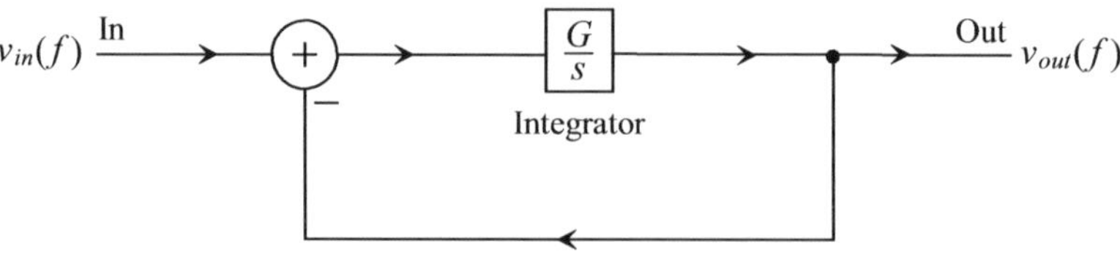

Figure 3.2 Block diagram of an integrator-based lowpass filter.

3.1.2 Active-RC Integrators

A continuous-time, fully-differential, analog integrator (CAI) is seen in Fig. 3.3. The CAI goes by other names, including the Miller integrator, the active-RC integrator, and when the resistors are replaced with MOSFETs operating in the triode region, the MOSFET-C integrators. The gain of the CAI can be written as

$$\frac{v_{out}}{v_{in}} = \frac{v_{out+} - v_{out-}}{v_{in+} - v_{in-}} = \frac{1}{s} \cdot \overbrace{\frac{1}{RC}}^{G} \tag{3.5}$$

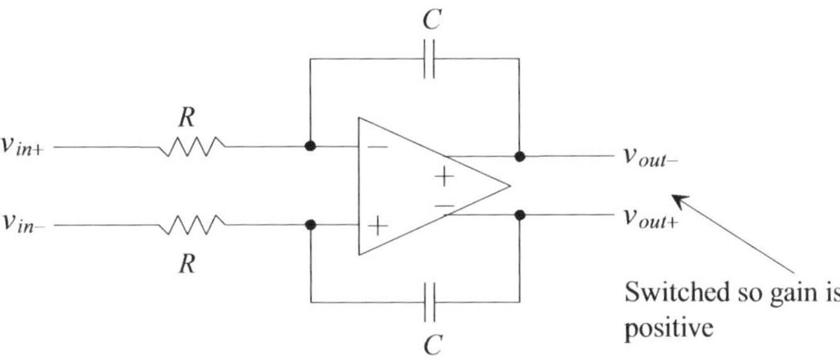

Figure 3.3 A continuous-time analog integrator (CAI).

Reviewing Fig. 3.2, we see that the CAI of Fig. 3.3 alone will implement the needed integration but not the summing (difference) block. By adding an additional feedback path, the entire block diagram of Fig. 3.2 can be implemented. Figure 3.4 shows the integrator-based implementation of the circuits in Figs. 3.1 and 3.2 (noting the op-amp must be able to drive a resistive load). This filter is called an *active-RC* filter because the RC is used with an active element (the op-amp). At this point there are several practical and useful modifications that we can make to this filter. However, let's work an example before moving on.

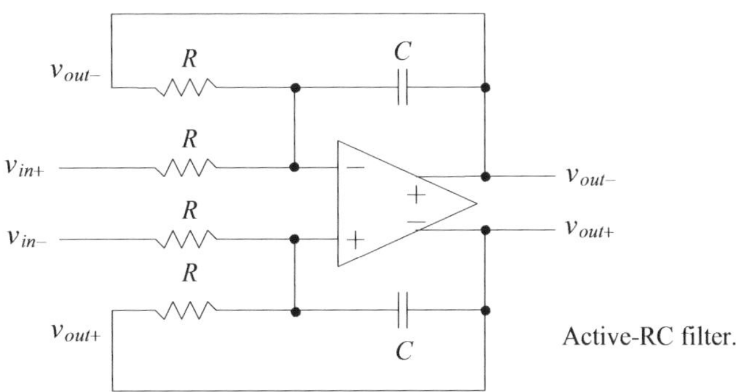

Figure 3.4 Implementation of a first-order lowpass filter using a CAI.

Example 3.1
Simulate the operation of the filter in Fig. 3.4 from DC to 100 MHz if $R = 10k$ and $C = 10$ pF. Show both the magnitude and phase responses of the filter. Assume the op-amp is ideal.

From Fig. 3.1 we know the 3 dB frequency of the filter is 1.59 MHz. The simulation results are shown in Fig. 3.5. The magnitude and phase response follow, as expected, the responses for the simple RC filter shown in Fig. 3.1. ■

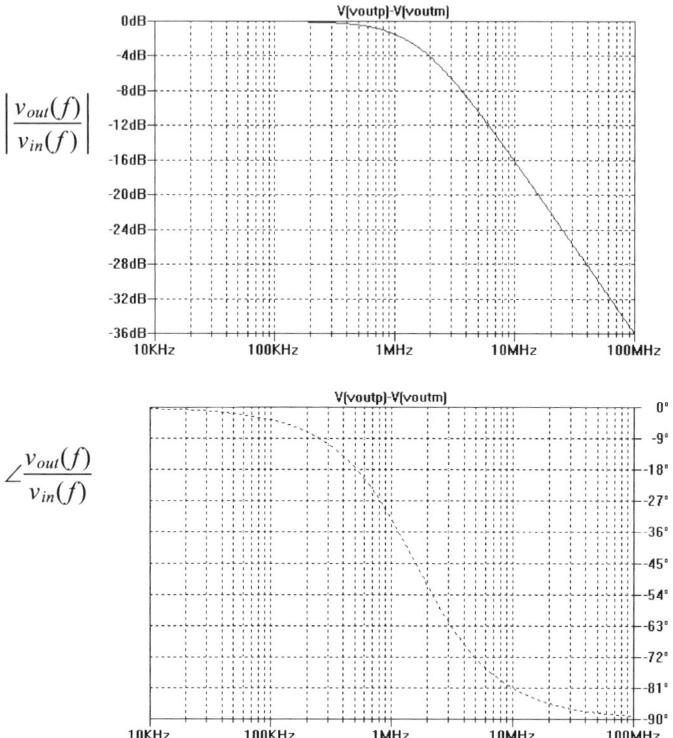

Figure 3.5 Magnitude and phase responses for the first-order filter in Fig. 3.4 if R = 10k and C = 10 pF.

What would happen if we switched what we define as v_{out+} and v_{out-} in the filter described in Ex. 3.1 without changing any other connections? Perhaps it is trivial, but the answer is that the output will be inverted. We can modify the block diagram of Fig. 3.2 by simply multiplying the output by -1, as seen in Fig. 3.6. The phase shift in Fig. 3.5 would shift up or down by 180 degrees. It would vary from 180 to 90 degrees, or from -180 to -270 (because $+180$ degrees is the same as -180 degrees) instead of from 0 to -90 degrees. If we allow the resistors used in the filter to have different values, as seen in Fig. 3.6, we can add a feedback gain to our block diagram. Assuming the outputs are labeled so that we don't have an inversion in the output of the filter (i.e., they are labeled as seen in Fig. 3.4), we can write

$$\frac{v_{in}}{R_I} - \frac{v_{out}}{R_F} = \frac{v_{out}}{1/sC} \tag{3.6}$$

It's important to notice (for later use) that in order to subtract the output from the input, the voltages are first changed to currents and then summed (or more correctly subtracted) on the inputs of the op-amp. This equation can be rewritten as

$$\frac{v_{out}}{v_{in}} = \frac{\frac{R_F}{R_I}}{1 + sR_F C} \tag{3.7}$$

Using the block diagram in Fig. 3.6, we can write

$$\frac{v_{out}}{v_{in}} = \frac{\frac{1}{G_2}}{1 + \frac{s}{G_1 G_2}} \text{ and } f_{3dB} = \frac{G_1 G_2}{2\pi} \tag{3.8}$$

Equating coefficients in these equations results in

$$G_2 = \frac{R_I}{R_F} \text{ and } G_1 = \frac{1}{R_I C} \tag{3.9}$$

Note that at DC where $s \to 0$, the block diagram in Fig. 3.6 becomes the classic feedback diagram with the forward gain approaching infinity and a feedback factor of G_2. Then from classic feedback theory, the closed-loop gain becomes $1/G_2$ or, for the filter in Fig. 3.6, R_F/R_I. Of course, analyzing this circuit (using loop equations) at DC when the capacitor is an open results in the same gain.

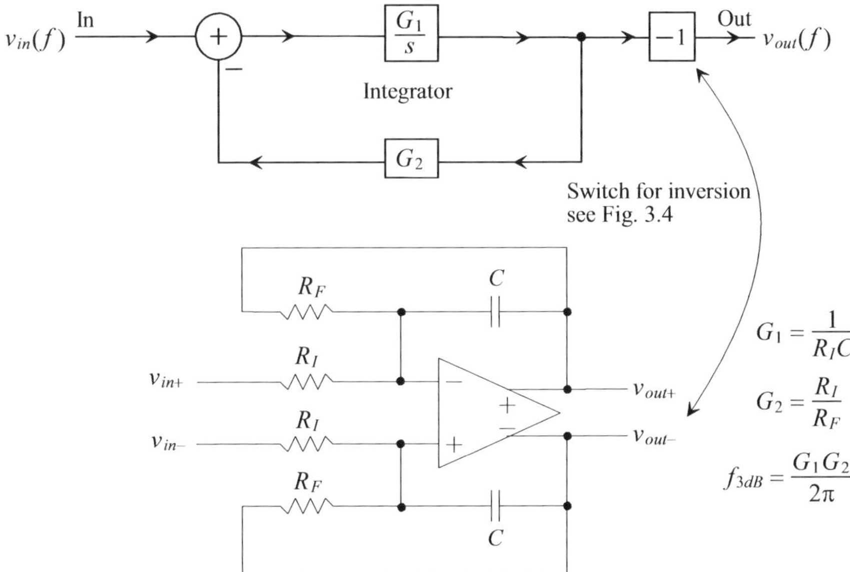

Figure 3.6 Integrator-based first-order filter.

Example 3.2
Modify the filter in Ex. 3.1 so that the low-frequency gain is 20 dB.

Using Eq. (3.7) or Eq. (3.8), we leave $C = 10$ pF and $R_F = 10$k. To get the gain of 10, we make $R_I = 1$k. The simulation results are shown in Fig. 3.7. ■

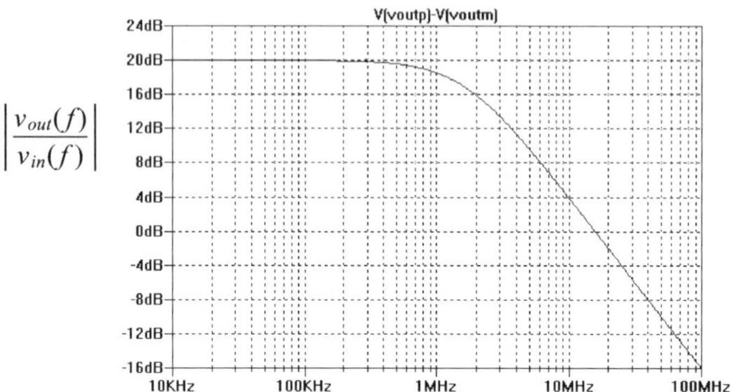

Figure 3.7 A first-order filter with gain, see Ex. 3.2.

Effects of Finite Op-Amp Gain Bandwidth Product, f_{un}

In the previous two examples we assumed a near-ideal op-amp. The open-loop gain of the op-amp can be written, assuming a dominant-pole compensated op-amp, by

$$A_{OL}(f) = \frac{v_{out}}{v_+ - v_-} = \frac{A_{OLDC}}{1 + j \cdot \frac{f}{f_{3dB}}} \tag{3.10}$$

where v_+ and v_- are the voltages on the noninverting and inverting op-amp input terminals, respectively. Note that we are using f_{3dB} in both Figs. 3.6 and Eq. (3.10) to indicate the 3 dB frequency of a frequency response. While the symbol is the same the actual values vary from circuit to circuit. When a practical op-amp is operating at frequencies above a few kHz, we can approximate the open-loop response (knowing the imaginary part of the denominator is much larger than the real part) as

$$A_{OL}(f) \approx \frac{A_{OLDC} \cdot f_{3dB}}{j \cdot f} = \frac{f_{un}}{j \cdot f} = \frac{2\pi f_{un}}{s} = \frac{\omega_{un}}{s} \tag{3.11}$$

where f_{un} is the frequency where the op-amp's open loop gain is unity (0 dB). Rewriting Eq. (3.6) to include the op-amp's finite gain bandwidth product (that is, f_{un}) and assuming, without the loss of generality, that the op-amp is operating with a single-ended output (v_+ tied to V_{CM} [AC ground]), results in

$$\frac{v_{in} - v_-}{R_I} + \frac{v_{out} - v_-}{R_F} + sC \cdot (v_{out} - v_-) = 0 \tag{3.12}$$

After some algebraic manipulation with $v_- = -v_{out}/A_{OL}(f)$, we get

$$\frac{v_{out}}{v_{in}} = \frac{-\frac{R_F}{R_I}}{\underbrace{1 + sCR_F}_{\text{Desired response}} + \frac{sCR_F}{A_{OL}(f)} + \frac{1}{A_{OL}(f)}\left(1 + \frac{R_F}{R_I}\right)} \tag{3.13}$$

or, with $A_{OL}(f) = \omega_{un}/s$

$$\frac{v_{out}}{v_{in}} = \frac{-\frac{R_F}{R_I}}{s^2\frac{CR_F}{\omega_{un}} + s \cdot \left[CR_F + \frac{1}{\omega_{un}}\left(1 + \frac{R_F}{R_I}\right)\right] + 1} \tag{3.14}$$

This equation is very revealing and shows just how significant a limitation the op-amp can be in a filter. For the moment, to simplify things, let's assume $\omega^2 << \omega_{un}/CR_F$ so that the s^2 term in Eq. (3.14) is negligible. We can then write the magnitude and phase responses as

$$\left|\frac{v_{out}}{v_{in}}\right| = \frac{\frac{R_F}{R_I}}{\sqrt{1 + \left[\omega CR_F + \frac{\omega}{\omega_{un}}\left(1 + \frac{R_F}{R_I}\right)\right]^2}} \text{ and } \angle\frac{v_{out}}{v_{in}} = -\tan^{-1}\left[\omega CR_F + \frac{\omega}{\omega_{un}}\left(1 + \frac{R_F}{R_I}\right)\right] \tag{3.15}$$

Example 3.3

Suppose a first-order filter is designed based on the topology seen in Fig. 3.6, where $\omega_{un} = 10/R_F C$ and $R_F/R_I = 10$. Assuming $\omega^2 << \omega_{un}/CR_F$, comment on how the magnitude and phase responses of the filter will be affected by the finite op-amp, f_{un}.

The op-amp's unity gain frequency is only 10 times larger than the bandwidth of the filter. This means the op-amp's closed-loop bandwidth (with a gain of 10) is equal to the desired bandwidth of the filter. The bandwidth of a gain of 10 op-amp circuit is $f_{un}/10$, which here is equal to the ideal filter 3 dB frequency of $1/2\pi R_F C$. The magnitude of the filter's response can be approximated as

$$\left|\frac{v_{out}}{v_{in}}\right| = \frac{\frac{R_F}{R_I}}{\sqrt{1 + [\omega \cdot 2CR_F]^2}} \text{ and } \angle\frac{v_{out}}{v_{in}} = -\tan^{-1}[\omega \cdot 2CR_F]$$

which shows the filter's 3 dB frequency is off by a factor of 2. ■

The point of the preceding example is, in general lowpass filter design, to minimize the effects of the op-amp's finite f_{un} a low value of closed-loop DC gain should be used. In the remaining discussion let's assume $R_F/R_I = 1$, so Eq. (3.14) can be approximated as

$$\left|\frac{v_{out}}{v_{in}}\right| \approx \frac{1}{\frac{CR_F}{\omega_{un}} \cdot s^2 + s \cdot CR_F + 1} \tag{3.16}$$

The poles of this transfer function are located at

$$s_{p1,p2} = \frac{-CR_F \pm \sqrt{(CR_F)^2 - 4\frac{CR_F}{\omega_{un}}}}{2 \cdot \frac{CR_F}{\omega_{un}}} \tag{3.17}$$

noting that if $\omega_{un} \to \infty$, then $s_{p1} = \infty$, and $s_{p2} = -1/CR_F$ (the ideal position of the pole, see Fig. 3.1).

To get some idea of the required op-amp f_{un} ($\omega_{un} = 2\pi f_{un}$), let's assume that we want the pole to vary no more than 1% from the ideal location due to finite op-amp bandwidth

$$\frac{-1}{CR_F} = \frac{99}{100} \cdot \frac{-CR_F - \sqrt{(CR_F)^2 - 4\frac{CR_F}{\omega_{un}}}}{2 \cdot \frac{CR_F}{\omega_{un}}} \tag{3.18}$$

This can be rewritten as

$$1.01 = \frac{\omega_{un} \cdot CR_F}{2} - \frac{1}{2}\sqrt{(\omega_{un} \cdot CR_F)^2 - 4(\omega_{un} \cdot CR_F)} \tag{3.19}$$

If we let $x = \omega_{un} \cdot CR_F$, then we need to solve

$$x - \sqrt{x^2 - 4x} = 2.02 \tag{3.20}$$

knowing x is positive and much larger than one ($\omega_{un} >> 1/CR_F$). Solving Eq. (3.20) for x, results in $x = 100$. This means the op-amp's unity gain frequency must be 100 times larger than the filter's f_{3dB} in order for the variation of this frequency (the pole) to deviate less than 1% from the ideal. If we can withstand a 10% decrease in the filter's cutoff frequency, then f_{un} need only be 10 times larger than the filter's f_{3dB}. Clearly, from Eq. (3.14), the first-order filter's frequency response is actually second-order when the op-amp's gain bandwidth product f_{un} is a factor. Therefore, the shapes of the magnitude and phase responses will deviate from the ideal first-order shapes seen in Fig. 3.1. *We can draw two very practical conclusions.* First, even if it were possible to fabricate precise resistor and capacitor values, the limitations of the op-amp's finite bandwidth may still require the use of tuning when filtering with active-RC integrator-based filters. Tuning would consist of adding or removing resistors and capacitors to adjust the precise filter cutoff frequency (adding/removing the elements using either fuses or, if possible, MOSFET switches). Second, the op-amp's f_{un} should be at least 10 times larger than the cutoff frequency (f_{3dB}) of the filter (again assuming a closed-loop DC gain of unity, i.e., $R_F/R_I = 1$). This is a general "rule-of-thumb." Precision filters (well-defined magnitude and phase responses) would require wider bandwidth op-amps. Consider the following example.

Example 3.4

Repeat Ex. 3.1 if an op-amp is used with a DC gain of 10,000 and an f_{un} of 10 MHz.

Because the op-amp's A_{OLDC} is 10,000 and $f_{un} = 10$ MHz, the op-amp's open loop f_{3dB} is 1 kHz (see Eq. [3.11]). We can use the circuit shown in Fig. 3.8 in our SPICE simulation to model an op-amp with finite f_{un}. The RC in Fig. 3.8 is selected to give an op-amp open loop f_{3dB} of 1 kHz.

The 3-dB frequency of the filter described in Ex. 3.1 is, under ideal conditions, 1.59 MHz. Because our op-amp's unity gain frequency is only 10 MHz, we would expect, from Eq. (3.16), the op-amp to affect the frequency response of the filter. Figure 3.9 shows the simulation results using the op-amp model of Fig. 3.8. Comparing Fig. 3.9 to Fig. 3.5, we see differences in both the magnitude and phase responses of the filters. The magnitude response of Fig. 3.9 initially rolls off at 20 dB/decade below the ideal 1.59 MHz. Around 10 MHz the response

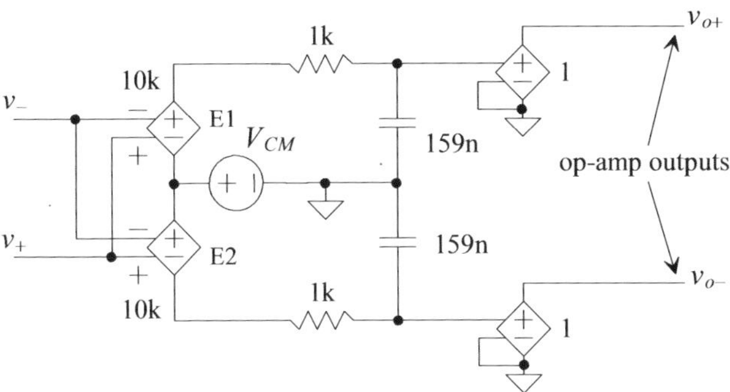

Figure 3.8 SPICE modeling of a differential input/output op-amp with finite bandwidth.

transitions to 40 dB/decade. Clearly this faster roll-off is the result of the op-amp's closed-loop pole coming into play. The limiting behavior of the op-amp, when looking only at the magnitude response, may be welcome (the filter's response rolls off faster) in a lowpass filter. However, it is not welcome in other filters (a highpass filter, for example). Figure 3.10 shows what happens if we decrease the filter's 3 dB frequency to 159 kHz by increasing the resistors used to 100k. What we are doing here is showing how making the op-amp's bandwidth much larger than the filter's affects the frequency response of the circuit. The magnitude response starts to fall off at –40 dB/decade at the op-amp's unity gain frequency, f_{un}, of 10 MHz. Also seen in Fig. 3.10 is the phase response of the filter. The op-amp's (closed-loop) phase response, which starts rolling off one decade below f_{un}, results in the final phase shift of the filter approaching –180 degrees. ■

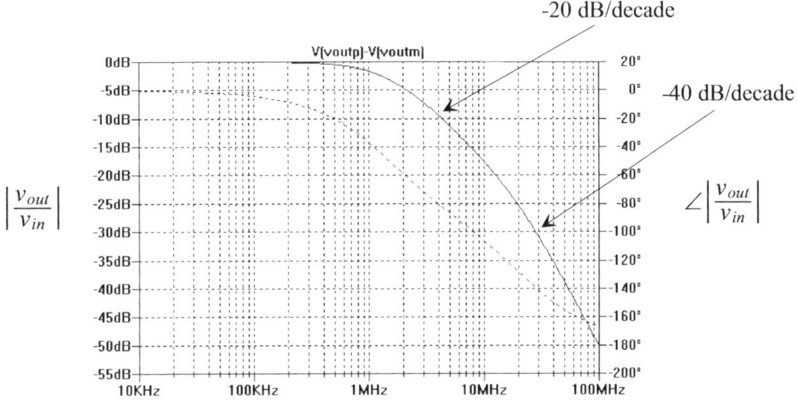

Figure 3.9 Magnitude and phase responses for the first-order filter in Fig. 3.4 if R=10k and C=10 pF using an op-amp with a 10 MHz unity-gain frequency.

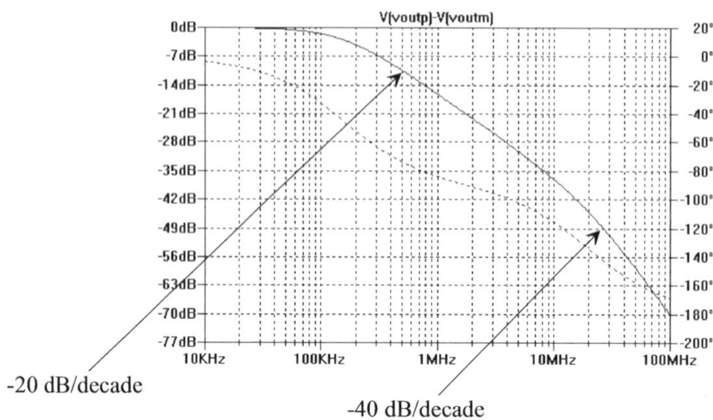

Figure 3.10 Increasing the resistance to 100k and replotting Fig. 35.9.

In order to model the effects of op-amp finite bandwidth on an active-RC filter's frequency response, we can add a pole to the ideal transfer function. Assuming unity gain in the passband (see Eq. [3.3]) results in

$$\frac{v_{out}(f)}{v_{in}(f)} = \frac{1}{1+s/G} \cdot \overbrace{\frac{1}{1+j\frac{f}{f_{un}}}}^{\text{Undesired}} \tag{3.21}$$

This result could have been used in the previous example to predict how the op-amp affects the filter's behavior. If the filter has gain (> 1) in the passband, see Eq. (3.8), we can modify this equation to read

$$\frac{v_{out}(f)}{v_{in}(f)} = \frac{\frac{1}{G_2}}{1+\frac{s}{G_1 G_2}} \cdot \overbrace{\frac{1}{1+j\frac{f}{G_2 \cdot f_{un}}}}^{\text{Undesired}} \tag{3.22}$$

For a higher order filter we would multiply the desired frequency response by the undesired term's (the op-amp's) response for each op-amp used in the circuit. Clearly, this limits the order of the filter (limits the number of op-amps used in a circuit; a first-order filter uses one op-amp, a second-order filter uses two op-amps, etc.). This is especially true if the filter has a passband approaching the f_{un} of the op-amps used.

Active-RC SNR

Consider the single-ended active-RC filter shown in Fig. 3.11. Let's assume an ideal op-amp with a maximum RMS output voltage of $VDD/(2\sqrt{2})$. The RMS input-referred noise of the filter, assuming thermal noise dominates over the bandwidth of interest, is simply $\sqrt{kT/C}$. The filter's *SNR* can then be written as

$$SNR = 20 \cdot \log \frac{VDD/\left(2\sqrt{2}\right)}{\sqrt{kT/C}} = 10 \cdot \log \frac{VDD^2/8}{kT/C} \tag{3.23}$$

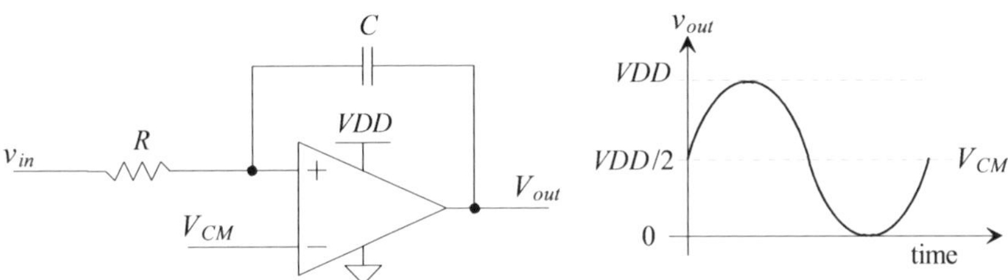

Figure 3.11 Estimating maximum possible SNR of an active-RC filter.

The size of the integrating capacitor fundamentally sets the *SNR* in integrator-based data converters or modulators. But consider the following: the maximum electrical energy stored in the capacitor used in an integrator is

$$\text{Maximum electrical energy} = \frac{1}{2}C \cdot \left(\frac{VDD}{2}\right)^2 \tag{3.24}$$

Equation (3.23) can then be rewritten as

$$SNR = 10 \cdot \log \frac{VDD^2/8}{kT/C} = 10 \cdot \log \frac{\frac{1}{2}C\left(\frac{VDD}{2}\right)^2}{kT} = 10 \cdot \log \frac{\text{Electrical energy}}{\text{Thermal energy}} \tag{3.25}$$

This equation can also be used to estimate the fundamental dynamic range, *DR*, of a filter. *Practically, DRs approaching 90 dB (*15 *bits)* can be attained using active-RC filters with good polysilicon resistors (to avoid the large nonlinear voltage coefficient associated with diffused or implanted resistors) and linear capacitors. Bandwidths approaching 50 MHz, assuming 500 MHz f_{un} op-amps are used, can be attained (at, of course, lower DRs).

3.1.3 MOSFET-C Integrators

Let's now look at a variation of the active-RC filter where the resistors are replaced with MOSFETs. Figure 3.12 shows a MOSFET-C filter. In order for the MOSFETs to behave as resistors they must remain in the triode region. Using long length devices helps ensure triode operation. Because the MOSFETs are operating as resistors, their speed is not governed directly by their gate-source voltage (overdrive voltage) or channel length. However, the linearity of the MOSFET resistors is still very important, as is the possibility that the MOSFETs will introduce a parasitic pole into the filter's frequency response because of the distributed resistance/capacitance of the channel (Fig. 3.12). For large input signals, the active MOSFET resistors become nonlinear, resulting in filters with *DRs of around only 40 dB*. The bandwidth of the MOSFET-C filters parallels that of the active-RC filters.

We might be questioning the usefulness of the MOSFET-C filter with a *DR* of only 40 dB. Clearly this filter will only find use in data conversion systems using six bits of resolution or less (36 dB *DR*) or in systems that process continuous-time signals. The big benefit of this filter over the active-RC filter is its ability to be tuned. Tuning the active-RC filter required adding or removing, via switches or fuses, resistors or capacitors in parallel or series with the existing resistors and capacitors. Tuning the MOSFET-C

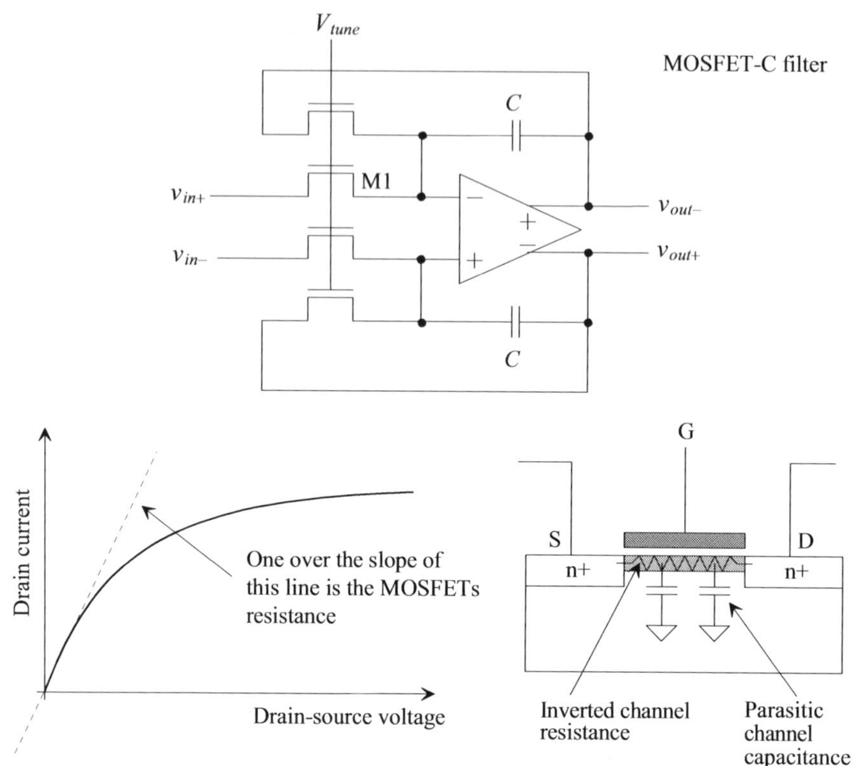

Figure 3.12 A first-order MOSFET-C filter.

filter shown in Fig. 3.12 can be accomplished by adjusting V_{tune}. If we assume long-channel behavior, we can write the resistance of the MOSFETs in terms of V_{tune} (assuming the input common-mode voltage of the op-amp is 0) as

$$R_n = \frac{1}{KP \cdot \frac{W}{L} \cdot \left(V_{tune} - V_{THN} - \underbrace{V_{DS}}_{=vin} \right)} \tag{3.26}$$

The current through M1 in Fig. 3.12 is

$$\frac{v_{in+}}{R_{n1}} = v_{in+} \cdot KP \cdot \frac{W}{L}(V_{tune} - V_{THN} - v_{in+}) \tag{3.27}$$

Some improvement in the linearity of the MOSFET resistors, say 10 dB (resulting in a *DR* of 50 dB), can be achieved by utilizing the fully-balanced signals available in the circuit. Consider replacing M1 in Fig. 3.12 with the pair of MOSFETs, M1A and M1B, shown in Fig. 3.13. The resulting current is now

$$\frac{v_{in+}}{R_{n1A}} + \frac{v_{in-}}{R_{n1B}} = KP \cdot \frac{W}{L}[v_{in+} \cdot (V_{tune+} - V_{THN} - v_{in+}) + v_{in-} \cdot (V_{tune-} - V_{THN} - v_{in-})] \tag{3.28}$$

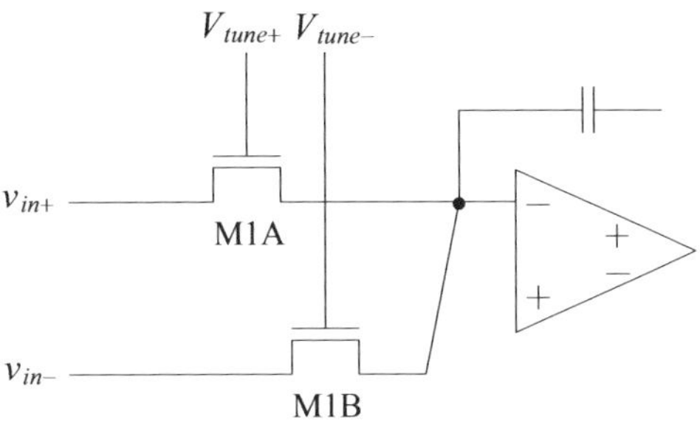

Figure 3.13 Linearizing MOSFET resistors.

Knowing $v_{in+} = -v_{in-}$ and letting $V_{tune} = V_{tune+} - V_{tune-}$, we can write the equivalent current through M1 as

$$\frac{v_{in+}}{R_{n1}} = v_{in+} \cdot KP \cdot \frac{W}{L} \cdot V_{tune} \tag{3.29}$$

The result is that the nonlinear behavior of the MOSFET's channel resistance due to the changing drain-source voltage cancels to a first order. Figure 3.14 shows the implementation of a first-order MOSFET-C filter using linearized MOSFETs.

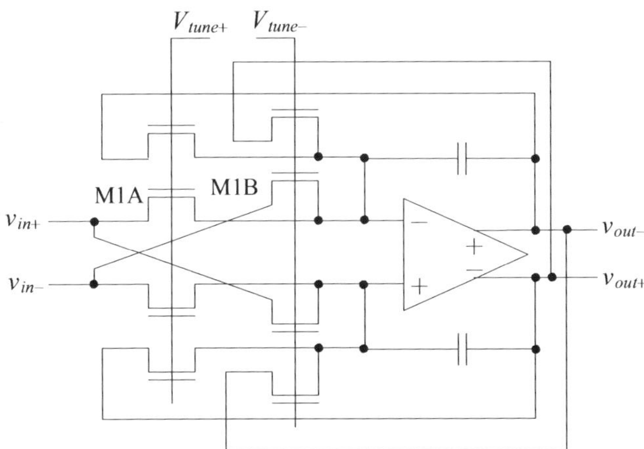

Figure 3.14 First-order MOSFET-C filter using linearized MOSFET resistors.

Why Use an Active Circuit (an Op-Amp)?

Before going any further, let's realize that we can get the exact same frequency performance using a simple resistor/capacitor or MOSFET/capacitor as we get when we use these elements with an op-amp. So, "Why use the op-amp?" The answer to this question comes when we realize that when a capacitive or resistive load is connected to the output of the filter (without an active element), the frequency behavior changes. Using the op-amp allows us to drive an arbitrary (within reason) capacitive or resistive load. Using an active element will also allow us to cascade first-order sections to implement higher order filters.

3.1.4 g_m-C (Transconductor-C) Integrators

The operational-transconductance amplifiers, OTA, is an amplifier that has only two high-impedance nodes: the amplifier's input and its output. Figure 3.15 shows a schematic symbol, transfer curves, and a possible implementation for an OTA (based on a fully-differential diff-amp). Transconductor-C, or g_m-C, filters use a circuit, a transconductor, that provides a linear voltage-current transfer curve. Our OTA in Fig. 3.15 does behave like a transconductor over a portion of the input voltage range but becomes nonlinear for large input voltage differences, $v_{in+} - v_{in-}$. By increasing the lengths of the NMOS diff-pairs used in the OTA, we can increase the linear common-mode range of the OTA, making it appear as though it were a transconductor. The fundamental problems with increasing the lengths of the diff-pair MOSFETs are the increase in the OTA's input capacitance (affecting the location of the filter's poles and zeroes) and, perhaps more fundamentally, the inherent reduction in the MOSFET's f_T.

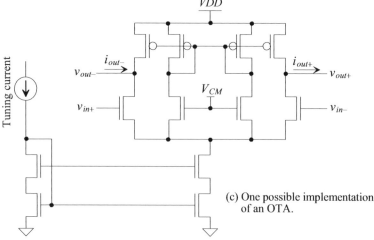

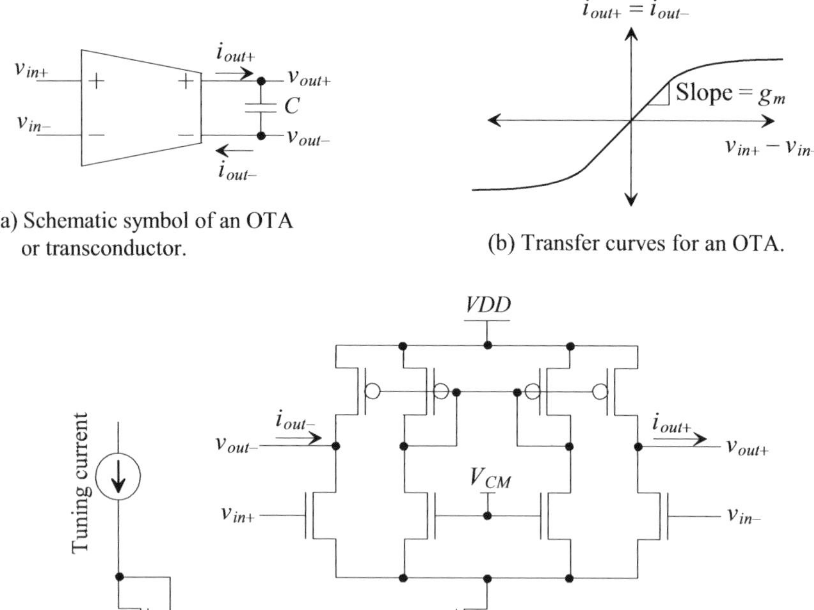

Figure 3.15 Showing an implementation of an OTA and transfer curves.

Before discussing these issues in more detail, let's look at an important limitation of g_m-C filters; namely, the fact that the transconductor's input voltage must vary. In any precision application, the input voltage must remain constant because of the roll-off associated with the amplifier's CMRR (unless, of course, the common-mode voltage can be held at a precise value). This *limits the DR of g_m-C filters to around 50 dB*. Again, this is not too useful if used as an antialiasing or reconstruction filter unless the system's resolution is less than eight bits (48 dB SNR). The g_m-C filter finds extensive use in continuous-time signal processing.

We can relate the input voltage difference to the output voltage difference for the circuit in Fig. 3.15a using

$$v_{out+} - v_{out-} = \frac{i_{out+}}{j\omega C} = \frac{i_{out-}}{j\omega C} = \frac{g_m(v_{in+} - v_{in-})}{j\omega C} \tag{3.30}$$

Comparing this result to Eq. (3.5), we can use the same design techniques if we require

$$G = \frac{1}{RC} = \frac{g_m}{C} \text{ or } f_{3dB} = \frac{G}{2\pi} \tag{3.31}$$

The big benefit of the g_m-C filter over the active-RC filter is the ability to tune the filter by adjusting the transconductor's g_m.

The circuit of Fig. 3.15a implements an integrator, as does the active-RC circuit of Fig. 3.3. However, as seen in Fig. 3.2, we also need to implement a summing block in a first-order filter. In order to move toward the goal of implementing the summing block, consider the transconductor circuit shown in Fig. 3.16a. The output current of a single transconductor is $g_m(v_{in+} - v_{in-})$. We can sum this current with the output current from the second transconductor to implement the summing block in Fig. 3.2. As seen in Fig. 3.16b, the outputs of the two OTAs are combined, so the output currents from each transconductor subtract. Assuming each transconductor has the same transconductance, we can write

$$g_m(v_{in+} - v_{in-}) - g_m(v_{out+} - v_{out-}) - j\omega C(v_{out+} - v_{out-}) = 0 \tag{3.32}$$

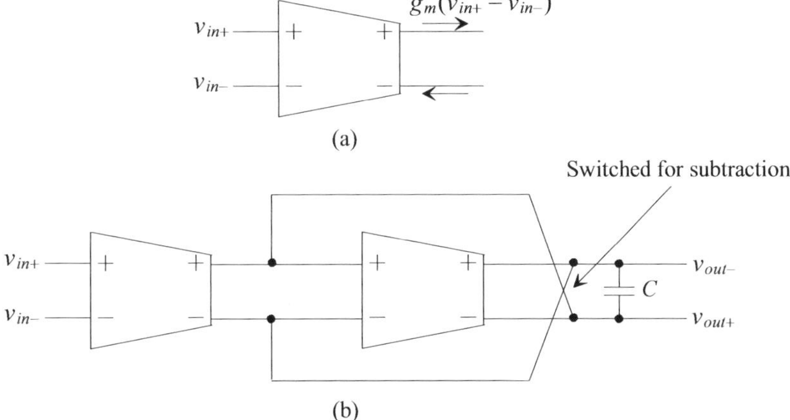

Figure 3.16 Implementing a first-order filter using transconductors.

or

$$\frac{v_{out+} - v_{out-}}{v_{in+} - v_{in-}} = \frac{1}{1 + j\omega C \cdot \frac{1}{g_m}} \tag{3.33}$$

If the transconductors have different g_ms, we can design a filter with a DC gain (see Problem 3.10), which can be characterized using Eq. (3.8).

Example 3.5

Repeat Ex. 3.1 using a g_m-C filter with a g_m of 100 μA/V.

To simulate a transconductor using SPICE, a voltage-controlled current source can be used as seen in Fig. 3.17. In order to set the output common-mode voltage to V_{CM} in the simulation, we add the large resistors (whose values can be changed to simulate the finite, nonideal, output resistance of the OTA) connected to V_{CM}. If we didn't use these resistors, after reviewing Fig. 3.16, it would result in an unknown common-mode voltage on the second transconductor input.

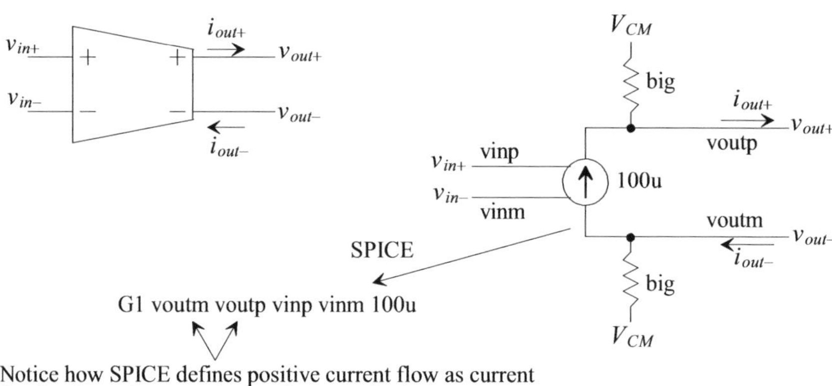

Figure 3.17 Modeling an ideal transconductor in SPICE using a voltage-controlled current source.

In order to have the same time constant, and thus pole location, as in Ex. 3.1, let's set the capacitor value, in the schematic of Fig. 3.16 to 10 pF. The value of the transconuctance, $1/g_m$, is 10k. The simulation results are shown in Fig. 3.18. As we would expect, the shape follows, exactly, that of the active-RC filter in Fig. 3.5. Also, although not shown, the phase response matches as well. ■

Common-Mode Feedback Considerations

In most OTAs the load capacitance is used for compensation. These capacitances compensate both the normal, forward, differential signal path as well as the CMFB path. Reviewing Fig. 3.15, we see that the capacitance in part (a) indeed does provide a load for differential signals. However, any signal that is common to both outputs (a common-mode signal) doesn't cause a displacement current to flow through the capacitor. Because both sides of the capacitor change at the same rate for common-mode signals, the change in voltage across the capacitor is zero. This can result in unstable CMFB

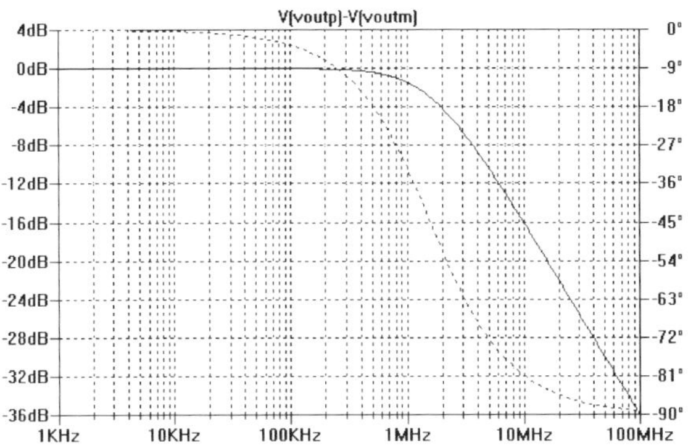

Figure 3.18 Simulation results for Ex. 35.5.

loops. Figure 3.19 shows how we would break the capacitor in Fig. 3.15 up into two components to provide the same loading for differential and common-mode signals. We can write

$$\frac{v_{out+}}{1/j\omega 2C} = i_{out+} = i_{out-} = \frac{-v_{out-}}{1/j\omega 2C} \tag{3.34}$$

$$i_{out+} = g_m(v_{in+} - v_{in-}) = i_{out-} \tag{3.35}$$

$$\frac{v_{out+} - v_{out-}}{v_{in+} - v_{in-}} = \frac{g_m}{j\omega C} \tag{3.36}$$

which is the same result as Eq. (3.30).

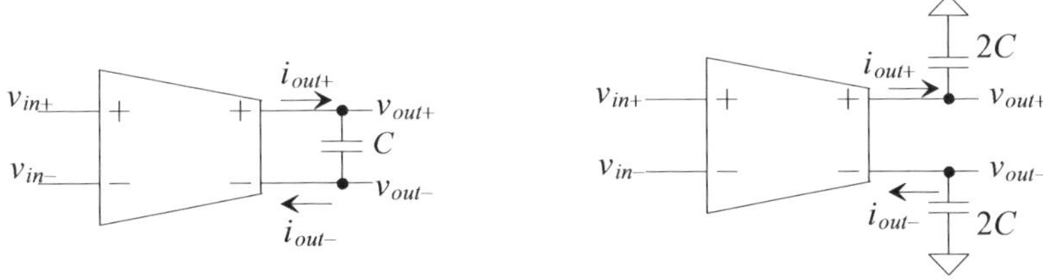

Figure 3.19 Showing how we break the capacitor up to provide a load for the CMFB circuit.

A High-Frequency Transconductor

A transconductor, Fig. 3.20, can be implemented using inverters. To increase the input common-mode range of the transconductor, the lengths of INV1 and INV2 can be increased as we discussed for the diff-amp seen in Fig. 3.15. This, again, lowers the location of the parasitic poles introduced into the transconductor's response. Note, in this circuit, that other than the power supplies and the tuning voltage, there are only two sets of nodes: the input and the output nodes. This allows the transconductor's capacitances to sum with the load capacitances and be tuned out.

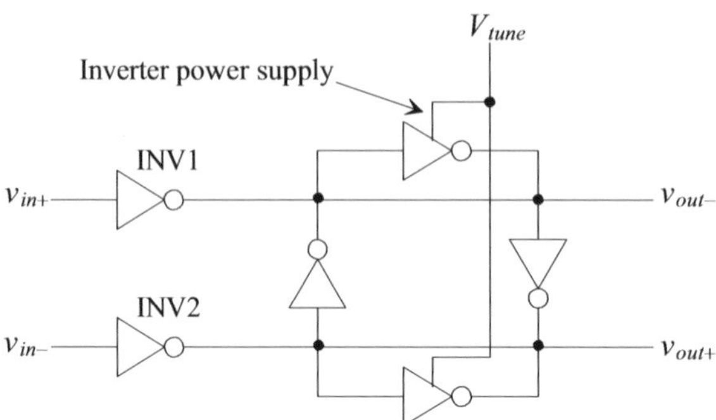

Figure 3.20 High-frequency transconductor.

3.1.5 Discrete-Time Integrators

Let's consider using a discrete-analog integrator, DAI, discussed in the last chapter to implement a first-order filter. The output of a DAI with two delaying inputs is

$$\frac{v_{out}(z)}{v_1(z)-v_2(z)}=\frac{C_I}{C_F}\cdot\frac{z^{-1}}{1-z^{-1}} \tag{3.37}$$

If we apply the filter's input to the noninverting input of the DAI and feed the output back to the subtracting input, see Fig. 3.21, we get

$$\frac{v_{out}(z)}{v_{in}(z)-v_{out}(z)}=\frac{C_I}{C_F}\cdot\frac{z^{-1}}{1-z^{-1}} \tag{3.38}$$

we can write, see Eqs. (1.62) and (1.66)

$$\left|\frac{v_{out}}{v_{in}-v_{out}}\right|=\frac{C_I}{C_F}\cdot\frac{1}{2\left|\sin\pi\frac{f}{f_s}\right|}\text{ and }\angle\frac{v_{out}}{v_{in}-v_{out}}=-\pi\frac{f}{f_s}-\frac{\pi}{2}\text{ for }0<f<f_s \tag{3.39}$$

The sampling frequency (the frequency the discrete-time filter is clocked at) is f_s , while the filter's input frequency is labeled f. If we require $f<<f_s$ (say at least sixteen times less so $\sin\pi\frac{f}{f_s}\approx\pi\frac{f}{f_s}$), then we can rewrite Eq. (3.39) as

$$\frac{v_{out}(z)}{v_{in}(z)-v_{out}(z)}=\frac{C_I}{C_F}\cdot\frac{z^{-1}}{1-z^{-1}}\approx\frac{C_I}{C_F}\cdot\frac{f_s}{j2\pi f}=G\cdot\frac{1}{s} \tag{3.40}$$

where

$$G=\frac{C_I}{C_F}\cdot f_s=\frac{1}{C_F R_{sc}}\text{ and }f_{3dB}=\frac{G}{2\pi} \tag{3.41}$$

placing the gain of the integrator in the same form as Eqs. (3.4). The variable R_{sc} is a switched capacitor resistor

$$R_{sc}=\frac{1}{C_I\cdot f_s}=\frac{T_s}{C_I} \tag{3.42}$$

The equivalent block diagrams for first-order filters using CAIs and using the DAIs (again assuming $f<<f_s$) are compared in Fig. 3.21.

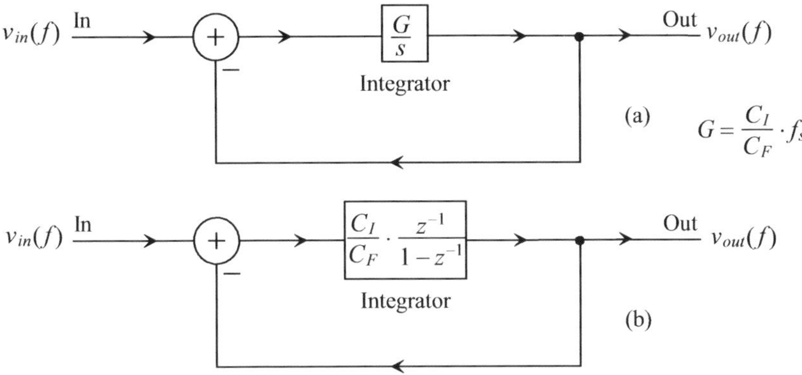

Figure 3.21 Block diagram of an integrator-based lowpass filter. (a) Continuous-time and (b) the discrete-time equivalent.

Example 3.6

Sketch the implementation of a DAI-based (switched-capacitor), first-order filter with characteristics like the one in Ex. 3.1. Use SPICE to simulate the design.

The schematic of the filter is shown in Fig. 3.22. Here we are assuming the clocking frequency of the filter is 100 MHz. If the feedback capacitance, C_F, is 10 pF, the size of the input capacitor, C_I, is then, from Eq. (3.42) and knowing R_{sc} is 10k, 1 pF. The 3 dB frequency of the filter is, once again, $1/2\pi R_{sc} C_F = 1.59$ MHz. Because of the time-domain (clock) component of the filter, we can't use an AC analysis for the SPICE simulation. Let's apply a 1 V peak-to-peak sinewave to the filter at 1.59 MHz and verify the output of the filter is 3 dB down (0.707 V peak-to-peak). The results are seen in Fig. 3.23.

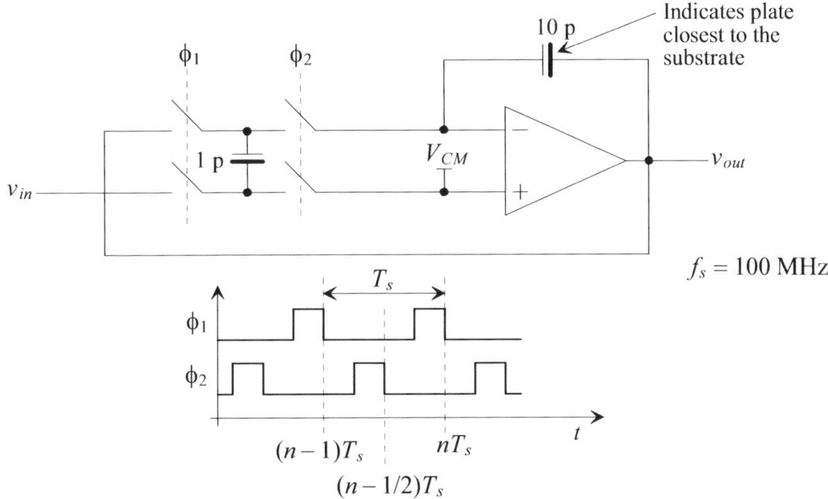

Figure 3.22 A switched capacitor, first-order filter similar to the one described in Ex. 3.1.

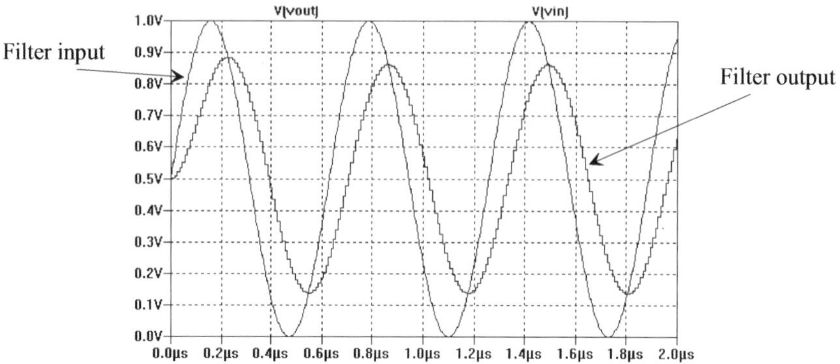

Figure 3.23 Output of the switched-capacitor circuit in Fig. 3.22.

Let's comment on the exact transfer function of the DAI in Fig. 3.22. We see in Fig. 3.22 that the output is indeed fed back to the input through a ϕ_1 controlled switch. The result is that the output, through the feedback loop, sees one clock cycle delay, z^{-1}. The output is assumed settled on the falling edge of ϕ_2 during each clock cycle. Because of this, the input, which is settled on the falling edge of ϕ_1, sees only a half clock cycle delay, $z^{-1/2}$. This means that the DAI in Fig. 3.22 really has a transfer function of

$$v_{out}(z) = \frac{C_I}{C_F} \cdot \frac{z^{-1}}{1-z^{-1}} \cdot [v_{in}(z) \cdot z^{1/2} - v_{out}(z)] \tag{3.43}$$

If we think of the input signal arriving a half-clock cycle earlier, then the only difference in the transfer function here, when compared to the continuous-time equivalent and assuming $f << f_s$, is a small phase difference. We can delay the input signal a full clock cycle by adding a ϕ_1 controlled switch on the output of the filter. However, because this switch may be part of the next filter section, we don't discuss this option further. ■

Example 3.7

Sketch the switched-capacitor implementation of the discrete-time lowpass (first-order) filter shown in Fig. 3.24.

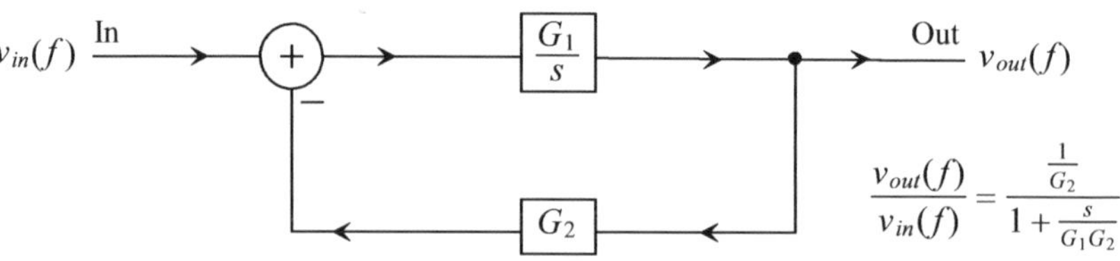

Figure 3.24 General implementation of a lowpass first-order filter.

The implementation is seen in Fig. 3.25. The coefficients are

$$G_1 = \frac{C_{I1}}{C_F} \cdot f_s \text{ and } G_2 = \frac{C_{I2}}{C_F} \cdot f_s \cdot \frac{1}{G_1} = \frac{C_{I2}}{C_{I1}} \quad (3.44)$$

The DC gain, as seen in Eq. (3.8) is set by the ratio of C_{I2} to C_{I1} $(1/G_2)$, and the filter's 3 dB frequency is

$$f_{3dB} = \frac{G_1 G_2}{2\pi} \quad (3.45)$$

Note that the DAI used in this filter has a transfer function of

$$v_{out}(z) = \frac{z^{-1}}{1 - z^{-1}} \cdot \left[\frac{C_{I1}}{C_F} \cdot v_{in}(z) \cdot z^{1/2} - \frac{C_{I2}}{C_F} \cdot v_{out}(z)\right] \quad (3.46)$$

■

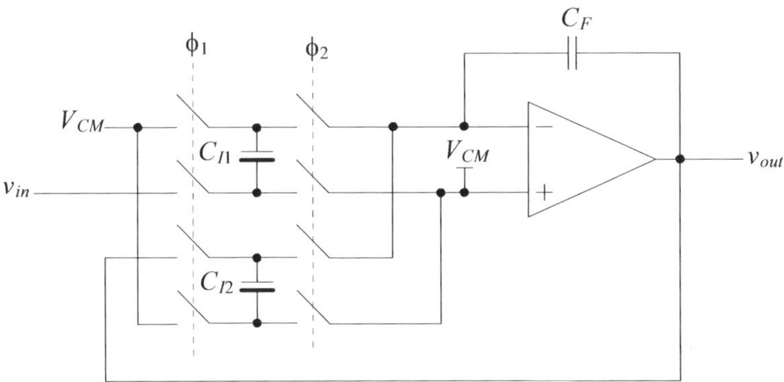

Figure 3.25 Implementation of the block diagram shown in Fig. 3.24.

The big benefit of switched-capacitor-based filters is the fact that the filters' poles and zeroes are determined by a ratio of capacitors and an external clock frequency (which is often a precise frequency set by a crystal oscillator). No tuning is needed. Varying the clocking frequency can precisely set the filter's characteristics for adaptive filtering (changing the filter's characteristics on the fly). Switched-capacitor filters with SNDRs in the 90 dB range have been attained at audio frequencies.

In the previous two examples we used ideal op-amps and didn't concern ourselves with the potential aliasing resulting from the analog sample-and-hold operation on the input of the filter. An analog antialiasing filter, AAF, (that is, not a discrete-time filter) must be used to remove the potential aliased signal prior to sampling. As with the noise-shaping modulator-based data converters we'll study later in the book, the fact that the sampling frequency, f_s, is much larger than the input frequencies of interest allows a relaxed AAF design. Indeed, the resistance of the MOSFET switches used combined with the switched input capacitance (C_I) form a lowpass filter. This filtering may serve as the switched-capacitor filter's AAF in many designs.

An Important Note

If we pause for a moment and think about the filters we have covered in this chapter we come to the realization that all require precise analog components. High-speed, wide-bandwidth op-amps and/or components with precise matching or absolute values are needed. We might argue that this would be a reason to focus our discussion on digital filtering (filters using only multipliers, delays, and adders) instead of filters using analog components. However, digital filters can't alone filter an analog waveform without first running the signal through an ADC. Further, traditional digital filters that use general-purpose multipliers at reasonable speeds can be very large (take up a significant layout area). They can be so large in fact as to not be practical in a general purpose filtering application. Special-purpose chips have been fabricated specifically for digital filtering (called digital signal processors, DSPs).

The noise-shaping topologies that use oversampling, discussed later in the book, can reduce the requirements placed on the analog components in the circuit. *Isn't it logical then to attempt to combine noise-shaping with purely digital filtering for the design of analog interfaces?* The answer is obviously, "yes"; however, as mentioned above, we have some caveats. While the resulting interface will place lower demands on the precision of the analog circuitry, we'll need to develop digital filters that don't rely on complex multipliers. The multiplications we do use should be simple, perhaps requiring an additional adder, or trivial (shift) multiplication. Also we'll need to use the digital filters to not only filter the input signal but to filter out the modulation noise present in the output of the NS modulator. Again these topics are discussed in much greater detail later in the book.

Exact Frequency Response of an Ideal Discrete-Time Filter

Figure 3.26 shows an equivalent diagram for Figs. 3.21a and 3.21b. We have replaced the ratio of capacitors, C_I/C_F, with the variable A in the figure. To determine the transfer function we can write

$$v_{out}(f) = A \cdot [v_{in}(f) - v_{out}(f)] \cdot \frac{z^{-1}}{1-z^{-1}} \tag{3.47}$$

or

$$\frac{v_{out}(f)}{v_{in}(f)} = \frac{Az^{-1}}{Az^{-1}+1-z^{-1}} = \frac{A}{z-(1-A)} \tag{3.48}$$

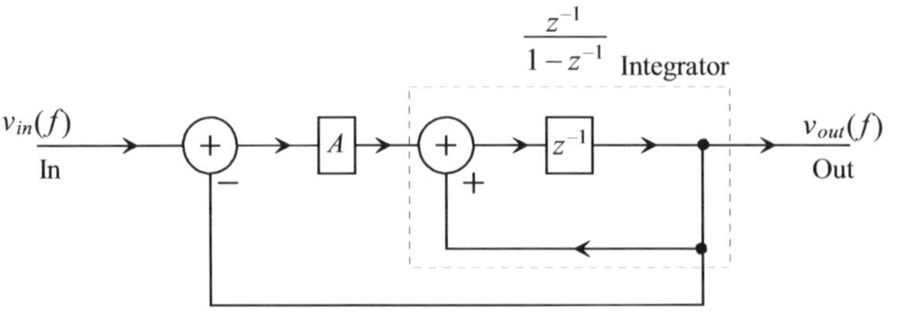

Figure 3.26 Digital block diagram of an integrator-based discrete-time lowpass filter.

Figure 3.27 shows the z-plane plot and magnitude response for this first-order filter. Remembering that all discrete-time filters have a periodic frequency response (a period of f_s or one complete revolution around the unit circle), we can compare this filter's response to the filters in the previous examples. To do so, let's assume $f_s = 100$ MHz and write, from Eq. (3.41),

$$f_{3dB} = \frac{Af_s}{2\pi} = 1.59\text{ MHz} = \frac{0.1 \cdot 100\text{ MHz}}{2\pi} \quad \text{that is, } A = 0.1 \qquad (3.49)$$

We can write the magnitude response of Eq. (3.48), with $z = \cos 2\pi f/f_s + j\sin 2\pi f/f_s$, as

$$\left|\frac{v_{out}}{v_{in}}\right| = \frac{A}{\sqrt{(\cos 2\pi f/f_s - 1 + A)^2 + \sin^2 2\pi f/f_s}} \qquad (3.50)$$

Figure 3.28 shows the responses of the first-order discrete-time filter. The maximum attenuation of the filter occurs at $f_s/2$ or 50 MHz here and is, from Fig. 3.27 with $A = 0.1$, 0.0526 or – 25.6 dB.

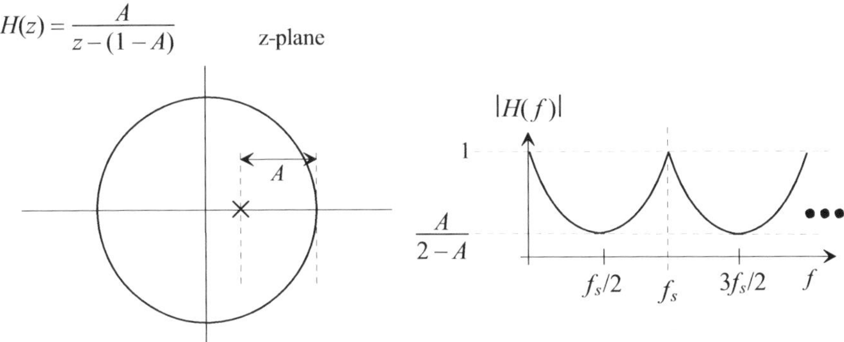

Figure 3.27 The z-plane representation and magnitude response of a first-order discrete-time filter.

3.2 Filtering Topologies

In this section we present some basic filter building blocks using integrators. Our focus is on continuous-time analog implementations of these filters.

3.2.1 The Bilinear Transfer Function

Consider the block diagram of the general first-order filter shown in Fig. 3.29. We can relate the filter's output to its input using

$$[v_{in}(f) \cdot [1 + sG_3] - G_2 \cdot v_{out}(f)] \cdot \frac{G_1}{s} = v_{out}(f) \qquad (3.51)$$

or

$$\frac{v_{out}(f)}{v_{in}(f)} = \frac{1}{G_2} \cdot \frac{1 + \frac{s}{1/G_3}}{1 + \frac{s}{G_1 G_2}} \qquad (3.52)$$

This filter's transfer function is termed "bilinear" because it is the ratio of two linear functions. Using this topology we can implement lowpass, allpass (used for phase

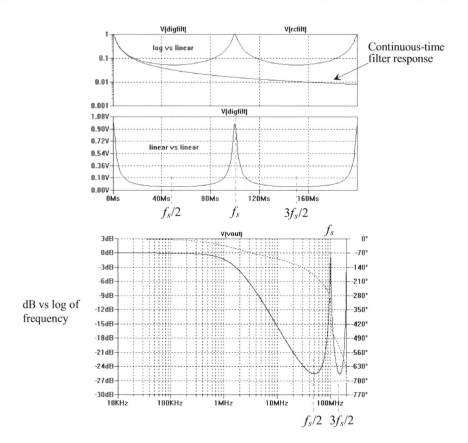

Figure 3.28 The magnitude response of the discrete-time first-order filter of Fig. 3.26 with an A of 0.1.

shifting), and highpass filters (keeping in mind that the highpass filter will ultimately change into a bandpass filter response because of the op-amp's or transconductor's high frequency rolloff). The location of the filter's pole is given by

$$f_{3dB,pole} = \frac{G_1 G_2}{2\pi} \tag{3.53}$$

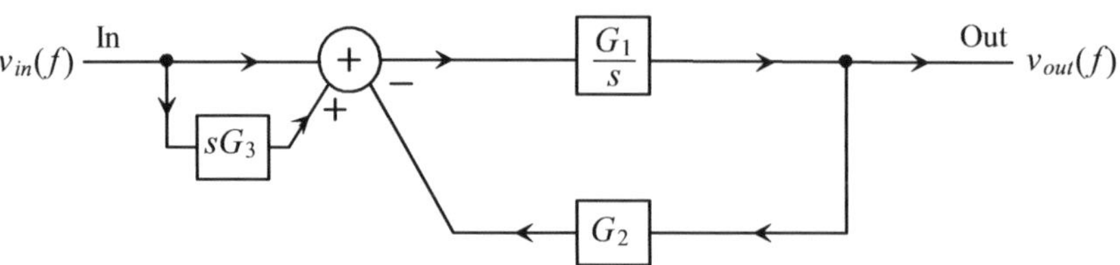

Figure 3.29 Implementation of a bilinear transfer function using an integrator.

while the filter's zero is located at

$$f_{3dB,zero} = \frac{1}{2\pi G_3} \tag{3.54}$$

The filter's gain at DC, in all cases, is

$$A_{DC} = \frac{1}{G_2} \tag{3.55}$$

Active-RC Implementation

The active-RC implementation of the bilinear transfer function is seen in Fig. 3.30. Again, as mentioned earlier, the resulting active-RC transfer function suffers from poor repeatability from one process run to the next. The RC time constants must be tuned, on-chip, with fuses (or switches) and adding/removing resistors or capacitors. Note how the summation is implemented by changing the input/output voltages to currents. The currents are summed at the inputs of the op-amp (which remain, ideally, at the common-mode voltage, V_{CM}). This is important to note in both the active-RC and switched-capacitor implementations.

We won't discuss the implementation of the MOSFET-C-based bilinear transfer function. It should be obvious that replacing the resistors in Fig. 3.30 with MOSFETs or linearized MOSFETs (see Figs. 3.12-3.14) provides a MOSFET-C implementation.

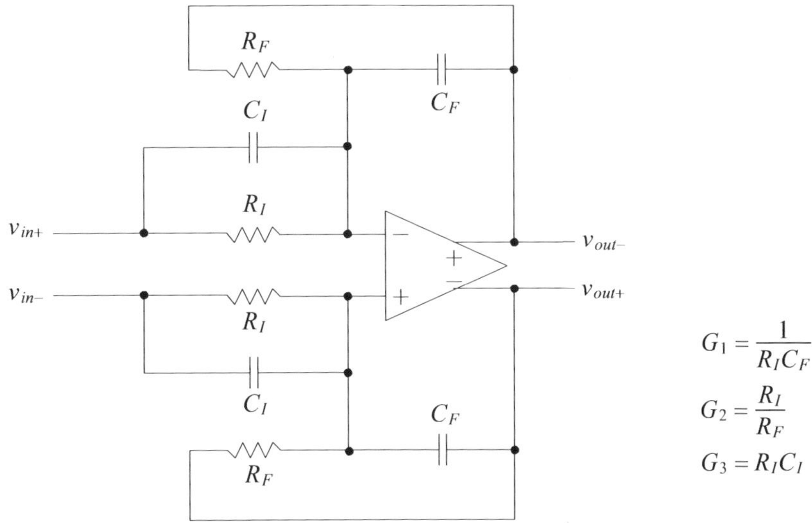

Figure 3.30 Implementation of an active-RC bilinear transfer function filter.

Transconductor-C Implementation

Figure 3.31 shows the implementation of the bilinear transfer function using transconductor stages. Again, as with the active-RC filter, signals are summed using currents. Summing the currents at the output nodes results in

$$g_{m1}(v_{in+} - v_{in-}) - g_{m2}(v_{out+} - v_{out-}) + \frac{v_{in+} - v_{out+}}{1/s2C_1} - \frac{v_{out+}}{1/s2C_2} = 0 \tag{3.56}$$

where we know $v_{out+} = -v_{out-}$ and $v_{in+} = -v_{in-}$. It will be helpful to write

$$\frac{v_{in+}}{1/s2C_1} = \frac{2v_{in+}}{1/sC_1} = \frac{v_{in+} - v_{in-}}{1/sC_1} \tag{3.57}$$

Using this expression, we can write Eq. (3.56) as

$$(v_{out+} - v_{out-}) \cdot (s(C_1 + C_2) + g_{m2}) = (v_{in+} - v_{in-}) \cdot (g_{m1} + sC_1) \tag{3.58}$$

or finally

$$\frac{v_{out+} - v_{out-}}{v_{in+} - v_{in-}} = \frac{g_{m1}}{g_{m2}} \cdot \frac{1 + \frac{s}{g_{m1}/C_1}}{1 + \frac{s}{g_{m2}/(C_1+C_2)}} \tag{3.59}$$

It's important to note that when looking at this equation, the location of the pole and zero can be adjusted by changing each transconductor's g_m independently. The ability to adjust one variable in a filter's transfer function and only change the position of a single pole or zero is called *orthogonal tuning*.

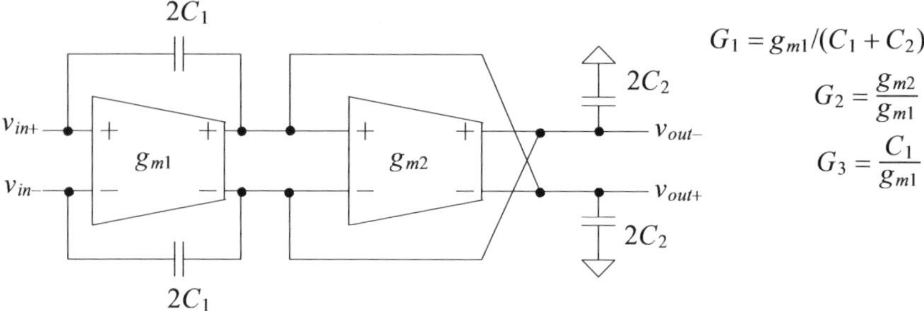

Figure 3.31 Implemention of a bilinear filter using transconductors.

Switched-Capacitor Implementation

The switched-capacitor, SC, implementation of the bilinear filter is seen in Fig. 3.32. This filter is directly derived from the active-RC filter of Fig. 3.30. From Eq. (3.42) we can write

$$R_I = \frac{1}{C_{I1} \cdot f_s} \text{ and } R_F = \frac{1}{C_{I2} \cdot f_s} \tag{3.60}$$

and so

$$G_1 = \frac{C_{I1}}{C_F} \cdot f_s,\ G_2 = \frac{C_{I2}}{C_{I1}}, \text{ and } G_3 = \frac{C_{I3}}{C_{I1} \cdot f_s} \tag{3.61}$$

Note how, in this discrete-time filter, the passband gain is C_{I3}/C_F when the filter is designed for a highpass response (and the filter no longer behaves like a discrete-time filter). The gain at DC in all situations is C_{I1}/C_{I2}.

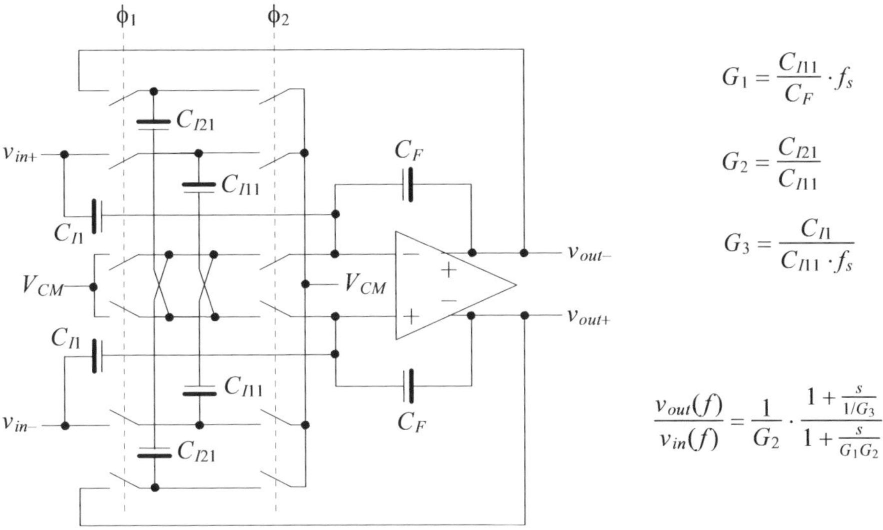

Figure 3.32 Implemention of a bilinear filter using switched capacitors.

3.2.2 The Biquadratic Transfer Function

As we briefly indicated in the last section, higher order filters can be implemented by cascading first-order sections. However, because the pole and zero locations in these first-order filters are restricted to real values, the performance of these cascades is poorer than filters with complex pole and zero locations. For example, cascading two identical lowpass filters having f_{3dB} frequencies of 1.59 MHz would result in a filter that has an attenuation of 6 dB at 1.59 MHz and a – 40 dB/decade roll off at higher frequencies. Using a second-order filter, we can design a filter to have a sharper transition at 1.59 MHz so that the attenuation is less than 6 dB at 1.59 MHz (however, the roll off remains –40 dB/decade). Further, we can use these sections to implement higher order filters using Butterworth, Chebyshev, Elliptic (Cauer), or Bessel responses.

The biquadratic, or "biquad" for short, filter transfer function (a ratio of two quadratic equations) is given by

$$\frac{v_{out}}{v_{in}} = \frac{a_2 s^2 + a_1 s + a_0}{s^2 + \left(\frac{2\pi f_0}{Q}\right)s + (2\pi f_0)^2} \tag{3.62}$$

where $2\pi f_0 = \omega_0$. The complex-conjugate poles are located at

$$p_1, p_2 = -\frac{\pi f_0}{Q} \pm \frac{1}{2}\sqrt{\left(\frac{2\pi f_0}{Q}\right)^2 - 4(2\pi f_0)^2} \tag{3.63}$$

or

$$p_1, p_2 = -\frac{\pi f_0}{Q} \pm j \cdot 2\pi f_0 \sqrt{1 - \left(\frac{1}{2Q}\right)^2} \tag{3.64}$$

In order to move toward the goal of implementing a biquad, consider the block diagram in Fig. 3.33. This block diagram is essentially the cascade of two first-order blocks (as seen in Fig. 3.29) except that instead of feeding the output of the second stage back to its input, we feed it back to the input of the first stage. We can determine the transfer function of this filter by writing

$$v_{out1} = (v_{in} + sG_3 v_{in} - G_5 v_{out} - G_2 v_{out1}) \cdot \frac{G_1}{s} \tag{3.65}$$

or

$$v_{out1}\left(\frac{s + G_1 G_2}{s}\right) = v_{in}\frac{(1+sG_3)G_1}{s} - v_{out}\frac{G_1 G_5}{s} \tag{3.66}$$

Further, we can relate v_{out1} to the output using

$$v_{out} = v_{out1}(1 + sG_6) \cdot \frac{G_4}{s} \tag{3.67}$$

Using Eq. (3.66) with Eq. (3.67) gives

$$v_{out} = \left(v_{in} \cdot \frac{(1+sG_3)G_1}{s + G_1 G_2} - v_{out}\frac{G_1 G_5}{s + G_1 G_2}\right) \cdot (1 + sG_6)\frac{G_4}{s} \tag{3.68}$$

or

$$v_{out}\left(1 + \frac{G_1 G_4 G_5 (1 + sG_6)}{s^2 + sG_1 G_2}\right) = v_{in}\left(\frac{(1+sG_3)G_1(1+sG_6)G_4}{s^2 + sG_1 G_2}\right) \tag{3.69}$$

Finally, the transfer function of the biquad is given by

$$\frac{v_{out}}{v_{in}} = \frac{s^2 G_1 G_3 G_4 G_6 + s(G_1 G_3 G_4 + G_1 G_4 G_6) + G_1 G_4}{s^2 + s(G_1 G_2 + G_1 G_4 G_5 G_6) + G_1 G_4 G_5} \tag{3.70}$$

Equating terms in Eqs. (3.62) and (3.70) gives

$$a_2 = G_1 G_3 G_4 G_6 \tag{3.71}$$

$$a_1 = G_1 G_3 G_4 + G_1 G_4 G_6 \tag{3.72}$$

$$a_0 = G_1 G_4 \tag{3.73}$$

$$\frac{2\pi f_0}{Q} = G_1 G_2 + G_1 G_4 G_5 G_6 \tag{3.74}$$

$$(2\pi f_0)^2 = G_1 G_4 G_5 \tag{3.75}$$

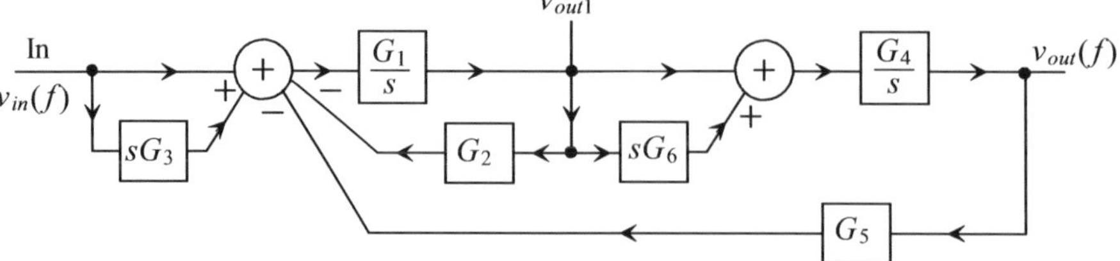

Figure 3.33 Implementation of a biquadratic transfer function using two integrators.

Active-RC Implementation

Figure 3.34 shows the active-*RC* implementation of the biquad filter along with the associated design equations. It should be noted that this is the general design schematic. If, for example, a lowpass filter needs to be implemented, the filter can be greatly simplified.

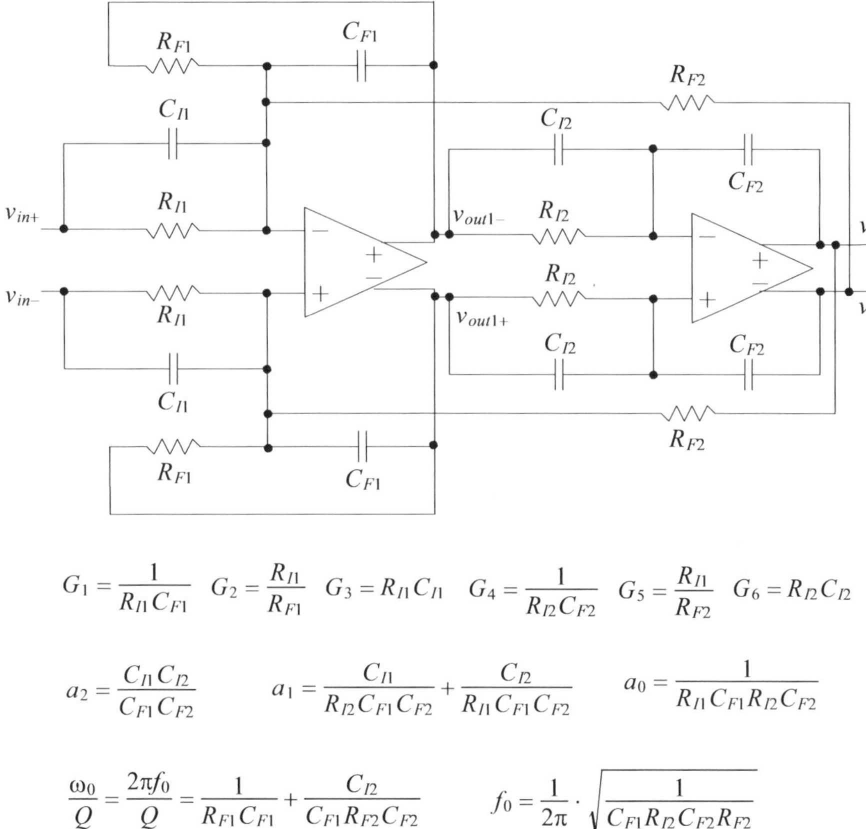

$$G_1 = \frac{1}{R_{I1}C_{F1}} \quad G_2 = \frac{R_{I1}}{R_{F1}} \quad G_3 = R_{I1}C_{I1} \quad G_4 = \frac{1}{R_{I2}C_{F2}} \quad G_5 = \frac{R_{I1}}{R_{F2}} \quad G_6 = R_{I2}C_{I2}$$

$$a_2 = \frac{C_{I1}C_{I2}}{C_{F1}C_{F2}} \qquad a_1 = \frac{C_{I1}}{R_{I2}C_{F1}C_{F2}} + \frac{C_{I2}}{R_{I1}C_{F1}C_{F2}} \qquad a_0 = \frac{1}{R_{I1}C_{F1}R_{I2}C_{F2}}$$

$$\frac{\omega_0}{Q} = \frac{2\pi f_0}{Q} = \frac{1}{R_{F1}C_{F1}} + \frac{C_{I2}}{C_{F1}R_{F2}C_{F2}} \qquad f_0 = \frac{1}{2\pi} \cdot \sqrt{\frac{1}{C_{F1}R_{I2}C_{F2}R_{F2}}}$$

Figure 3.34 Implementation of the active-RC biquadratic transfer function filter.

Figure 3.35 shows the frequency response, pole-zero locations in the s-plane, and transfer function for a second-order lowpass circuit made using an inductor (*L*), capacitor (*C*), and resistor (*R*). This LRC circuit has the same frequency response shape as a lowpass biquad filter. However, the DC gain of the LRC circuit must be unity while the DC gain of the biquad filter can be set to $a_0/(2\pi f_0)^2$. Note that if the pole quality factor, *Q*, is greater than $1/\sqrt{2}$ the response will show peaking. Setting *Q* to 0.707 results in the Butterworth or maximally flat response.

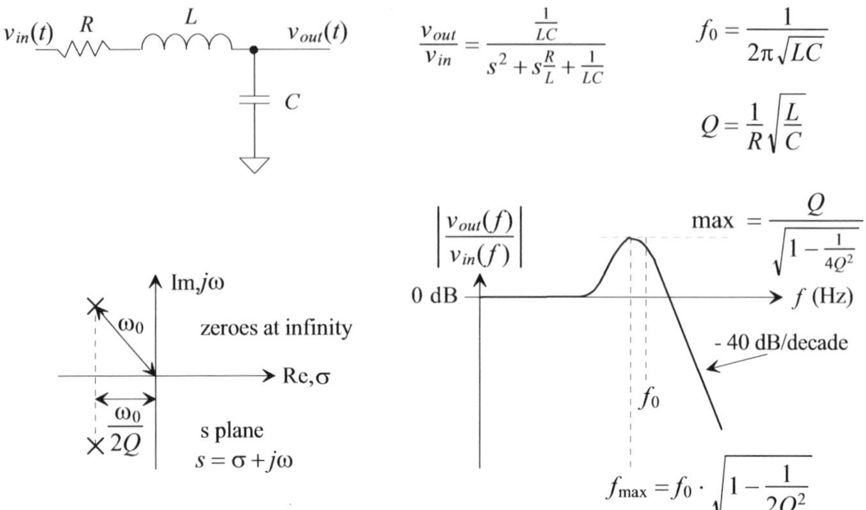

Figure 3.35 Second-order lowpass filter.

Example 3.8
Design an LRC circuit with a Q of 0.707 and a cutoff frequency (f_0) of 1.59 MHz.

From Fig. 3.35 we have two equations we need to solve

$$f_0 = \frac{1}{2\pi\sqrt{LC}} = 1.59\ MHz \rightarrow LC = 10 \times 10^{-15} \text{ and } Q = \frac{1}{\sqrt{2}} = \frac{1}{R}\sqrt{\frac{L}{C}}$$

We can set $C = 100$ pF, then $L = 100$ μH, and $R = 1.414$k (definitely not practical values if the circuit is going to be purely integrated). The response of the resulting LRC circuit is shown in Fig. 3.36. Note how the cutoff frequency is set by the

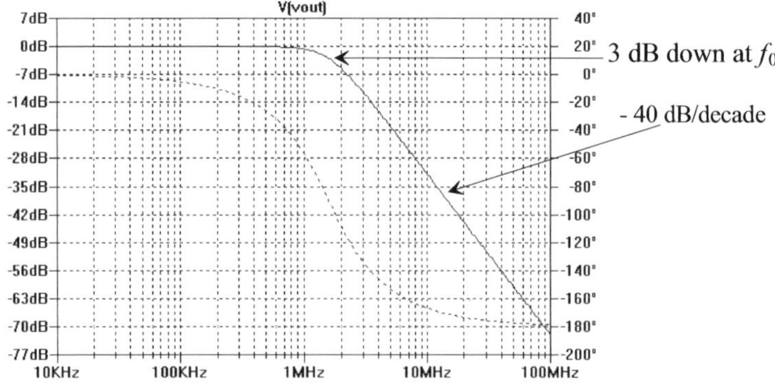

Figure 3.36 Second-order magnitude response for the circuit described in Ex. 3.8.

inductor and capacitor, while the Q of the circuit is set by all three elements (variations in the resistance having the largest effect on the circuit's Q). Higher Q indicates the poles are moving toward the imaginary axis (keeping in mind that a system with right-half plane poles is unstable [oscillates]) and more peaking, Fig. 3.37. ■

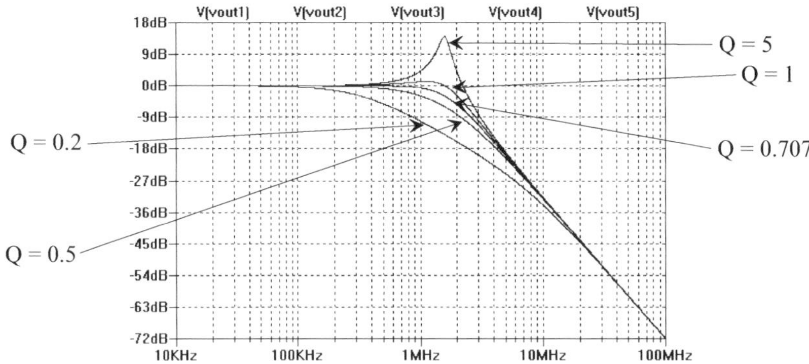

Figure 3.37 The effect of Q on the frequency response of a second-order lowpass filter.

Example 3.9
Simulate the design of an active-RC filter that has frequency characteristics similar to Fig. 3.36.

Using the basic topology of Fig. 3.34, we see that for a lowpass filter, $C_{I1} = C_{I2} = 0$ and therefore $G_3 = G_6 = 0$. Further

$$f_0 = 1.59\ MHz = \frac{1}{2\pi}\sqrt{\frac{1}{C_{F1}R_{I2}C_{F2}R_{F2}}}$$

which we shall use to set $R_{I2} = R_{F2} = 10k$ and $C_{F1} = C_{F2} = 10\,pF$. The Q of the filter is given by

$$Q = \frac{1}{\sqrt{2}} = 2\pi f_0 \cdot R_{F1}C_{F1} \rightarrow R_{F1} = 7.07\,k\Omega$$

Knowing $a_2 = a_1 = 0$, the gain at DC is $a_0/(2\pi f_0)^2$ or R_{F2}/R_{I1} $(1/G_5)$, which is 1 here. The simulation results are shown in Fig. 3.38. ■

Notice that at DC, when used in the lowpass configuration, the outputs of the first integrator, v_{out1+} and v_{out1-}, must be equal. If not, the difference is integrated by the second section. As the frequency increases, so does the difference in these voltages.

Figure 3.39 shows the second-order bandpass response. Again, as with the second-order lowpass response, the center frequency (resonant frequency) is set by the values of the inductor and capacitor. The Q of the filter indicates how narrow the bandpass response is; higher Q indicates a narrower response. Note how the response eventually rolls off at –20 dB/decade. At low frequencies the capacitor can be thought of as an open (resulting in a first-order RL circuit response), while at high frequencies the inductor can be thought of as an open (resulting in a first-order RC circuit response).

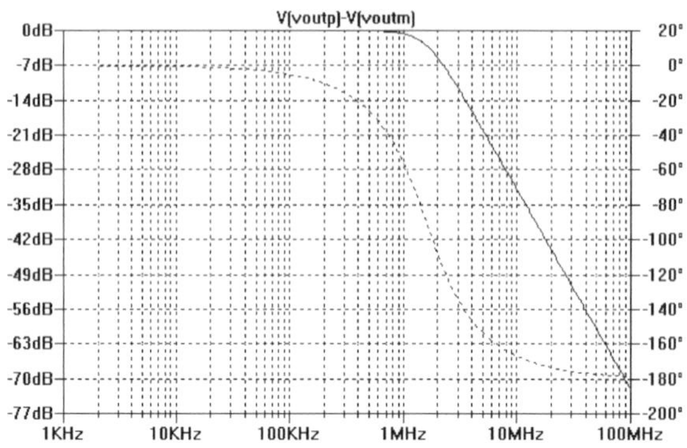

Figure 3.38 Magnitude and phase responses for the active-RC filter of Ex. 3.9.

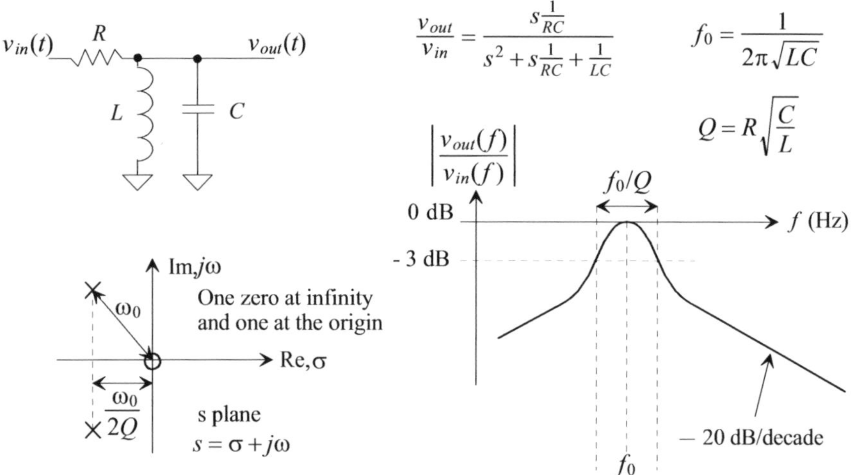

Figure 3.39 Second-order bandpass filter.

Example 3.10
Repeat Ex. 3.8 for the bandpass *LRC* circuit.

Again, we can set $C = 100$ pF and $L = 100$ μF. Solving for Q using the equation in Fig. 3.39 results in

$$Q = R\sqrt{\frac{C}{L}} = 0.707 = R\sqrt{\frac{100p}{100\mu}} \rightarrow R = 707$$

The simulation results are seen in Fig. 3.40. ■

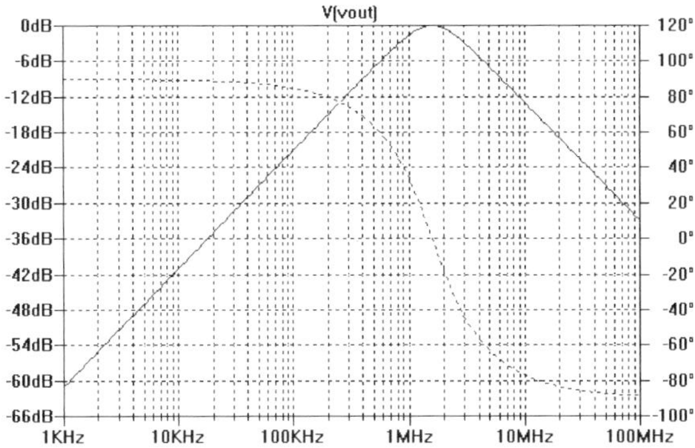

Figure 3.40 Bandpass response of a second-order circuit with a Q of 0.707.

Example 3.11
Repeat Ex. 3.10 if the Q is increased to 20.

Figure 3.41 shows the simulation results. To attain a Q of 20, we use a resistor of 20k with the inductor and capacitor values remaining unchanged. ■

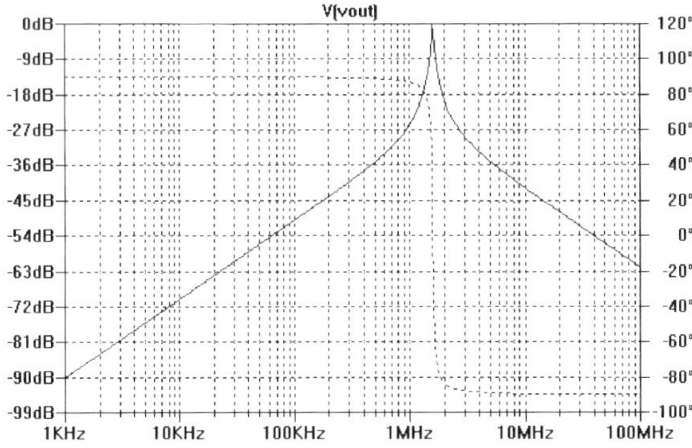

Figure 3.41 Bandpass response of a second-order circuit with a Q of 20.

Example 3.12
Use an active-RC filter to implement a filter with the response shown in Fig. 3.41.

Let's begin by writing the filter's transfer function

$$\frac{v_{out}}{v_{in}} = \frac{a_1 s}{s^2 + \left(\frac{2\pi f_0}{Q}\right)s + (2\pi f_0)^2} = \frac{a_1 s}{s^2 + \left(\frac{10\times 10^6}{20}\right)s + (10\times 10^6)^2}$$

Looking at this equation, Eq. (3.62), and Fig. 3.34 we see that $a_2 = a_0 = 0$ and so $C_{I2} = 0$ and $R_{I1} = \infty$. Further then

$$a_1 = \frac{C_{I1}}{R_{I2}C_{F1}C_{F2}} \qquad f_0 = \frac{1}{2\pi}\sqrt{\frac{1}{C_{F1}R_{I2}C_{F2}R_{F2}}} = 1.59 \text{ MHz}$$

$$\frac{2\pi f_0}{Q} = \frac{2\pi \cdot 1.59 \text{ MHz}}{20} = \frac{1}{R_{F1}C_{F1}}$$

The passband gain (the maximum gain) occurs at f_0 and is calculated by replacing s in the transfer function above with $j2\pi f_0$. It is given by

$$A_{passband} = \frac{a_1 Q}{2\pi f_0}$$

If, again, we set $C_{F1} = C_{F2} = 10$ pF and $R_{I2} = R_{F2} = 10$k, we get an f_0 of 1.59 MHz. Further then, with a Q of 20, we can set R_{F1} to 200k. Finally, setting the passband gain to unity results in

$$a_1 = \frac{2\pi f_0}{Q} = \frac{C_{I1}}{R_{I2}C_{F1}C_{F2}} \rightarrow C_{I1} = 0.5 \text{ pF}$$

While these values do result in a biquad with the shape seen in Fig. 3.41, the values are not practical. Redoing the calculations while trying to minimize the component spread gives another possible solution: R_{I2} = 100k, C_{F1} = 20p, R_{F1} = 100k, C_{F2} = 5p, C_{I1} = 5p, and R_{F2} = 1k. The simulation results are seen in Fig. 3.42. ■

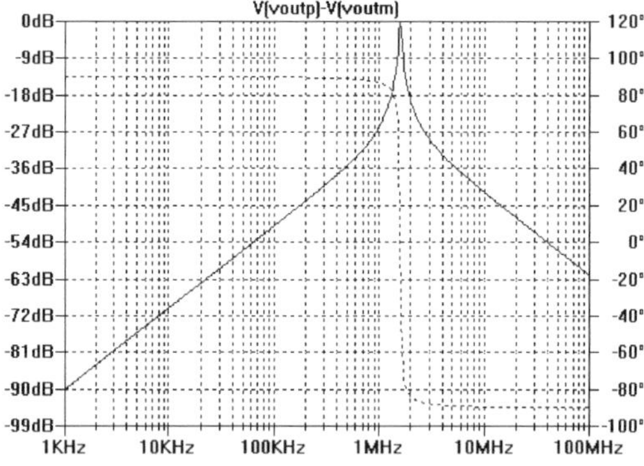

Figure 3.42 Outputs of the biquad of Ex. 35.12 using active-RC elements.

Switched-Capacitor Implementation

The switched-capacitor implementation of the biquad circuit is shown in Fig. 3.43. Note how this circuit is a simple translation of the active-RC circuit of Fig. 3.34. Again, if the filter designed using this section has a lowpass or bandpass response, it can be simplified. For example, from Figs. 3.35 and 3.39 (the implementation of lowpass and bandpass filters), we see that a_2 is zero. This indicates that G_6 *can be set to zero* (removing C_{I2} in Figs. 3.34 or 3.43). The resulting second-order filter response can be written as

$$\frac{v_{out}}{v_{in}} = \frac{a_1 s + a_0}{s^2 + \left(\frac{2\pi f_0}{Q}\right)s + (2\pi f_0)} = \frac{sG_1G_3G_4 + G_1G_4}{s^2 + sG_1G_2 + G_1G_4G_5} \tag{3.76}$$

Technically, the filter is no longer biquadratic, so we will refer to it as a second-order filter.

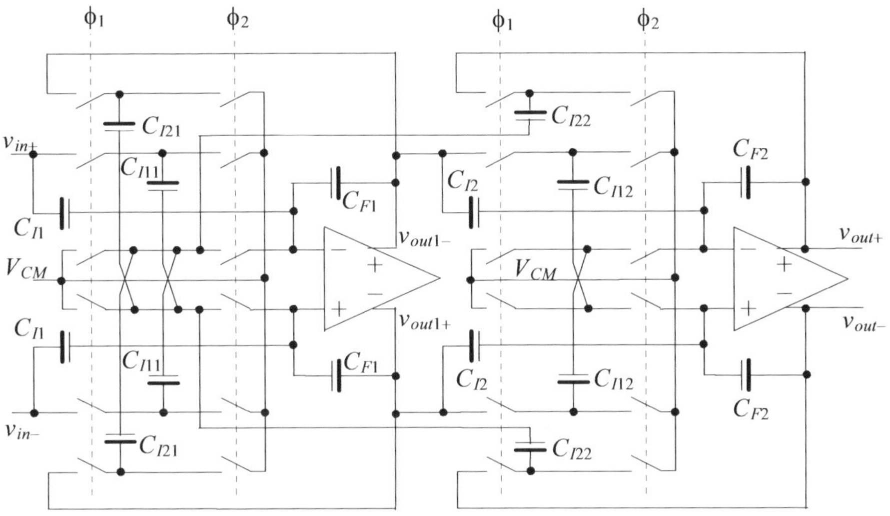

$$G_1 = \frac{C_{I11}}{C_{F1}} \cdot f_s \quad G_2 = \frac{C_{I21}}{C_{I11}} \quad G_3 = \frac{C_{I1}}{C_{I11} \cdot f_s} \quad G_4 = \frac{C_{I12}}{C_{F2}} \cdot f_s \quad G_5 = \frac{C_{I22}}{C_{I11}} \quad G_6 = \frac{C_{I2}}{C_{I12} \cdot f_s}$$

Figure 3.43 Implementing a biquad filter using switched capacitors.

The biquad in Fig. 3.43 can look confusing until we start to dissect it. If we understand the topology of Fig. 3.32, we see that the switched-capacitor biquad is nothing more than two bilinear filters connected in cascade. The only difference is that the switched capacitance, C_{I22}, is fed back to the input of the first op-amp to simulate the feedback resistance, R_{F2}, in Fig. 3.34. This circuit, for the general lowpass or bandpass filter implementation, gets much simpler when the unused components are removed.

High Q

We have a major concern alluded to in Ex. 3.12 when using either of the topologies of Fig. 3.34 or 3.43 with a large Q. As we saw in this example the capacitor values were within a factor of 4 of each other (20p and 5p) but the resistors used were two orders of magnitude different (100k and 1k). This large difference can be traced to, again assuming $G_6 = 0$,

$$\frac{2\pi f_0}{Q} = G_1 G_2 \text{ and } 2\pi f_0 = \sqrt{G_1 G_4 G_5} \text{ or } Q = \frac{\sqrt{G_4 G_5}}{\sqrt{G_1}\, G_2} = \sqrt{\frac{C_{F1}}{R_{I2} C_{F2} R_{F2}}} \cdot R_{F1} \tag{3.77}$$

This equation shows that R_{F1} has the largest direct dependence on Q. Using a large value of R_{F1} results in a smaller feedback signal (a smaller amount of current is fed back to the input of the first op-amp). In other words G_2 in Fig. 3.33 is small.

In order to minimize the amount of signal, v_{out1}, fed back and summed with the input signal, while at the same time forcing the components to have similar values, consider the modified, from Fig. 3.33, biquad block diagram shown in Fig. 3.44. All we have done here is added a separate signal path in parallel with the G_2 path. Instead of subtracting, though, we are now adding the signal to the input summing block. Equation (3.70) can be rewritten, assuming G_6 is zero (a bandpass or lowpass response), as

$$\frac{v_{out}}{v_{in}} = \frac{s(G_1 G_3 G_4) + G_1 G_4}{s^2 + sG_1(G_2 - G_{2Q}) + G_1 G_4 G_5} \tag{3.78}$$

or, equating the coefficient of s in the denominator of this equation with the coefficient of s in the denominator of Eq. (3.62), results in

$$\frac{2\pi f_0}{Q} = G_1(G_2 - G_{2Q}) \tag{3.79}$$

The implementation of the "high-Q" biquad is seen in Fig. 3.45 (with G_6 included). The additional gain from the figure is

$$G_{2Q} = \frac{R_{I1}}{R_{F1}\frac{Q}{Q-1}} = G_2 \cdot \frac{Q-1}{Q} \tag{3.80}$$

Rewriting Eq. (3.79) results in

$$\frac{2\pi f_0}{Q} = G_1 G_2\left(1 - \frac{Q-1}{Q}\right) = \frac{G_1 G_2}{Q} \tag{3.81}$$

Let's use this result in the following example.

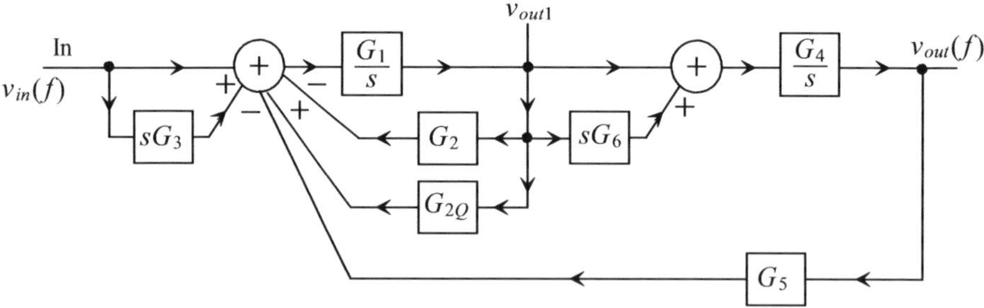

Figure 3.44 Implementation of a "high-Q" biquadratic transfer function.

Example 3.13
Repeat Ex. 3.12 using the high-Q circuit of Fig. 3.45.

The passband gain is 1 so we know that

$$a_1 = \frac{2\pi f_0}{Q} = \overbrace{G_1 G_3 G_4}^{a_1} = G_1(G_2 - G_{2Q}) = \frac{G_1 G_2}{Q}$$

or $2\pi f_0 = 10 \times 10^6 = G_1 G_2 = 1/R_{F1}C_{F1}$. We also know that

$$G_1 G_3 G_4 = \frac{C_{I1}}{C_{F1} R_{I2} C_{F2}} = G_1(G_2 - G_{2Q}) = \frac{1}{C_{F1} R_{F1}} \cdot \frac{1}{Q}$$

and

$$2\pi f_0 = 10 \times 10^6 = \sqrt{\frac{1}{C_{F1} R_{I2} C_{F2} R_{F2}}}$$

In an attempt to minimize component spread let's set R_{F1} to 5k, R_{I2} to 20k, C_{I1} to 4p, C_{F2} to 20p, C_{F1} = 20p, R_{F2} = 1.25k, and finally R_{F1Q} = 5.25k (roughly). The simulation results and schematic are shown in Fig. 3.46. ■

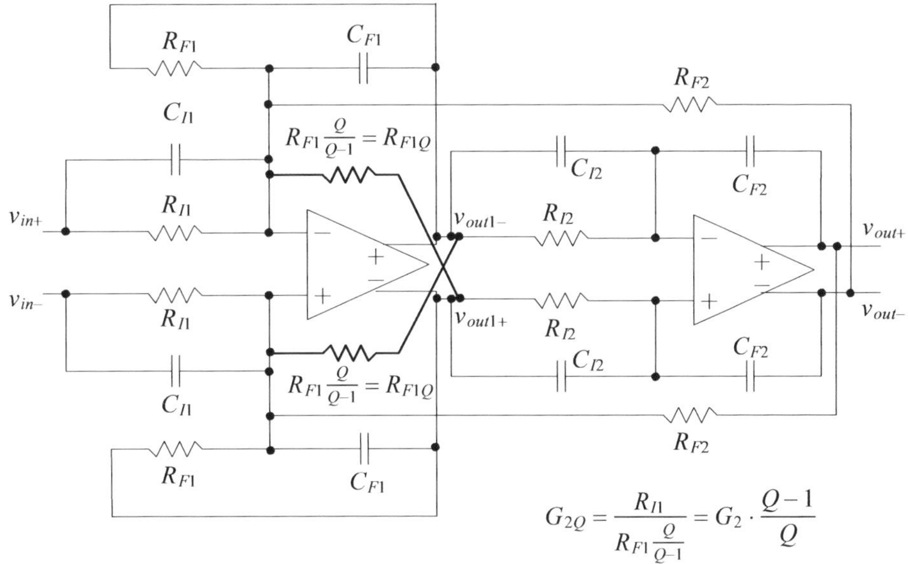

Figure 3.45 Implementation of the "high-Q" active-RC biquadratic transfer function filter. The bold lines indicate the added components.

Example 3.14

Repeat Ex. 3.13 using a switched-capacitor implementation. Assume that the filter is clocked at 100 MHz.

To implement the filter, we need to replace the resistors in Fig. 3.46 with switched capacitors. However, we notice in the gain equations that the resistors are all ratios of capacitors. This means we can reduce the size of the filter by scaling the values in Fig. 3.46. In order to do this let's divide all capacitors by 10 and multiply all resistors by 10. Therefore, we can write C_{I1} = 0.4p, C_{F1} = 2p, and C_{F2} = 2p. The resistors can be calculated using

$$R_{F1} = \frac{1}{C_{I21} \cdot f_s} = 50k \rightarrow C_{I21} = 0.2p$$

$$R_{F1Q} = \frac{1}{C_{I21Q} \cdot f_s} = 52.5k \rightarrow C_{I21Q} = 0.190p$$

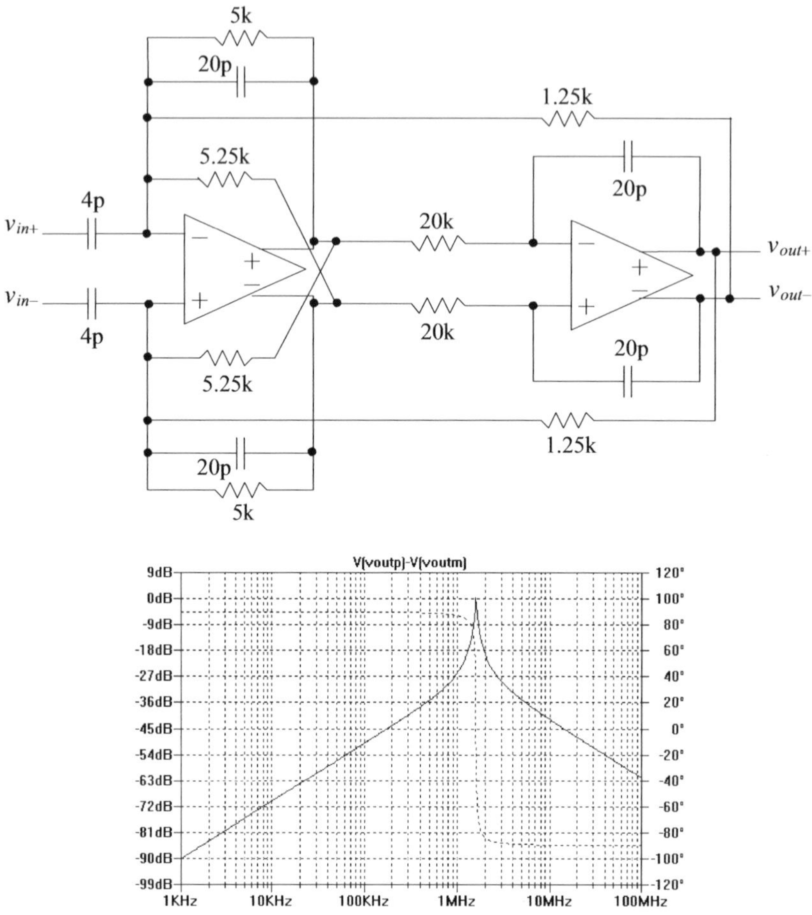

Figure 3.46 Bandpass filter discussed in Ex. 3.13.

$$R_{I2} = \frac{1}{C_{I12} \cdot f_s} = 200k \rightarrow C_{I12} = 0.05p \text{ (!)}$$

$$R_{F2} = \frac{1}{C_{I22} \cdot f_s} = 12.5k \rightarrow C_{I22} = 0.8p$$

Looking at the value of C_{I12}, we see it may be too small. Let's change its scale factor from 10 to 4. This means

$$R_{I2} = \frac{1}{C_{I12} \cdot f_s} = 80k \rightarrow C_{I12} = 0.125p$$

We have to scale C_{F2} as well (so that G_4 remains constant). Now $C_{F2} = 5$ pF.

Figure 3.47 shows the implementation of the filter. Note how easy it was to implement the high-Q circuitry. All we did was add two capacitors to the circuit. Also note how the circuit is simplified after removing the unused components (R_{f1} and C_{f2}).

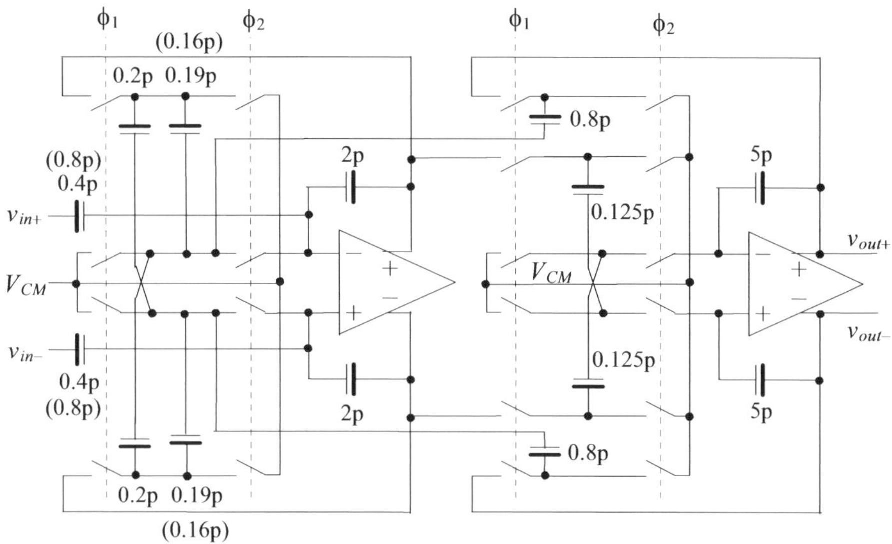

Figure 3.47 Switched-capacitor implementation of a high-Q filter; see Ex. 35.14.

Again, as in Ex. 3.6, because this circuit can only be simulated using a transient analysis, we will input a sinewave at a known frequency and verify that we get the correct output. Looking at Fig. 3.46, we see that if we apply a 1 V signal to the filter at 1.59 MHz we should get a 1 V signal out at 1.59 MHz. However, as seen in Fig. 3.48, the filter is unstable and oscillates. Using simulations it's easy to show that even if we ground the inputs of the filter, the outputs oscillate at f_0 (1.59 MHz). To understand why, remember in Fig. 3.39 that as the Q of the bandpass filter is increased, the poles move closer to the right-half plane. If, for some reason, the poles move into the right-half plane, the filter will become an oscillator (unstable). It's important to remember that when we designed the filter we approximated our discrete-time variable z as $1+\frac{s}{f_s}=1+j\frac{2\pi f}{f_s}$ (when f $<<f_s$ $\cos\frac{2\pi f}{f_s}\approx 1$ and $\sin\frac{2\pi f}{f_s}\approx\frac{2\pi f}{f_s}$). We could be more exact and write

$$z=e^{j\frac{2\pi f}{f_s}}=\cos\frac{2\pi f}{f_s}+j\sin\frac{2\pi f}{f_s} \tag{3.82}$$

which clearly will not follow $1+\frac{2\pi f}{f_s}$ for frequencies f approaching the sampling frequency f_s. As we discussed in Ch. 2, sampled signals will have spectral content in excess of the sampling frequency. Practically, the spectral content is limited by the combination of the switches "on" resistance and the capacitors in the circuit.

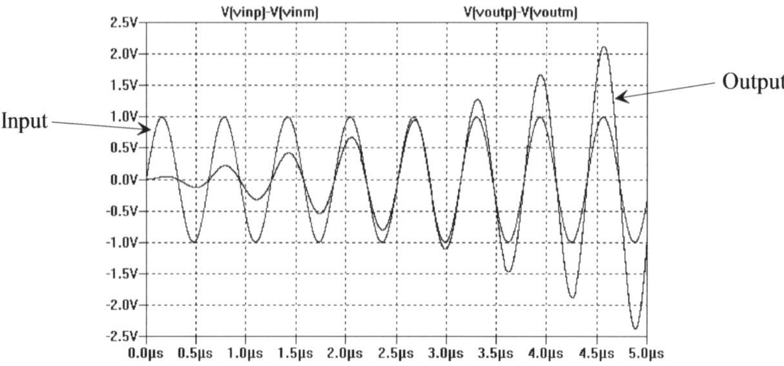

Figure 3.48 Output of the filter in Fig. 3.47 showing instability.

As f gets larger, the cosine term will decrease from one, causing the real portion to get smaller. A decrease in the real component, as seen in the complex plane in Fig. 3.39, causes the pole to move closer to the right-half plane (causing the Q to increase).

Practically, the maximum Q we can design for (but not attain) is in the neighborhood of 5. If we redesign the biquad of Fig. 3.47 to have a Q of 5, we see that all we need to change is C_{I1} (from 0.4p to 0.8p) and C_{I21Q} (from 0.19p to 0.16p). The simulation results are seen in Fig. 3.49. In this figure, we adjusted the input frequency until we reached a 3 dB point (the filter's center frequency was 1.59 MHz, as expected). This occurred at 1.52 MHz and 1.66 MHz. The Q of the circuit is not 5, but is, from Fig. 3.39, 1.59/(1.66 – 1.52) or 11.36. ■

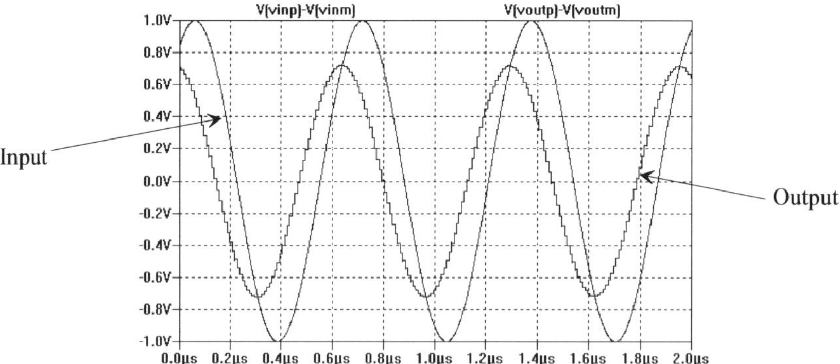

Figure 3.49 Output of the filter in Fig. 3.47 after lowering the Q to maintain stability.

Q Peaking and Instability

While it would appear the active-RC circuit is the best choice for high-Q filter implementations, we must remember that the discussion neglected the effects of the finite gain-bandwidth product (f_{un}) of the op-amps. We can model these effects by replacing the ideal integrator gain of $1/s$ with

$$\frac{1}{s} \to \frac{1}{s\left(1+\frac{s}{2\pi f_{un}}\right)} \tag{3.83}$$

Using this result, we can rewrite the pole locations of Eq. (3.62) (see Eqs. [3.63] and [3.64]) as

$$\frac{1}{\left(s\left(1+\frac{s}{2\pi f_{un}}\right)+p_1\right)} \cdot \frac{1}{\left(s\left(1+\frac{s}{2\pi f_{un}}\right)+p_2\right)} \tag{3.84}$$

or, looking at a single term,

$$s\left(1+\frac{s}{2\pi f_{un}}\right)+p_1 = s+p_1-\overbrace{\frac{(2\pi f)^2}{2\pi f_{un}}}^{\text{Unwanted}} \tag{3.85}$$

This subtraction results in a shift in the pole toward the right-half plane, increasing the Q of the circuit; Fig. 3.50. Reviewing Eq. (3.64), we can subtract the unwanted term in Eq. (3.85) to estimate the shift in the Q or

$$\frac{\pi f_0}{Q} - \frac{(2\pi f)^2}{2\pi f_{un}} \text{ or at } f = f_0 \text{ we can write } \pi f_0\left(\frac{1}{Q} - \frac{2f_0}{f_{un}}\right) \tag{3.86}$$

The shifted Q is then

$$\frac{1}{Q_{shift}} = \frac{1}{Q} - \frac{2f_0}{f_{un}} \to Q_{shift} = \frac{Q}{1-\frac{Q2f_0}{f_{un}}} \tag{3.87}$$

So, for the filter Q to remain finite, we require

$$\frac{Q2f_0}{f_{un}} << 1 \tag{3.88}$$

Let's use this result in the following example.

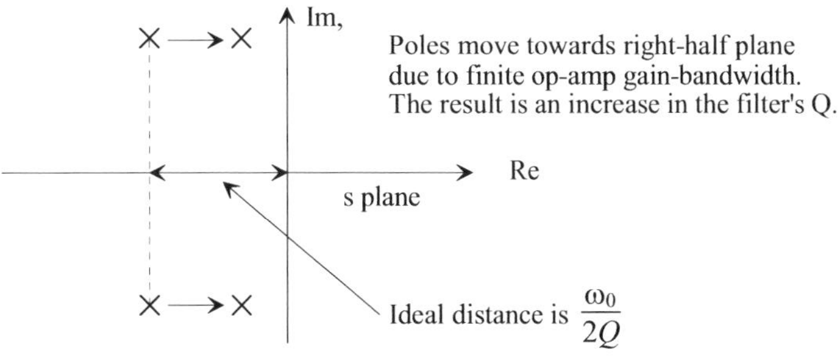

Figure 3.50 Showing Q peaking resulting from the op-amp finite gain bandwidth product.

Example 3.15

Resimulate the filter in Ex. 3.13 using op-amps that have an f_{un} of 100 MHz.

The center frequency, f_0, of this filter is 1.59 MHz and the Q is 20. Using Eqs. (3.87) and (3.88), we can estimate the increase in Q due to op-amp finite gain-bandwidth product as

$$\frac{Q2f_0}{f_{un}} = \frac{20 \cdot 2 \cdot 1.59}{100} = 0.636 \rightarrow Q_{shift} = 55$$

Practically, this is too high of a Q (the poles are too close to the right-half plane), and the filter will be unstable (noise in the circuit, or simulation noise in the simulation, will push the poles into the right-half plane). Figure 3.51 shows the simulation results (see also Ex. 3.4). The inputs to the filter are grounded. The unstable oscillation frequency is close to the ideal, f_0, but is shifted by a small amount. ■

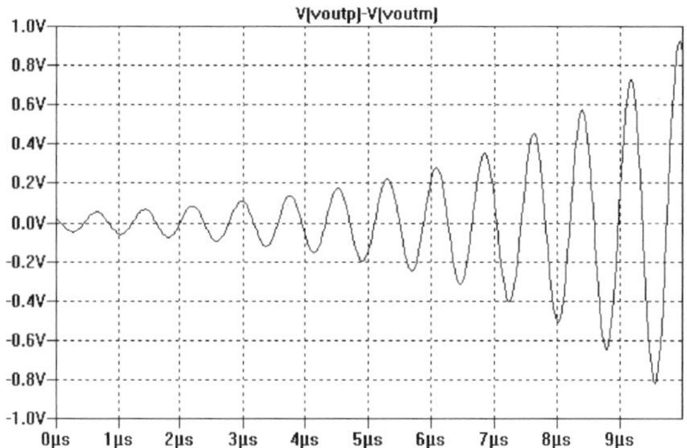

Figure 3.51 Showing how the filter of Ex. 3.13 becomes unstable due to finite op-amp bandwidth.

Transconductor-C Implementation

Let's redraw the bilinear filter in Fig. 3.31 as seen in Fig. 3.52. We redraw it like this to show how the feedback gain, G_2, is implemented. In the block diagram of the biquad filter shown in Fig. 3.43, we used a similar scheme to implement the feedback gain, G_5. Figure 3.53 shows the implementation of a biquad filter using transconductors where we have drawn it so that the transconductors appear to be connected in series. This topology can be redrawn so that it looks similar to Fig. 3.52 (showing a direct correspondence between it and Fig. 3.43). Note that we could have drawn the schematic without the crossing wires if we switched the output polarity of two of the transconductors (that is, put the minus output on the top of the output instead of the plus output).

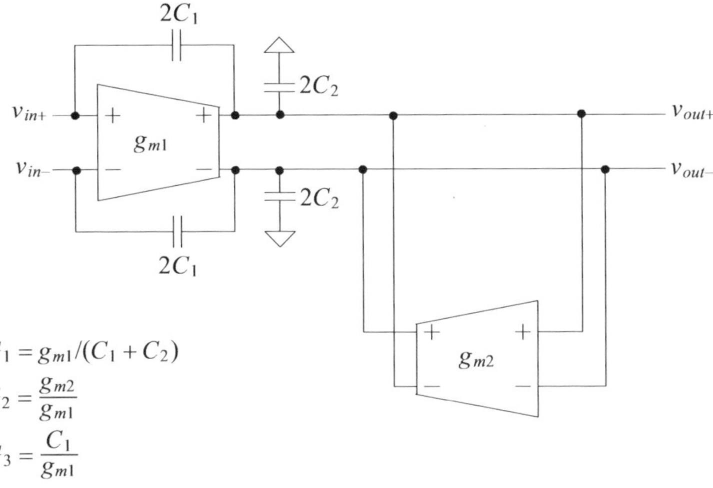

Figure 3.52 Redrawing the bilinear filter shown in Fig. 3.31.

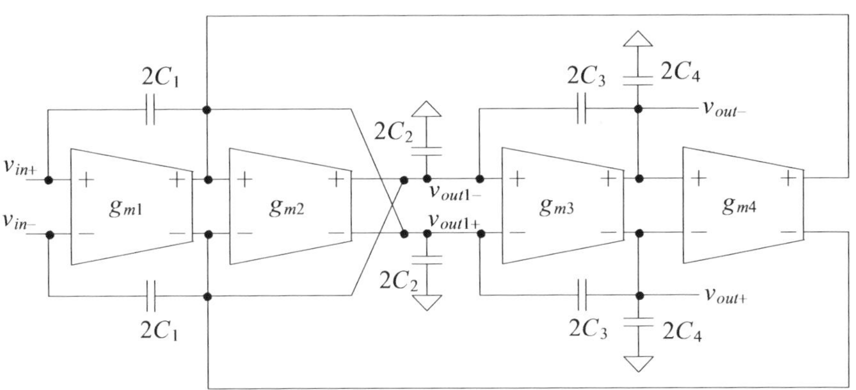

$$G_1 = g_{m1}/(C_1 + C_2) \; G_2 = \frac{g_{m2}}{g_{m1}} \quad G_3 = \frac{C_1}{g_{m1}} \quad G_4 = g_{m3}/(C_3 + C_4) \; G_5 = \frac{g_{m4}}{g_{m1}} \quad G_6 = \frac{C_3}{g_{m3}}$$

Figure 3.53 Implementing a biquadratic filter using transconductors.

ADDITIONAL READING

[1] S. Franco, *Design with Operational Amplifiers and Analog Integrated Circuits*, Third Edition, McGraw-Hill, 2003. ISBN 978-0071207034

[2] R. Schaumann and M. E. Van Valkenburg, *Design of Analog Filters*, Oxford University Press, 2001. ISBN 978-0195118773

[3] P. R. Gray, D. A. Hodges, and R. W. Brodersen (eds.), *Analog MOS Integrated Circuits*, Wiley-IEEE, 1980. ISBN 0-471-08964-8

[4] P. R. Gray, B. A. Wooley, and R. W. Brodersen (eds.), *Analog MOS Integrated Circuits II*, Wiley-IEEE, 1989. ISBN 0-87942-246-7

[5] P. A. Lynn, *An Introduction to the Analysis and Processing of Signals*, Hemisphere Publishing Corporation, 1989. ISBN 0-89116-981-4

[6] Y. P. Tsividis and J. O. Voorman (eds.), *Integrated Continuous-Time Filters: Principles, Design, and Applications*, Wiley-IEEE, 1993. ISBN 0-7803-0425-X

[7] B. Nauta, *Analog CMOS Filters for Very High Frequencies*, Kluwer Academic Publishers, 1993. ISBN 0-7923-9272-8

QUESTIONS

3.1 Resketch Fig. 3.2 for the following circuit.

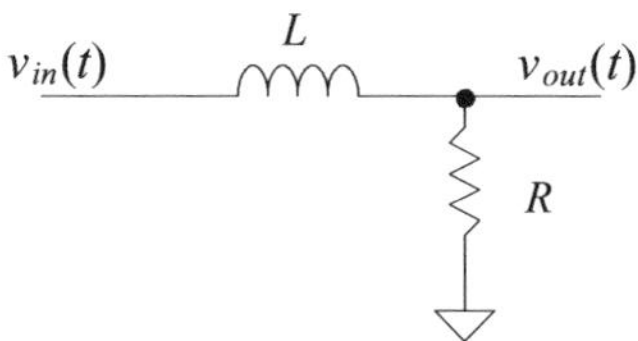

Figure 3.54 First-order lowpass filter using an inductor and a resistor.

3.2 Show that Eq. (3.6) is still valid if the circuit's inputs and outputs are referenced to the common-mode voltage, V_{CM}. (The op-amp inputs should also be at V_{CM}.)

3.3 Sketch the implementation of a first-order lowpass filter using a CAI with a 3 dB frequency of 10 MHz and a DC gain of 6 dB. Simulate your design to verify it works as expected.

3.4 Plot, in the complex plane, the ideal pole location and the actual pole locations due to finite op-amp unity gain frequency for the filter described in Ex. 3.4.

3.5 Plot Eq. (2.59) of the last chapter using SPICE and the op-amp model shown in Fig. 3.8.

3.6 Suppose an antialiasing filter was required for a 12-bit data converter. Further assume the filter is to be implemented using an active-RC topology. If $VDD = 1.0$ V, estimate the minimum value of the integration capacitor that should be used, assuming the filter's noise performance is dominated by thermal noise. Is it wise, for 12-bit system performance, to design the filter so that its SNR is equal to the SNR of the data converter?

3.7 Repeat question 3.6 if the op-amp used in the filter has a linear output swing of 80% of the power supply voltage.

3.8 Derive the transfer function for the filter shown in Fig. 3.16 if the transconductors have different g_ms. Sketch the block diagram, similar to the one seen in Fig. 3.6, for the filter.

3.9 Derive the transfer function for the following first-order transconductor filter.

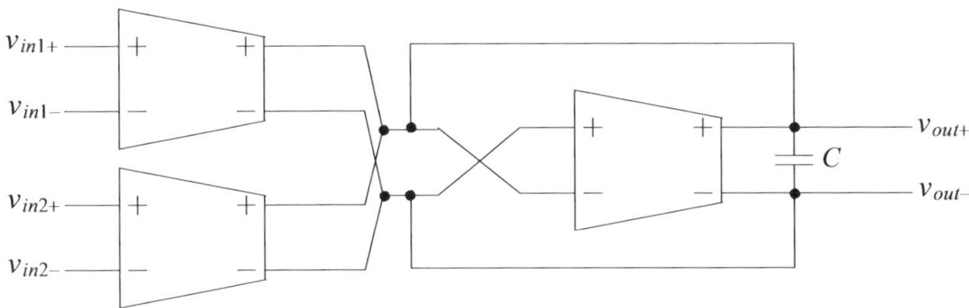

Figure 3.55 A first-order filter with two inputs.

3.10 Show the derivation details that result in Eqs. (3.44) and (3.46).

3.11 Show the details of how the gains, G, are derived in Fig. 3.30.

3.12 Is it possible to tune the gain, Q, and cutoff frequency of the lowpass biquad independently? If so, how? Give examples using the simulation netlist used to generate Fig. 3.38.

3.13 What happens to the poles in the biquadratic equation, Eq. (3.62), if the Q is less than 0.5? (Hint: The filter behaves like the cascade of two first-order filters.) Is the f_{max} equation in Fig. 3.35 valid?

3.14 Compare the size of the elements used in Exs. 3.8 and 3.9. Is there a benefit to using an active element for monolithic implementation?

3.15 Show, using the simulations from Ex. 3.14, that increasing the switch resistance, and thus the spectral content present in a switched capacitor circuit, can help to stabilize high-Q switched-capacitor bandpass filters.

3.16 Redesign and simulate the operation of the filter discussed in Ex. 3.14, with a Q of 5, while trying to minimize the difference between C_{I1} and C_{F2}. Suggest a possible modification to the filter topology (similar to how we add G_{2Q} in Fig. 3.45) to reduce this component spread.

3.17 Show how to derive the transfer function of the transconductor-C biquad filter seen in Fig. 3.53. Can this filter be orthogonally tuned? If so, how?

3.18 Repeat Ex. 3.9 using the transconductor-based biquad.

3.19 How would a "high-Q" biquad be implemented using transconductors? Repeat Ex. 3.12 using the transconductor-based biquad.

3.20 Show, using biquad sections, how the lowpass ladder filter seen in Fig. 3.56 can be implemented.

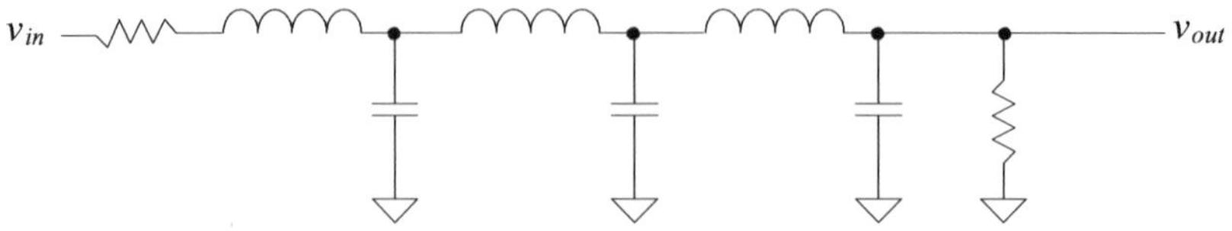

Figure 3.56 Implementing a ladder filter using biquads, see problem 3.20.

Chapter

4

Digital Filters

In this chapter we study the implementation of digital filters specifically useful for mixed-signal circuit design. As in the last chapter, our filters will be based on the integrator and, to a lesser extent, the differentiator. Prior to studying the material in this chapter the reader may want to review Sec. 1.2.

4.1 SPICE Models for DACs and ADCs

In order to verify, using SPICE simulations, the theory for digital filters discussed in this chapter, we start out by developing simulation models for ideal digital-to-analog converters (DACs) and analog-to-digital converters (ADCs). Our goal is to generate ideal DACs and ADCs that we can place in a mixed-signal simulation to either generate a digital word based on an analog input or look at the spectrum of a digital signal.

4.1.1 The Ideal DAC

Consider the ideal transfer characteristics of a 3-bit DAC shown in Fig. 4.1. Notice in this figure that we have drawn two reference voltages, V_{REF+} and V_{REF-}, and are assuming that $V_{REF+} > V_{REF-}$. When a digital input of 000 is applied to the DAC, the output voltage becomes V_{REF-}. When the input code is increased to 001, the output of the DAC (an analog voltage defined at discrete amplitude levels) increases by one least significant bit (LSB). If the DAC has an input code with a number of bits, N, then we can define an LSB as

$$1 \text{ LSB} = \frac{V_{REF+} - V_{REF-}}{2^N} = V_{LSB} \text{ for } N \geq 2 \tag{4.1}$$

If, for example, $V_{REF+} = 1.25$ V and $V_{REF-} = 0.25$ V and $N = 3$, then our LSB, the vertical distance between adjacent points in Fig. 4.1, is 0.125 V. Note that in our discussion of an ideal DAC we are assuming that the output of the DAC ranges from V_{REF-} up to $V_{REF+} - 1$ LSB. We could just as easily have assumed that the output ranged from $V_{REF-} + 1$ LSB up to V_{REF+}. The important thing to notice is that the DAC output range is 1 LSB smaller than the difference between the positive and negative reference voltages. For the DAC developed in this chapter, we will assume $V_{REF+} = VDD = 1.0$ V and $V_{REF-} = 0$ V. Selection of the power supply rails, which are noise free in a SPICE

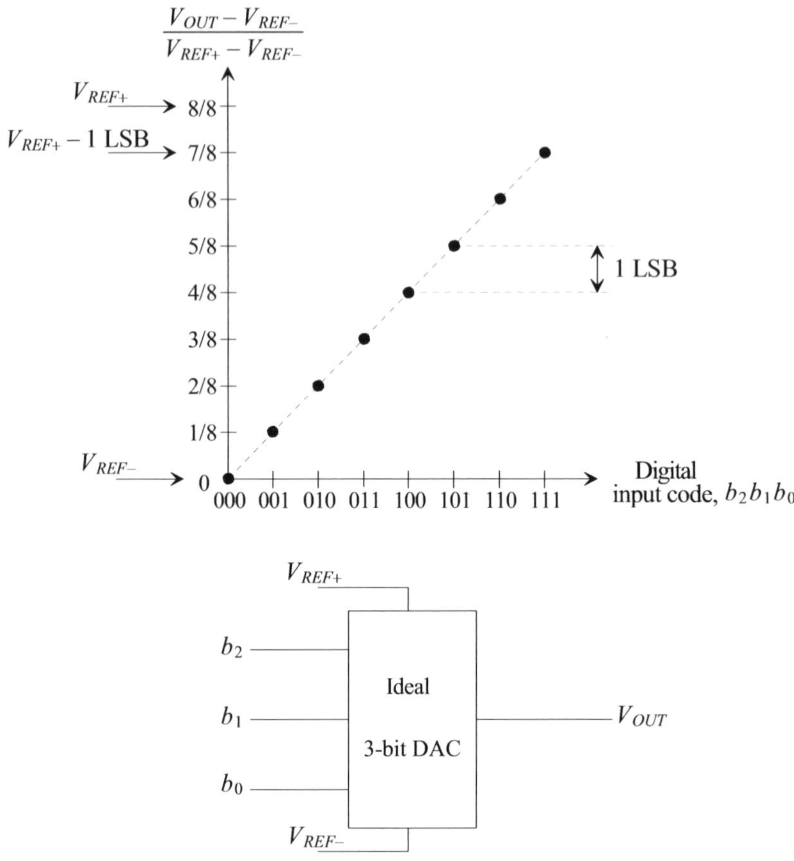

Figure 4.1 An ideal 3-bit DAC.

simulation, allow the maximum output range for the DAC (assuming the reference voltages are indeed the maximum and minimum voltages in the system, i.e., no charge pumps or external, larger, power supply voltages). If we need more resolution when using our ideal DAC, we will simply increase the number of bits, N, used and thus decrease the value of the DAC's LSB.

SPICE Modeling the Ideal DAC

We can write the output of the ideal DAC in terms of the reference voltages and digital input codes b_N (which are logic "0" or "1"), and assuming that an input code of all zeroes results in an output voltage of V_{REF-}, as

$$V_{OUT} = (V_{REF+} - V_{REF-}) \cdot \left(\frac{b_{N-1}}{2^1} + \frac{b_{N-2}}{2^2} + \ldots + \frac{b_1}{2^{N-1}} + \frac{b_0}{2^N} \right) + V_{REF-} \qquad (4.2)$$

or

$$V_{OUT} = (V_{REF+} - V_{REF-}) \cdot \frac{1}{2^N} \cdot (b_{N-1}2^{N-1} + b_{N-2}2^{N-2} + ... + b_1 \cdot 2^1 + b_0) + V_{REF-} \qquad (4.3)$$

We can implement this equation, in SPICE, using a nonlinear dependent source (a B source). For a 3-bit, ideal DAC, the statement that implements this equation may look like

```
*Nonlinear dependent source, B, for generating the 3-bit DAC output
Bout Vout 0 V=((v(vrefp)-v(vrefm))/8)*(v(B2L)*4+v(B1L)*2+v(B0L))+v(vrefm)
```

The terms BXL correspond to logic signals that have a value of 1 V or 0 V.

Example 4.1
Write the nonlinear dependent SPICE source statement for an ideal 12-bit DAC.

The statement follows:

```
Bout Vout 0 V=((v(vrefp)-v(vrefm))/4096)*
+(v(B11L)*2048)+v(B10L)*1024+v(B9L)*512+v(B8L)*256+
+v(B7L)*128+v(B6L)*64+v(B5L)*32+v(B4L)*16+v(B3L)*8+
+v(B2L)*4+v(B1L)*2+v(B0L))+v(vrefm)
```

remembering that a "+" in the first column of a line indicates that the text on the remainder of the line behaves as if it were typed at the end of the previous line. It doesn't indicate addition. ■

The next thing we need to concern ourselves with is the digital logic levels. We want to use our ideal DAC with real circuits where the logic voltage levels may not be well defined. We need to determine and use a switching-point voltage based on the power-supply voltage *VDD*. We will assume the input logic code is a valid logic "1" if its amplitude is greater than *VDD*/2 and a logic "0" if its amplitude is less than *VDD*/2. The switch implementation used to generate the logic signals (1 V and 0 V) used in our dependent source from real signals is shown in Fig. 4.2.

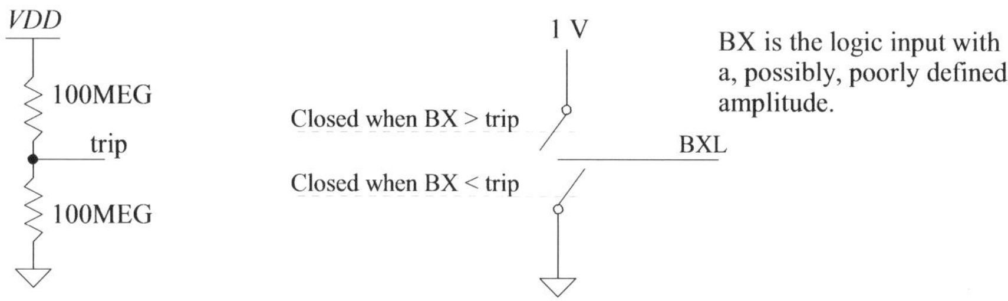

Figure 4.2 Generating logic levels using voltage-controlled switches.

4.1.2 The Ideal ADC

The characteristics of our ideal ADC are shown in Fig. 4.3. Notice that the transfer curve is shifted to the left. If we were to flip the curve on its side and mark, with black dots, the intersection of the analog input voltage with the ADC transfer curve, we would have the DAC transfer curve of Fig. 4.1. Again, 1 LSB is given by Eq. (4.1). Notice how converting a (normalized) input voltage of 0.1 V will result in an output code of 000 which is the same output code resulting from converting 0 V. Unlike the ideal DAC, the ideal ADC quantizes its input with the practical result of adding noise to the input signal. This noise is called *quantization noise*.

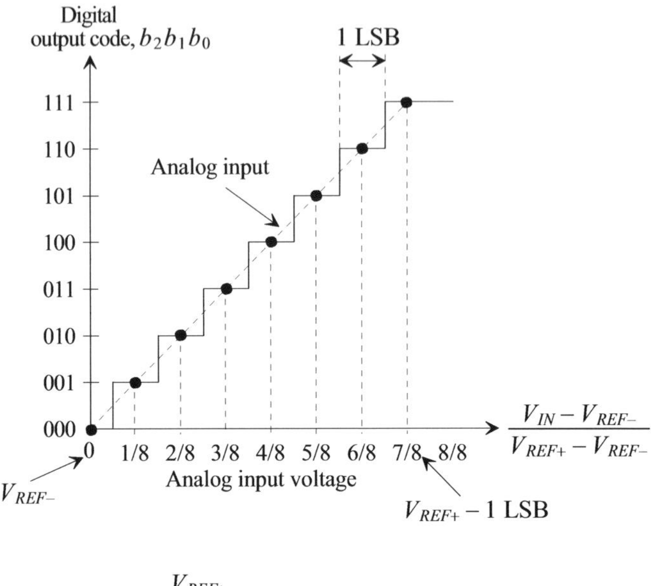

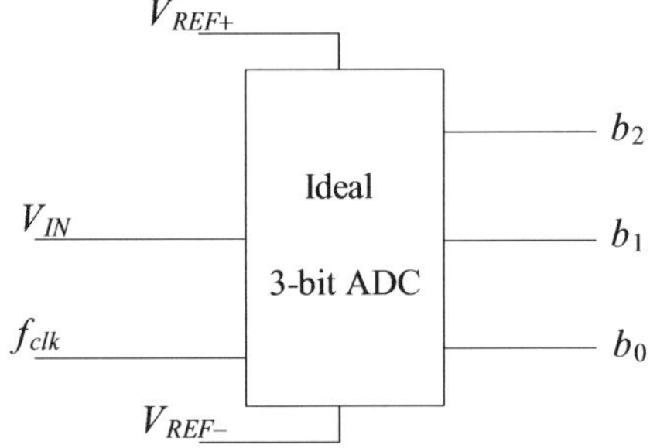

Figure 4.3 An ideal 3-bit ADC.

The implementation of the ideal ADC consists of an ideal S/H followed by passing the output of the S/H (the held signal) through an algorithm to generate the output bits. The algorithm we use is based on a pipeline ADC and follows:

1. The input signal is sampled and held.
2. This held signal is input to a comparator that compares the input value to a reference voltage.
3. If the input signal is greater than the reference voltage, the output bit is set to a high, and the reference signal is subtracted from the input. The difference is multiplied by two and passed to the output of the stage.

4. If the input signal is less than the reference voltage, the output bit is set low. The input signal is multiplied by two and passed to the output of the stage.

5. This output is used as the input to the next stage and steps 2, 3, and 4 above are repeated. This continues for N stages (where N is the number of bits in the ADC).

The reference voltage, or common mode voltage V_{CM}, can be determined by calculating the midpoint between V_{REF+} and V_{REF-} followed by subtracting V_{REF-} so that the V_{CM} is referenced to 0 V. This can be written as

$$V_{CM} = \frac{V_{REF+} + V_{REF-}}{2} \rightarrow V_{CM0} = \frac{V_{REF+} + V_{REF-}}{2} - V_{REF-} = \frac{V_{REF+} - V_{REF-}}{2} \quad (4.4)$$

We also want to level-shift the input signal so that it is referenced to 0 V. In addition, we want to shift the transfer curves to the left by 1/2 LSB as seen in Fig. 4.3. In order to do this we use the following SPICE statement (for an 8-bit ADC where V(OUTSH) is the output voltage of the ideal S/H [the input to the pipeline algorithm above])

```
* Level shift by VREFM and 1/2LSB
BPIP PIPIN 0 V=V(OUTSH)-V(VREFM)+((V(VREFP)-V(VREFM))/2^9)
```

The last term in this statement is 1/2 LSB, which is given by

$$1/2 \text{ LSB} = \frac{V_{REF+} - V_{REF-}}{2^{N+1}} \text{ assuming } V_{REF+} > V_{REF-} \geq 0 \quad (4.5)$$

We are level-shifting the input and common-mode voltage because we want to make the model as flexible as possible. For example, we want the ADC model to function if $V_{REF+} = 0.5$ V and $V_{REF-} = 0.25$ V. Note that if $V_{REF-} = 0$ and $V_{REF+} = VDD$, the model can be simplified.

4.1.3 Number Representation

Suppose we have an ideal ADC and DAC each with a resolution of 8-bits, a $V_{REF+} = 1$ V and a V_{REF-} of ground. Using Eq. (4.1) the data converter's LSB is 3.906 mV. The minimum output of the DAC, Fig. 4.1, is ground and the maximum output is VDD – 1 LSB or 0.9961 V. An input sinewave swinging from ground to VDD and centered around the common-mode voltage, V_{CM}, of 500 mV is seen in Fig. 4.4. Also seen are the corresponding digital codes and voltages. This number format is referred to as *binary offset* format. The output of our ideal ADC and the input of our ideal DAC are in this format. A more useful format, for adding and subtracting digital numbers, is the *two's complement* format, Fig. 4.5. We get this format by complementing the MSB of a word in

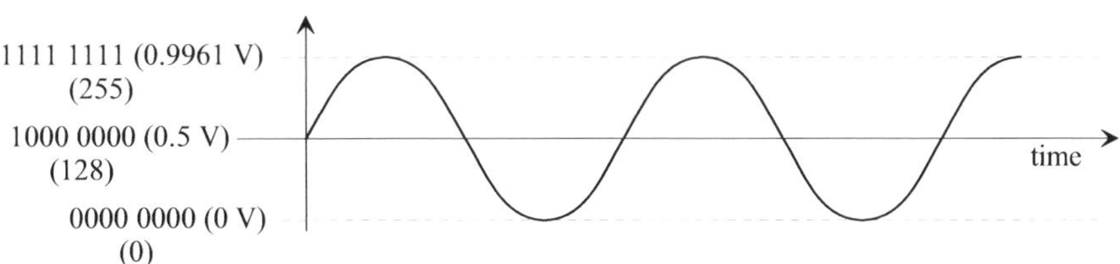

Figure 4.4 Representing a sinusoid in binary offset format.

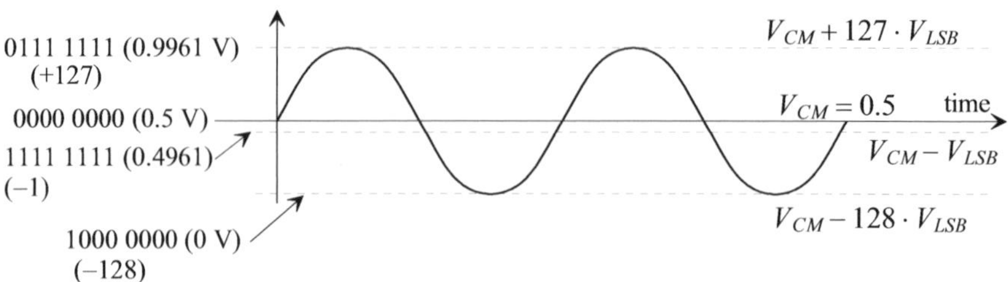

Figure 4.5 Representing a sinusoid in two's complement format.

binary offset (and thus to go back-and-forth between two's complement and binary offset we simply complement the MSB). Note, in Fig. 4.5, that −1 is represented using 1111 1111, −2 is represented using 1111 1110, etc. Also note that the MSB corresponds to the word's sign-bit. An MSB of 0 indicates a positive value (or the common-mode voltage, V_{CM}, 0000 0000) while an MSB of 1 indicates a negative value (a code or voltage below V_{CM}).

Increasing Word Size (Extending the Sign-Bit)

Often as we design digital filters we'll need to increase the word size to increase the resolution of the signal or to avoid register overflow. Figure 4.6 shows how we would increase the binary offset representation of the output of a 3-bit ADC to 8-bits while changing the format to two's complement representation. The MSB of the ADC's output is inverted and extended. Notice what would happen if our ADC's input signal were V_{CM}. The ADC's output is then 100. This is changed into 0000 0000. The common-mode signal, as seen in Fig. 4.5, can be thought of as a reference, or zero, level. Since we'll be using two's complement numbers throughout the rest of the book the reader should spend some time reviewing, and ensuring they understand, Figs. 4.4, 4.5, and 4.6.

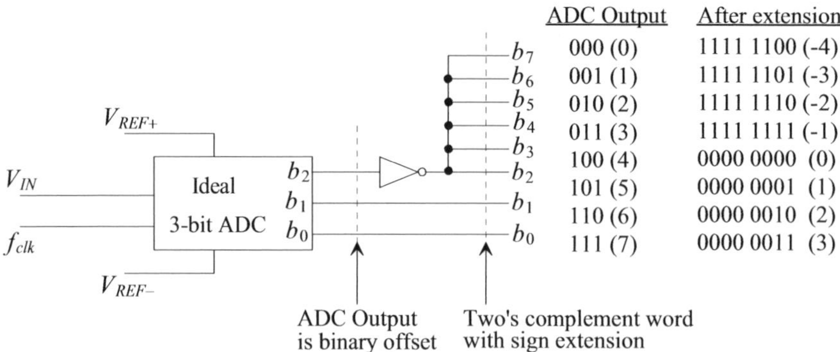

ADC Output	After extension
000 (0)	1111 1100 (-4)
001 (1)	1111 1101 (-3)
010 (2)	1111 1110 (-2)
011 (3)	1111 1111 (-1)
100 (4)	0000 0000 (0)
101 (5)	0000 0001 (1)
110 (6)	0000 0010 (2)
111 (7)	0000 0011 (3)

Figure 4.6 Showing how to change the output of an ADC to two's complement and how to extend the sign bit.

Example 4.2
Show how to convert a 1-bit binary offset number into 2-bit and 4-bit two's complement numbers.

A 1-bit binary offset number has values of 0 or 1 (common for the output of a noise-shaping ADC). To change this into a 2-bit two's complement number we'll represent 0 as −1 and 1 as +1 or

$$0 \rightarrow 11\ (-1)$$

$$1 \rightarrow 01\ (+1)$$

For 4-bit representations we can write $0 \rightarrow 1111\ (-1)$ and $1 \rightarrow 0001\ (+1)$. We'll use these results frequently when discussing noise-shaping (delta-sigma) data converters. ■

A comparator is an example of an ADC (or quantizer) that will generate the 1-bit number used in this example. Reviewing Eq. (4.1) we see that this equation doesn't work for determining the LSB of a 1-bit ADC. We can modify Eq. (4.1) for this situation

$$1\ \text{LSB} = V_{REF+} - V_{REF-} = V_{LSB} \text{ for } N = 1 \tag{4.6}$$

Adding Numbers and Overflow

Figure 4.7 shows how two's complement numbers are added. Note that the first thing we do, when adding two digital signals, is to extend our sign bit to avoid harmful overflow (see the allowable overflow examples in the figure). This increases the word size and allows the final output word size to always be large enough to accommodate the sum of the inputs. *Note that this is important.* In filters employing feedback, like the digital integrator seen in Fig. 1.26, we may increase the word size even more to avoid, or delay, harmful overflow. For an integrator a DC input will always, after some time, result in overflow (unless the integrator's input is always zeroes). Also note that harmful overflow is easy to detect since it will only occur when the two inputs are the same polarity (positive or negative). If the MSB of the two input words is a 0 (1) then the output word's MSB must be 0 (1) . If it's a 1 (0) then we know overflow occurred. In simpler terms, if the two inputs are positive (negative) then the output must be positive (negative).

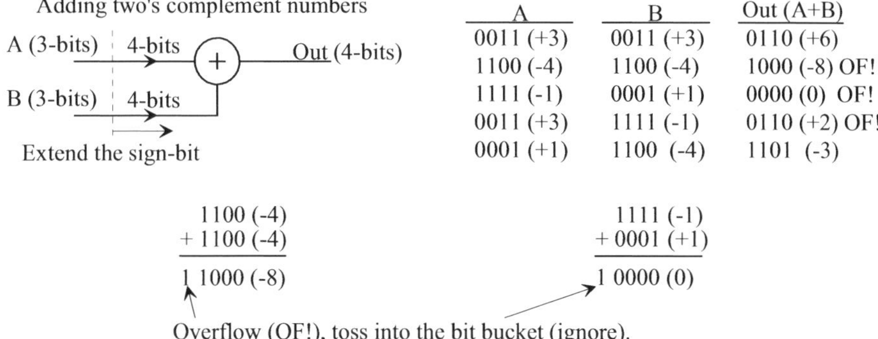

A	B	Out (A+B)
0011 (+3)	0011 (+3)	0110 (+6)
1100 (-4)	1100 (-4)	1000 (-8) OF!
1111 (-1)	0001 (+1)	0000 (0) OF!
0011 (+3)	1111 (-1)	0110 (+2) OF!
0001 (+1)	1100 (-4)	1101 (-3)

Figure 4.7 Showing how two's complement numbers are added.

Subtracting Numbers in Two's Complement Format

Figure 4.8 shows how two's complement numbers are subtracted. Here we are subtracting the input B from the input A. In order to do this we complement B (run the N-bit word through N inverters). We then tie the carry input of the adder high. Shown in Fig. 4.8 are several examples of subtraction. Note, again, how we extended the sign bit to ensure no harmful overflow occurs.

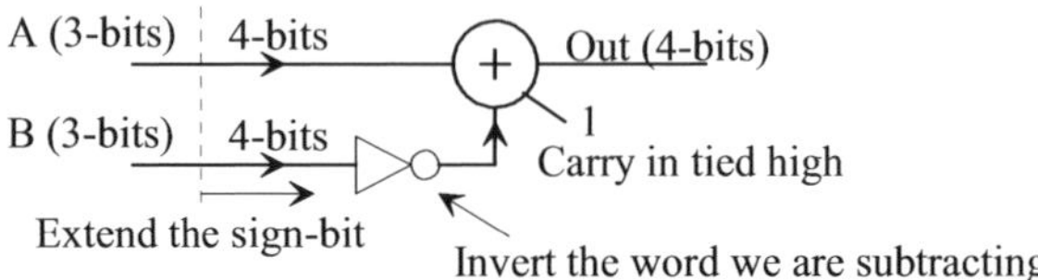

A	B	$\overline{B}+1$	Out (A-B)
0011 (+3)	1100 (-4)	0100 (+4)	0111 (+7)
1100 (-4)	1100 (-4)	0100 (+4)	0000 (0) OF!
1111 (-1)	0001 (+1)	1111 (-1)	1110 (-2) OF!
0011 (+3)	1111 (-1)	0001 (+1)	0100 (+4)
0010 (+2)	0001 (+1)	1111 (-1)	0001 (+1) OF!

Figure 4.8 Showing how two's complement numbers are subtracted.

4.2 Sinc-Shaped Digital Filters

Perhaps the most useful digital filters that we'll encounter when doing mixed-signal circuit design have Sinc-shaped responses. Let's start this section by discussing one of the simplest of these filters.

4.2.1 The Counter

Figure 4.9 shows how a counter can be used as a digital filter with a 1-bit input. If we assume the clock frequency is f_s (= $1/T_s$) and the counter is read out and reset every KT_s seconds then a constant input of 0s (−1 in two's complement, see Fig. 6.3)) will result in a counter output of 0 (−K) at the end of the time interval KT_s seconds. A constant input of logic 1s results in a counter output of K. An input of alternating 1s and 0s (the input is a 50% duty cycle squarewave with a frequency of $f_s/2$) results in an output of $K/2$ (0 for two's complement numbers). The output of the counter is related to its inputs using

$$y[Ki \cdot T_s] = \sum_{n=K(i-1)}^{K \cdot i-1} x[n \cdot T_s] \tag{4.7}$$

This equation simply indicates that we are taking K inputs, adding them together, and the result is the output of the filter (so we can think of this filter, **like any lowpass filter**, as an *averaging filter*). Rewriting Eq. (4.7) in the z-domain,

$$H(z) = \frac{Y(z)}{X(z)} = \sum_{n=0}^{K-1} z^{-n} = 1 + z^{-1} + z^{-2} + \ldots + z^{1-K} \tag{4.8}$$

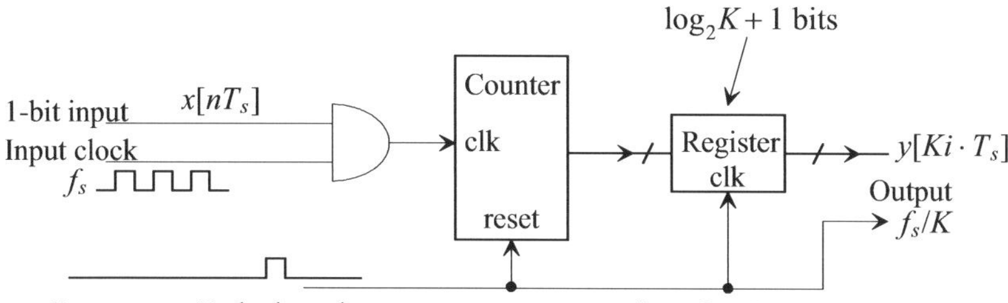

Figure 4.9 Using the counter as a digital filter.

or

$$H(z) = \frac{1-z^{-1}}{1-z^{-1}} \cdot (1 + z^{-1} + z^{-2} + ... + z^{1-K}) \tag{4.9}$$

The transfer function for the counter is then

$$H(z) = \frac{1-z^{-K}}{1-z^{-1}} \tag{4.10}$$

We'll see this equation frequently so let's spend some time on it. Note that the counter employs decimation, see Sec. 2.1.2, since the input word rate is K times faster than the output word rate. More on this in a moment (since aliasing will be a concern). The magnitude of the frequency response of Eq. (4.10) is given by

$$|H(f)| = \left| \frac{1 - e^{-j \cdot 2\pi K \frac{f}{f_s}}}{1 - e^{-j \cdot 2\pi \cdot \frac{f}{f_s}}} \right| = \frac{\sqrt{2(1 - \cos 2\pi K \cdot \frac{f}{f_s})}}{\sqrt{2(1 - \cos 2\pi \cdot \frac{f}{f_s})}} = \left| \frac{\sin\left(\pi K \cdot \frac{f}{f_s}\right)}{\sin\left(\pi \cdot \frac{f}{f_s}\right)} \right| \tag{4.11}$$

or

$$|H(f)| = K \cdot \left| \frac{Sinc\left(\pi \frac{K \cdot f}{f_s}\right)}{Sinc\left(\pi \frac{f}{f_s}\right)} \right| \tag{4.12}$$

Figure 4.10 shows the frequency response of the counter (see, also, Fig. 2.29). Note that for large K and small frequencies this equation can be approximated using

$$|H(f)| \approx K \cdot \left| Sinc\left(\pi \frac{K \cdot f}{f_s}\right) \right| \tag{4.13}$$

Figures 2.30 and 2.31 in Ch. 2 relate the amount of attenuation and droop we can expect from a Sinc response filter for various values of K. Again, these equations can be used to characterize all of the Sinc filters in this section.

Aliasing

Note, again, if we input a DC value of constant 1s then the counter's output is K (see Fig. 4.10). If our input is 101010... or 11001100... or 111000111000 etc. then the counter output is $K/2$. However, looking at the filter response seen in Fig. 4.10 and knowing an input of 101010 is a squarewave with a frequency of $f_s/2$, we would expect the counter's

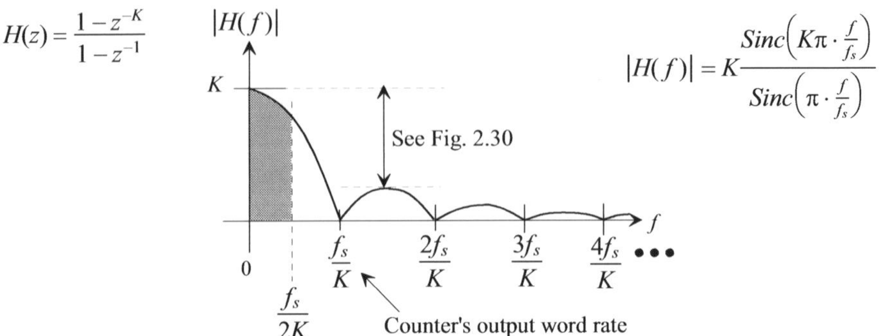

Figure 4.10 Frequency response of a Sinc-shaped digital filter.

output to be zero (and it is if we think in terms of two's complement numbers). However, let's look at the operation in terms of binary offset numbers.

Let's assume K is 8 and the counter is clocked at a frequency of 100 MHz. The 1-bit word rate coming into the counter is at 100 MHz while the counter is read out every 12.5 MHz (decimation by 8). Figure 4.11a shows the frequency response of the counter and the first-harmonic of the input of 1010101... or a squarewave at 50 MHz. We only concern ourselves with the first harmonic of the squarewave to keep the figures from getting too cluttered. A sinewave at 50 MHz, see Eqs. (1.85), having a peak-to-peak amplitude of 1, or a peak amplitude of 0.5, is represented at ± 50 MHz with dirac-delta functions having amplitudes of 0.25. Figure 4.11b shows the input signal spectrum after resampling at the output rate of 12.5 MHz (note the aliasing of the sampled signal at DC). In (c) we see that after including the counter's Sinc-shape response all of the tones disappear except for the aliased tone at DC (and tones at 100 MHz, 200 MHz, etc.) The output of the counter, which is read out as the counter is reset every KT_S seconds, (that is we don't look at the output of the counter while it's changing) is a constant value of 4 (that never changes) even though the input is a 50 MHz squarewave.

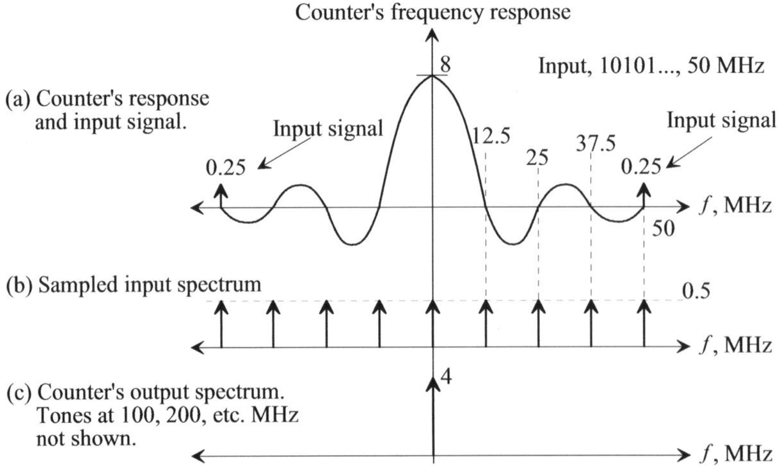

Figure 4.11 Showing how decimating using a counter results in aliasing.

Using a counter as a digital filter, as we've just seen, has limited uses because of the significant amount of aliasing that occurs with the inherent decimation. Note that we didn't discuss problems that occur when the input signal contains noise. One of the counter's uses, however, is in applications where the desired signal is constant (DC) such as some sensing applications. By using really large values of K the bandwidth can be shrunk down to limit the effects of noise while at the same time amplifying signals at DC (review Fig. 4.10).

The Accumulate-and-Dump

The counter's input was a 1-bit word. For N-bit input words the accumulate-and-dump can be used, Fig. 4.12, in place of a counter. The input signal is summed in a register for K clock cycles (note that this is the non-delaying integrator, also known as an *accumulator*, seen in Fig. 1.26). At the end of this time the sum is clocked into an output hold register and the summing register is reset (the integrator is reset). The equations and filtering behavior of the accumulate-and-dump are exactly the same as the counter. The maximum input word's value is $2^N - 1$ (all N bits high). In order to accommodate the summation of K, N-bit, input words the register size used must be at least $N + \log_2 K$ bits wide (rounding up to the nearest integer). The frequency response of the accumulate-and-dump filter or the counter at DC is K.

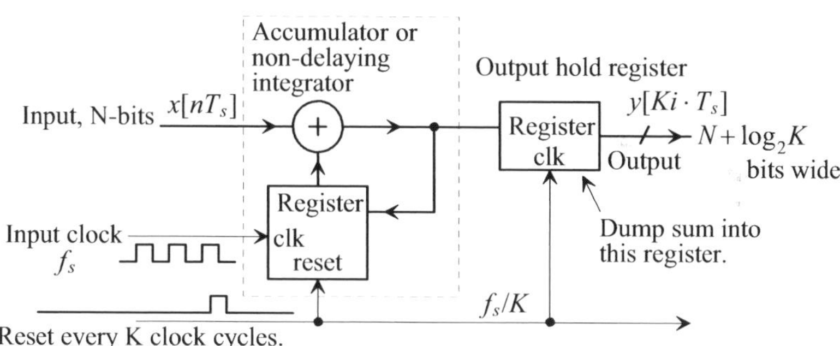

Figure 4.12 The accumulate and dump.

4.2.2 Lowpass Sinc Filters

Reviewing Eq. (4.10) one may wonder if we could implement this lowpass Sinc-response filter without the decimation inherent in a counter or an accumulate-and-dump. Figure 4.13 shows how cascading a comb filter with an integrator implements Eq. (4.10). Also seen in this figure are several z-plane plots and frequency responses for varying values of K. Note how, to get the lowpass response, we simply *cancel the zero* at DC (if this doesn't intuitively make sense review Sec. 1.2.3). We'll show that to implement bandpass or highpass Sinc-response filters all we do is cancel a zero using a pole at some other point on the unit circle. Notice in Fig. 4.13 that we've defined the bandwidth of the desired signal (where the droop is –3.9 dB, see Figs. 2.31 or 4.10) using

$$B = \frac{f_s}{2K} \tag{4.14}$$

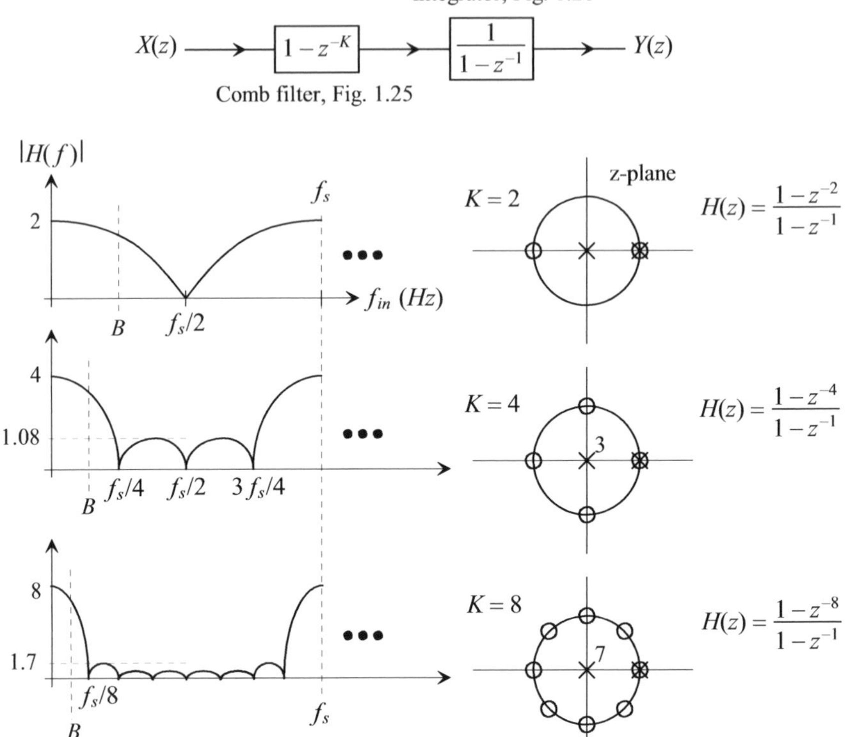

Figure 4.13 Lowpass Sinc-response filters with varying values of K.

Example 4.3

Assuming an 8-bit input word sketch the implementation of the Sinc-shaped filter

$$\frac{1-z^{-8}}{1-z^{-1}}$$

What is the output word size? Using SPICE verify that the filter's frequency response is given by Eq. (4.13) with $K = 8$. Assume $f_s = 100$ MHz.

The output of the filter is the sum of K, N-bit, words so the filter's output word size must grow to

$$\text{Output word size} = N + \log_2 K \tag{4.15}$$

For this example where $K = 8$ and the input word size is 8-bits the final output word must grow to 11-bits. The filter is sketched in Fig. 4.14.

Figure 4.15 shows the filter's input and output when the input frequency is 6.25 MHz ($=f_s/2K$). As seen in Figs. 4.10 and 2.31 the filter's attenuation is −3.9 dB or 0.638. The phase shift is – 78.75 degrees (see Eqs. [1.54] and [1.63]). ■

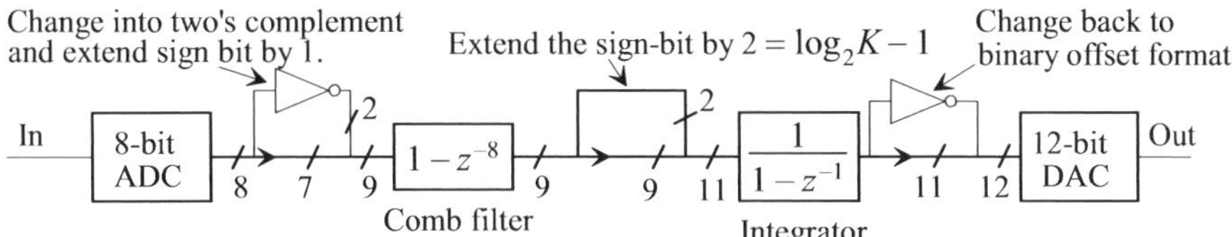

Figure 4.14 Digital filter sketch for Ex. 4.3.

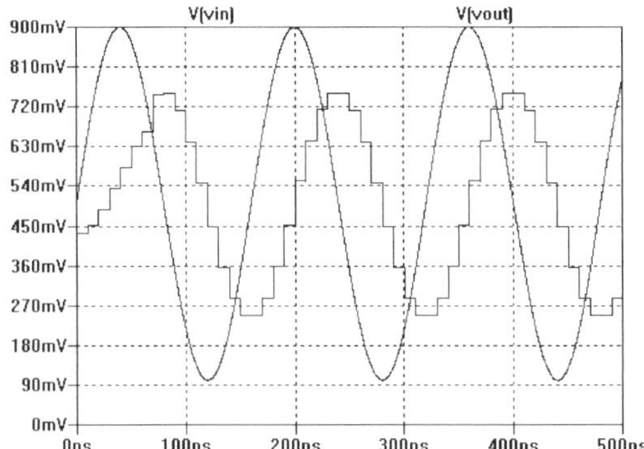

Figure 4.15 Input and output of the filter in Fig. 4.14 at a frequency of 6.25 MHz.

Example 4.4

Determine, and sketch, the time-domain impulse response of an averaging filter with $K = 8$.

The transfer function of the filter is given by Eq. (4.10) or

$$H(z) = \frac{1 - z^{-8}}{1 - z^{-1}} = 1 + z^{-1} + z^{-2} + z^{-3} + z^{-4} + z^{-5} + z^{-6} + z^{-7}$$

The time domain relationship between the input and the output is then

$$y[nT_s] = x[nT_s] + x[(n-1)T_s] + x[(n-2)T_s] + \ldots + x[(n-7)T_s]$$

Note that the output is simply a sum of the inputs over time, KT_s, as seen in Eq. (4.7). The time-domain impulse response of the first-order averaging filter is shown in Fig. 4.16. Note the rectangular shape. ■

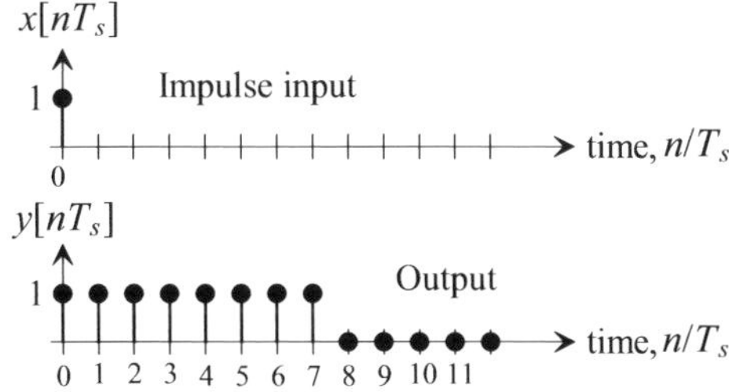

Figure 4.16 Impulse response of a K = 8 averaging filter.

Averaging without Decimation: A Review

The counter and the accumulate-and-dump discussed in Sec. 4.2.1 performed averaging and decimation in one stage. In other words, for example with $K = 4$, it summed four input samples, as shown in the sequence below, and passed the result to the output:

$$\overbrace{x(1)+x(2)+x(3)+x(4)}^{\text{First output}}+\overbrace{x(5)+x(6)+x(7)+x(8)}^{\text{Second output}}+x(9)+\ldots \tag{4.16}$$

where the output clocking frequency is, as seen in Fig. 4.9, f_s/K. For the lowpass Sinc-response (averaging) filters discussed in this section

$$\overbrace{x(1)+x(2)+x(3)+x(4)}^{\text{First output}}+x(5)+x(6)+x(7)+x(8)+x(9)+\ldots$$

$$x(1)+\overbrace{x(2)+x(3)+x(4)+x(5)}^{\text{Second output}}+x(6)+x(7)+x(8)+x(9)+\ldots$$

$$x(1)+x(2)+\overbrace{x(3)+x(4)+x(5)+x(6)}^{\text{Third output}}+x(7)+x(8)+x(9)+\ldots$$

$$x(1)+x(2)+x(3)+\overbrace{x(4)+x(5)+x(6)+x(7)}^{\text{Fourth output}}+x(8)+x(9)+\ldots$$

(4.17)

where the outputs of the averaging filter occur at the same rate as the inputs, f_s. The z-domain representation of the Sinc filter in Fig. 4.13 is the same as the counter or the accumulate-and-dump's transfer function

$$y(nT_s)=x[nT_s]+x[(n-1)T_s]+x[(n-2)T_s]+\ldots\rightarrow Y(z)=X(z)(1+z^{-1}+z^{-2}+\ldots+z^{1-K}) \tag{4.18}$$

or, again, a filter response given by Eq. (4.10)

Cascading Sinc Filters

The transfer function of a cascade of L of these Sinc-response filters can be written as

$$H(z)=\left[\frac{1-z^{-K}}{1-z^{-1}}\right]^L \tag{4.19}$$

$$|H(f)|=K^L\cdot\left[\frac{Sinc\left(K\pi\frac{f}{f_s}\right)}{Sinc\left(\pi\frac{f}{f_s}\right)}\right]^L \tag{4.20}$$

The attenuation through a cascade of L Sinc lowpass filters, see Eqs. (2.38) and (2.39) is

$$\left|\frac{\text{Main lobe}}{\text{First sidelobe}}\right|=K^L\cdot\sin^L\left(\frac{1.5\pi}{K}\right)\approx L\cdot 13\text{ dB for }K\geq 8 \tag{4.21}$$

while the droop at $f_s/2K$, see Fig. 4.17, is

$$\text{Droop}=\left[K\cdot\sin\left(\frac{\pi}{2K}\right)\right]^{-L}\approx L\cdot(-3.9)\text{ dB for }K\geq 8 \tag{4.22}$$

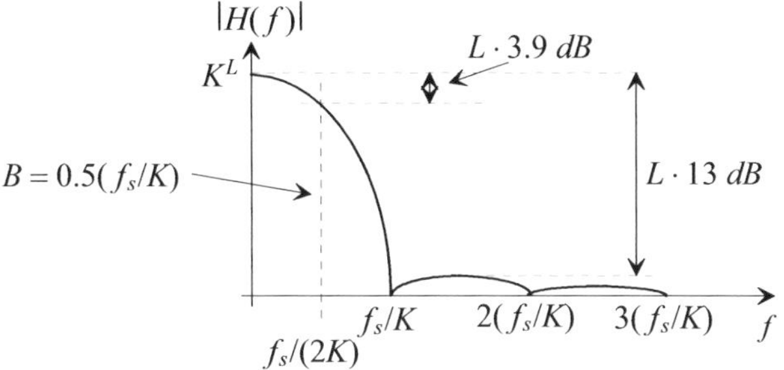

Figure 4.17 General frequency response of a lowpass Sinc (averaging) filter.

Example 4.5
Repeat Ex. 4.4 if an $L = 2$ averaging filter is used.

The transfer function of the filter is

$$H(z) = \left[\frac{1 - z^{-8}}{1 - z^{-1}}\right]^2 = 1 + 2z^{-1} + 3z^{-2} + 4z^{-3} + 5z^{-4} + \ldots + 3z^{-12} + 2z^{-13} + z^{-14}$$

The time domain relationship is

$$y[nT_s] = x[nT_s] + 2x[(n-1)T_s] + 3x[(n-2)T_s] + \ldots + 2x[(n-13)T_s] + x[(n-14)T_s]$$

The impulse response of the $L = 2$ lowpass Sinc filter is shown in Fig. 4.18. Note the triangular shape of the curve and how the impulse response of the $L = 2$ filter lasts twice as long as the $L = 1$ filter's response in Fig. 4.16. ■

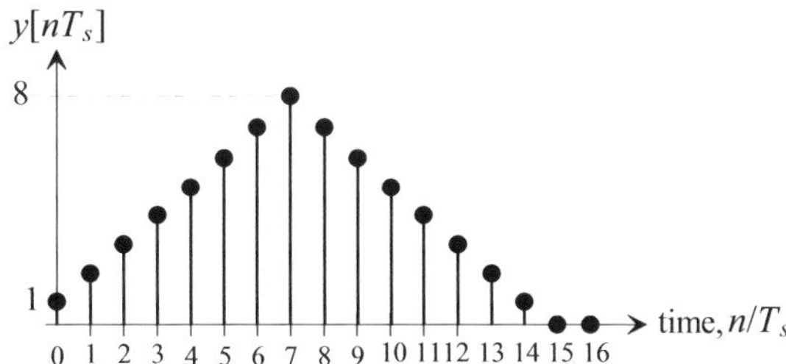

Figure 4.18 Impulse response of an L = 2 averaging filter with K = 8.

Finite and Infinite Impulse Response Filters

At this point a short note concerning digital filter names is in order. The comb filter seen in Fig. 1.24 (or Fig. 4.14), or the differentiator in Fig. 1.19, is an example of a *finite impulse response* (FIR) digital filter. Applying a unit amplitude impulse to the input of the comb filter causes the output of the comb filter to go to a 1 at the moment the impulse is applied and a –1 KT_s seconds later. The output is zero at all other times. In other words, the impulse response of the filter has a finite duration. The comb filter and differentiator are also called *non-recursive* filters since their outputs are only a function of their inputs (the output isn't fed back and used in the filter).

The integrator (also sometimes called an accumulator) shown in Figs. 1.26, 1.28, and 4.14 is an example of an *infinite impulse response* (IIR) digital filter. Applying a unit amplitude impulse to the input of the digital integrator, with zeroes the remaining times, causes the output of the integrator to increase to one and remain at one indefinitely. In other words, the output response of the integrator is of infinite duration. The integrator is also called a *recursive* filter since its current output value is a function of previous output values. Any digital filter with a denominator is a recursive filter.

While the averaging filters seen in Figs. 4.13 and 4.14 use an integrator (IIR section) and a comb filter (FIR section), overall, as seen in the previous two examples, it exhibits FIR behavior. Further, since the averaging filter's output is only a function of its inputs, as seen in Eq. (4.17), it is a non-recursive filter.

4.2.3 Bandpass and Highpass Sinc Filters

So far we've focused on lowpass filters. There are many situations, especially in communication systems, where we may want to filter or perform data conversion on a range of frequencies that doesn't extend from DC to B $(=f_s/2K)$. Bandpass ADCs and DACs, for example, are popular in communication systems. In this section, we introduce bandpass and highpass sinc-shaped filters.

Canceling Zeroes to Create Highpass and Bandpass Filters

In Fig. 4.13 we saw that we can generate a lowpass filter by canceling the zero at DC in a comb filter. We can generate a highpass, or differencing, filter by canceling a comb filter zero at $f_s/2$, as seen in the example shown in Fig. 4.19 with $K = 8$. The same equations, Eqs. (4.21) and (4.22), can be used to describe the behavior of this filter where, in the highpass response, the main lobe has shifted to $f_s/2$. Also, when looking at Fig. 4.19, remember that the frequency response of a digital filter is periodic with period f_s.

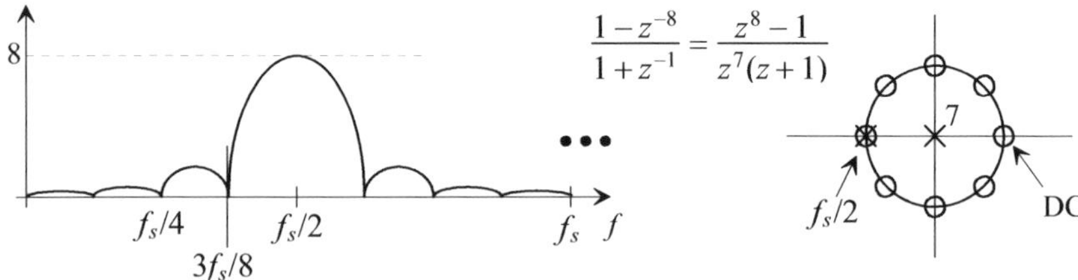

Figure 4.19 A highpass filter implementation using a comb filter.

We can generate a bandpass filter by canceling the zeroes at $f_s/4$ and $3f_s/4$, or some other frequencies, using a *digital resonator*. The general topology of the bandpass digital filter is shown in Fig. 4.20. Keeping in mind that the digital resonator is used to cancel the zeroes of the comb filter, we can write

$$H_D(z) = \frac{1}{1-2\cos\left[2\pi\frac{f}{f_s}\right]\cdot z^{-1}+z^{-2}} = \frac{z^2}{z^2-2\cos\left[2\pi\frac{f}{f_s}\right]\cdot z^{1}+1} = \frac{z^2}{\left(z-e^{+j\cdot 2\pi\frac{f}{f_s}}\right)\left(z-e^{-j\cdot 2\pi\frac{f}{f_s}}\right)} \tag{4.23}$$

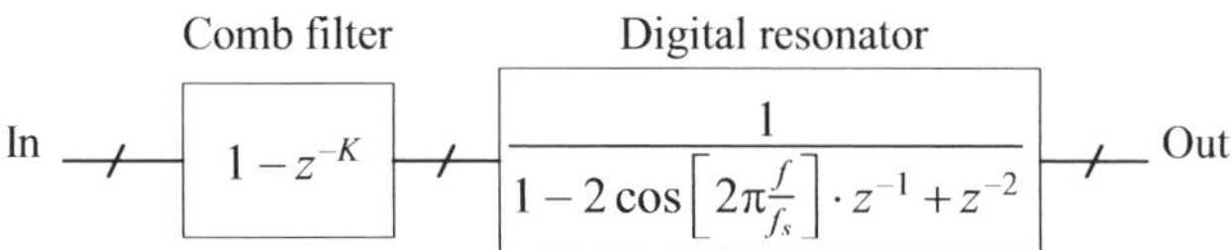

Figure 4.20 Implementing a sinc-shaped bandpass filter.

The time-domain representation of this equation is

$$y[nT_s] = 2\cos\left[2\pi\frac{f}{f_s}\right]\cdot y[(n-1)T_s] - y[(n-2)T_s] + x[nT_s] \tag{4.24}$$

It's desirable to determine at which frequencies the cosine term is an integer, zero, or a value that results in a trivial multiplication, that is, a shift so that we can implement the bandpass filter with trivial multiplications. In other words, we want a filter that uses only delays and additions so that its implementation is simple. The first frequency we will investigate is $f_s/4$. At this frequency the cosine term is zero and the digital-resonator/comb filter transfer function (the bandpass transfer function in Fig. 4.20) reduces to

$$H(z) = \frac{1 - z^{-K}}{1 + z^{-2}} \tag{4.25}$$

The magnitude and z-plane response of this filter, for $K = 8$, is shown in Fig. 4.21.

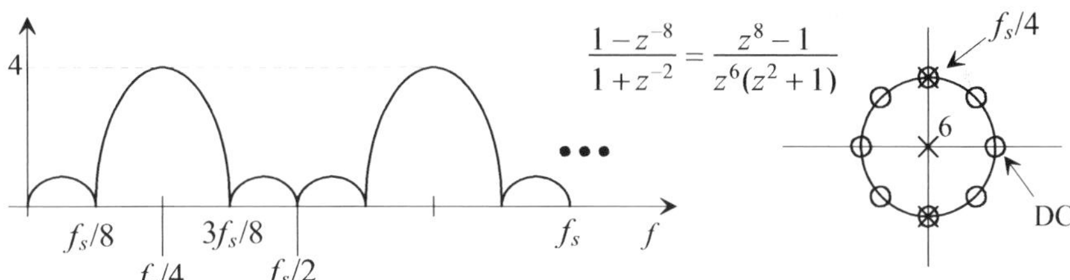

Figure 4.21 A bandpass filter implementation using a comb filter and digital resonator.

We can determine the magnitude response of Eq. (4.25) following the same procedure used to determine Eq. (2.38). The result, for the $f_s/4$ resonator, is

$$|H(f)| = \frac{\sqrt{2\left(1 - \cos K2\pi\frac{f}{f_s}\right)}}{\sqrt{2\left(1 + \cos 4\pi\frac{f}{f_s}\right)}} = \left|\frac{\sin\left(K\pi\frac{f}{f_s}\right)}{\cos\left(2\pi\frac{f}{f_s}\right)}\right| \tag{4.26}$$

At the center of the passband, that is $f_s/4$, $|H(f)| = K/2$. The ratio of the main lobe to the first side lobe, on either side, is plotted in Fig. 4.22 along with the lowpass Sinc filter response and is calculated using

$$\left|\frac{\text{Main lobe}}{\text{First side lobe}}\right| = \frac{K}{2}\cdot\sin\frac{3\pi}{K} \tag{4.27}$$

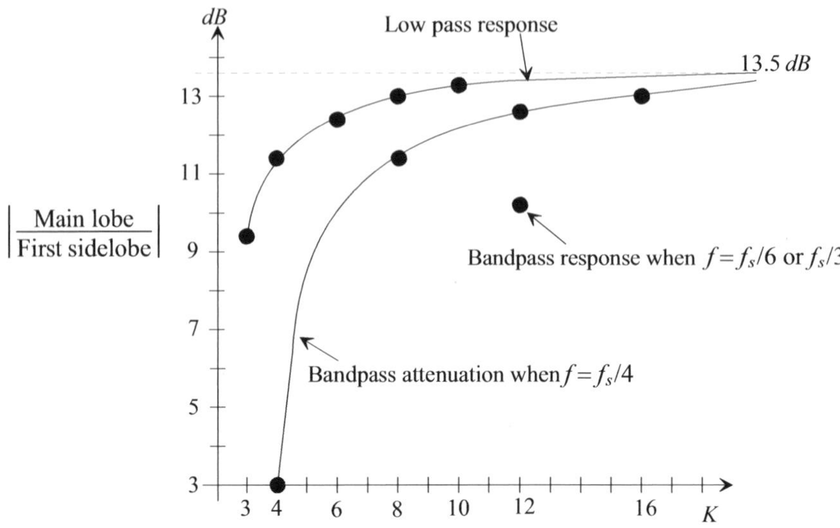

Figure 4.22 Lowpass and bandpass filter attenuation versus number of comb filter zeroes, K.

The cosine term in Eq. (4.24) can be set to ±1 when $f=f_s/6$ or $f_s/3$ resulting in a bandpass filter. It should be clear that with the appropriate choice of sampling frequency, number of zeroes K used in the comb filter, and value of the cosine term, many different combinations of simple bandpass filters can be implemented using these techniques. There are, however, only a few resonators that can be implemented using binary numbers. The resonators' coefficients must be such that the poles perfectly lie on the unit circle and cancel the comb filter's zeroes on the unit circle.

The ratio of the main lobe to the first sidelobe for $f=f_s/6$ or $f_s/3$ is given, assuming $K = 12, 24, \ldots$, by

$$\left|\frac{\text{Main lobe}}{\text{First side lobe}}\right| = \frac{K\sin\left(\frac{3\pi}{2K}\right)\sin\left(\frac{\pi}{3}-\frac{3\pi}{2K}\right)}{\sin\frac{\pi}{3}} = 1.15K\sin\left(\frac{3\pi}{2K}\right)\sin\left(\frac{\pi}{3}-\frac{3\pi}{2K}\right) \quad (4.28)$$

which is approximately 13.5 dB for $K = 24, 36, 48 \ldots$ and 10.15 dB for $K = 12$, Fig. 4.22.

In order to increase the amount of attenuation between the main lobe and the first side lobe in a bandpass filter implementation, we can cascade filter sections (as we did in the lowpass filter implementations discussed earlier). For example, cascading five $f_s/4$ bandpass filters with $K = 8$ will result in an attenuation of 57 dB. Also, note that by changing the sampling, or filter clock frequency f_s, we can easily change the bandpass filter's center frequency. A change in the clock frequency, and its selection, can easily be implemented using a counter and some control logic.

Example 4.6
Sketch the block level circuit diagram for an $f_s/4$ digital resonator.

From Eq. (4.24) the time domain representation of the $f_s/4$ resonator can be written as

$$y[nT_s] = x[nT_s] - y[(n-2)T_s]$$

The implementation is shown in Fig. 4.23. ■

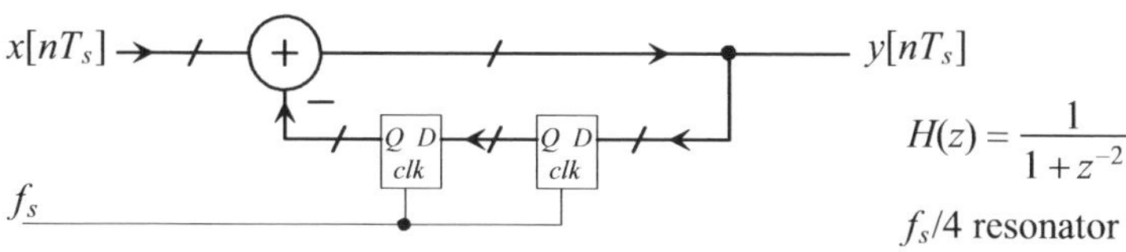

Figure 4.23 Implementation of a digital resonator.

Example 4.7

Assuming an 8-bit input word, sketch the implementation of the Sinc-shaped bandpass filter centered at $f_s/4$

$$\frac{1-z^{-8}}{1+z^{-2}}$$

What is the output word size? Using SPICE verify that the filter's frequency response is given by Eq. (4.26) with $K = 8$. Assume $f_s = 100$ MHz.

The sketch of the filter is seen in Fig. 4.24 (note the similarity to Fig. 4.14). The word size grows by $\log_2 K - 1$ from the input to the output. Here, where K is 8, the word grows 2-bits, to an output word size of 10-bits, (1-bit in the comb filter and 1-bit in the resonator for a gain of 4, see Fig. 4.21). Figure 4.25 shows the output of the resonator when the input is 25 MHz ($=f_s/4$). It can be useful to vary the input frequency and verify that Eq. (4.26) is correct. For example, if we change the input frequency to 12.5 MHz the filter's output is zero (this is the first adjacent zero point in the frequency response). When the input frequency is 15 MHz, the output of the filter is

$$|H(f)| = \left| \frac{\sin\left(K\pi\frac{f}{f_s}\right)}{\cos\left(2\pi\frac{f}{f_s}\right)} \right| = \left| \frac{\sin\left(8\pi\frac{15}{100}\right)}{\cos\left(2\pi\frac{15}{100}\right)} \right| = \left| \frac{-0.588}{0.588} \right| = 1 \rightarrow 0.25 \text{ (scaled by 4)}$$

where, since we normalize the filter's output so that passband gain is 1 (the input equals the output), the output of the filter is 0.25 the input. In the next chapter we'll see that we can throw away some of the lower bits in the output word since they don't contribute to an increase in the SNR. ■

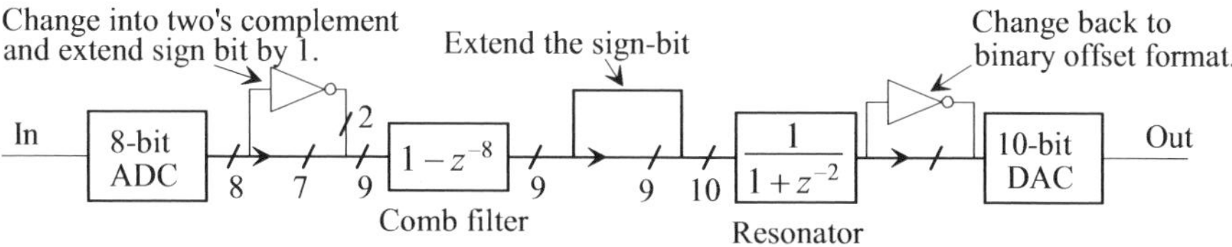

Figure 4.24 Digital filter sketch for Ex. 4.7.

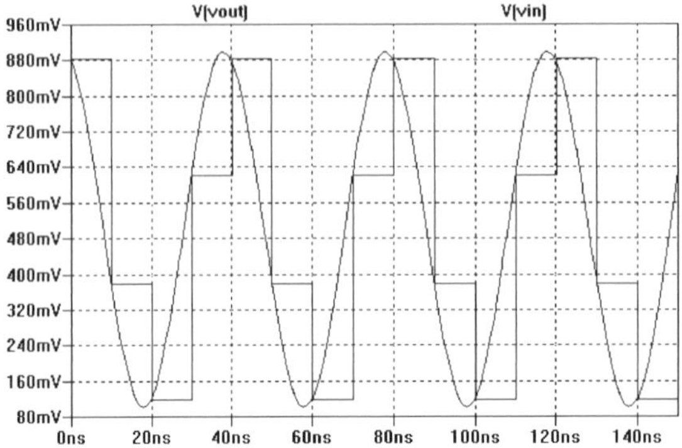

Figure 4.25 Input and output of the filter in Ex. 4.7 at $f_s/4$.

Frequency Sampling Filters

Consider the topology of a comb filter and resonators shown in Fig. 4.26. We are feeding the output of the comb filter through the resonators (with different center frequencies) and then using the combined sum of the resulting bandpass filter responses (the Sinc shapes) to build a bandpass filter. This is exactly the same as reconstructing a waveform in the time domain using an ideal RCF, as discussed in Ch. 2, except now we are using the

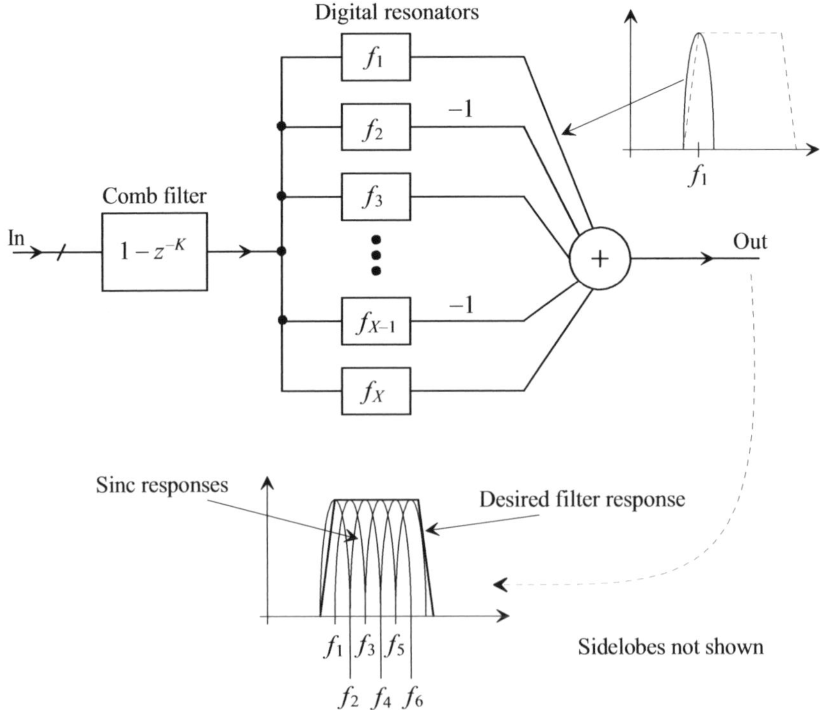

Figure 4.26 A frequency sampling filter.

summation of the frequency domain Sinc responses to generate a bandpass filter with a variable width. Note that every other digital resonator is subtracted rather than added to the final result. This is to account for the phase reversal between adjacent resonator outputs.

4.2.4 Interpolation using Sinc Filters

We saw back in Sec. 2.1.5 that an interpolator is a circuit that increases the clocking rate of the digital data. This was useful for, among other things, reducing the requirements placed on the reconstruction filter. The simplest interpolator was the input hold register, Fig. 2.28. Before we discuss interpolation using Sinc filters let's answer the question "Why can't we use an input hold register for interpolation here?"

Examine the block diagram in Fig. 4.27. The ADC is clocked at a rate of f_s so the anti-aliasing filter (AAF) limits the ADC's input spectrum to $f_s/2$. On the input of the ADC is a sample-and-hold (S/H) with a Sinc-shaped spectral response (–3.9 dB attenuation at the Nyquist frequency of $f_s/2$, see Fig. 2.17). We might expect, based on Fig. 2.29 and the associated discussion, that the hold register seen in Fig. 4.27 would provide an additional Sinc-shaped response in the signal path. However, remember (see Ex. 2.2) repetitively sampling and holding a signal results in only one S/H attenuation. Since the input signal in Fig. 4.27 sees this response when it passes through the S/H it isn't affected by another Sinc-shaped response as it passes through the hold register.

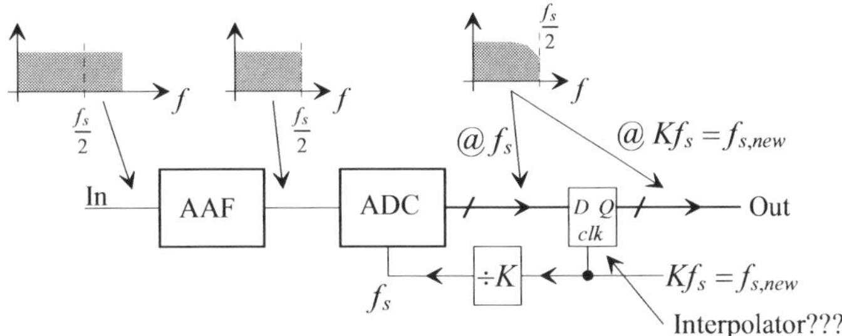

Figure 4.27 Interpolation using a hold register (see Sec. 2.1.5).

Note that because of this assumption that the data has passed through a S/H and experienced the associated Sinc-shaped filtering, we won't use zeroes padding in the discussions in this section. It should be straightforward to extend the discussions to designs that do use zeroes padding. Also note that if the digital data hasn't passed through a Sinc-response filter then passing it through a hold register, as discussed in Sec. 2.1.5, does introduce a Sinc-response (which, as we'll find out now, may often be desirable when performing interpolation).

The next question is "Why would we want the signal to see additional filtering?" The answer to this comes from reviewing Fig. 2.26. The image removal filter is an important component of an interpolator. We can implement this filter using

$$H(z) = \left[\frac{1 - z^{-K}}{1 - z^{-1}} \right]^L \tag{4.29}$$

The benefit, for an interpolator, comes from realizing that if we clock the comb filter with a slower clock, f_s/K, then one clock delay is KT_s so we can simplify the comb sections using

$$1 - z^{-K} \rightarrow 1 - z^{-1} \tag{4.30}$$

Figure 4.28 shows an interpolator implemented using a cascade of Sinc Filters. The attenuation through the interpolator is given, see Eq. (2.37), by

$$|H(f)| = \left| \frac{Sinc\left(K\pi \frac{f}{f_{s,new}}\right)}{Sinc\left(\pi \frac{f}{f_{s,new}}\right)} \right|^L \tag{4.31}$$

Larger attenuation can be achieved by cascading more stages, L. Again, the word size grows by 1-bit through each comb filter and by $\log_2 K - 1$ through each integrator. Also note that if we were to use zeroes padding instead of the inherent hold register present in the topology (see discussion on previous page) a selector, see Fig. 2.26, would be added in between the comb filters and integrators to introduce $K - 1$ zeroes.

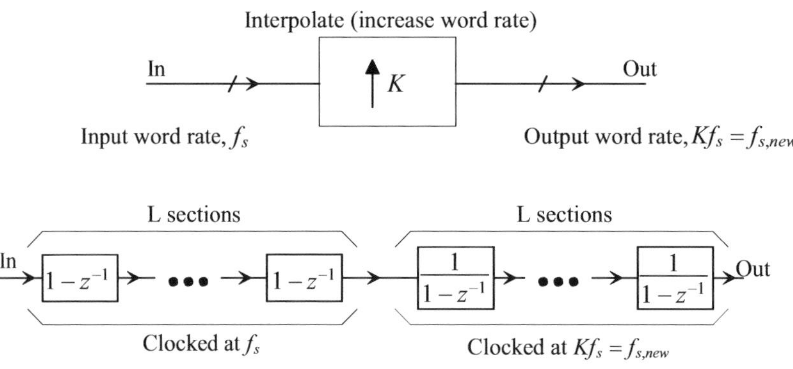

Figure 4.28 Interpolating using Sinc filters for image removal.

Example 4.8a

Using an 8-bit word, an input clock rate of 12.5 MHz, an input frequency of 1.1 MHz, and $K = 8$ determine the output word size, magnitude, and output clock rate for interpolators with $L = 1$ and 2. Also determine the attenuation between the passband and the first sidelobe, see Fig. 2.29. Verify your results with SPICE.

The Nyquist frequency is 6.25 MHz (the spectral content on the input of the ADC should be < 6.25 MHz to avoid aliasing). The output clock rate is 100 MHz. For the case when $L = 1$ the word grows 1-bit through the comb filter and 2-bits through the integrator so the output word size is 11-bits. When $L = 2$ the output word size is 14 bits. We'll see in the next chapter that the signal-to-noise ratio sets the number of bits we keep on the output of the filter or in between adjacent stages.

The attenuation through the filter is

$$|H(f)| = \left| \frac{Sinc\left(K\pi \frac{f}{f_{s,new}}\right)}{Sinc\left(\pi \frac{f}{f_{s,new}}\right)} \right|^L = \left| \frac{Sinc\left(8\pi \frac{1.1}{100}\right)}{Sinc\left(\pi \frac{1.1}{100}\right)} \right|^L \approx \left| \frac{\sin\left(8\pi \frac{1.1}{100}\right)}{8\pi \frac{1.1}{100}} \right|^L = (0.988)^L$$

or, for $L = 1$, $-$ 0.107 dB and for $L = 2$, $-$ 0.214 dB. We can estimate the attenuation between the passband and first sidelobe using information seen in Fig. 2.30. For $L = 1$ the attenuation is roughly 13 dB and for $L = 2$ the attenuation is 26 dB.

Figure 4.29 shows the simulation results. Looking closely we see that for $L = 1$ the interpolator's output is simply a linear change between adjacent output points. For $L = 2$ we see additional phase shift (more delay through the filter) and a more "rounded" behavior indicating a better representation of the original signal. It can be instructive to vary the input frequency in these simulations and look at the resulting output signals (especially showing how the output signal amplitude drops as the input signal moves closer to the Nyquist frequency of 6.25 MHz). ■

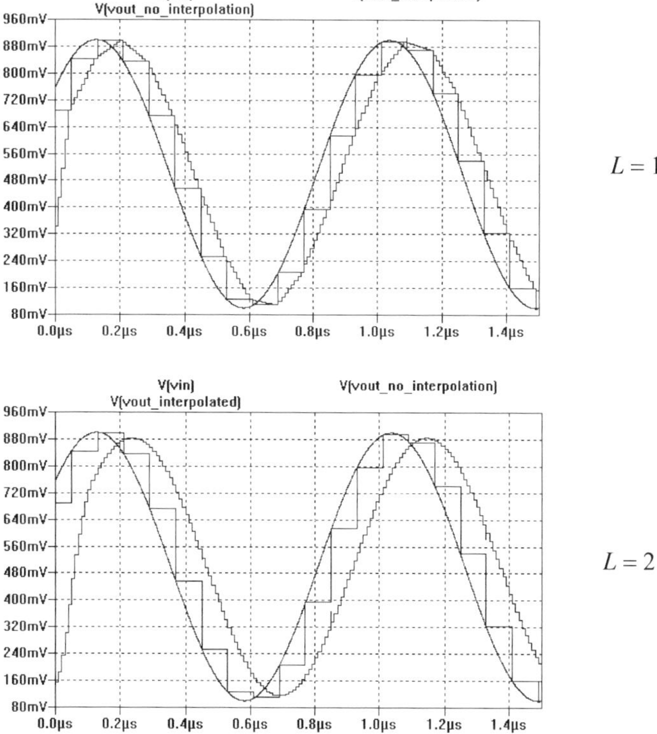

Figure 4.29 Interpolation using one- and two-stage Sinc filters.

Additional Control

In order to have an additional parameter for adjusting the image removal filter characteristics, Eqs. (4.29) and (4.31), consider adding delay to our comb filter sections as seen in Fig. 4.30.

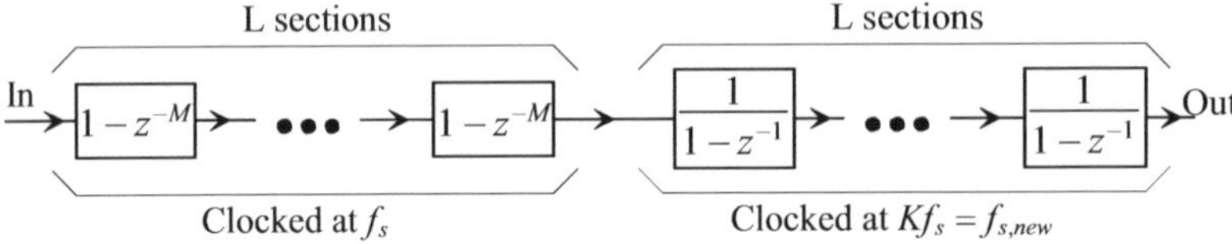

Figure 4.30 General interpolation using Sinc filters.

Using this topology we can describe the filter's characteristics using

$$H(z) = \left[\frac{1-z^{-KM}}{1-z^{-1}}\right]^L \tag{4.32}$$

and

$$|H(f)| = (KM)^L \left|\frac{Sinc\left(KM \cdot \pi\frac{f}{f_{s,new}}\right)}{Sinc\left(\pi\frac{f}{f_{s,new}}\right)}\right|^L \tag{4.33}$$

Figure 4.31 plots this equation. Note that the Nyquist frequency ($= f_s/2$) remains unchanged since it's set by the sampling in the ADC. By using the factor M we can achieve a narrower bandwidth but at the cost of more droop at the Nyquist frequency. The word size, in this filter, still grows by 1-bit in every comb section and by $\log_2(K \cdot M) - 1$ bits in the integrator sections.

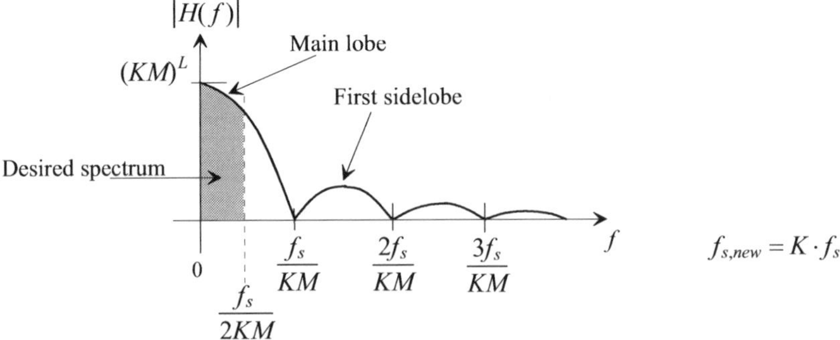

Figure 4.31 Frequency response image removal filter using a Sinc interpolator, Fig. 4.30.

Cascade of Integrators and Combs

A quick note before leaving this section. The filters we presented are often labeled CIC filters since they are formed by cascading integrators and comb filters. We'll continue to avoid using this term for reasons that will become clear in the next section.

4.2.5 Decimation using Sinc Filters

As seen in Fig. 2.12, resampling digital data at a lower frequency (decimation) consists of passing the data through a digital anti-aliasing filter followed by resampling. In this section we discuss using a Sinc filter, Eqs. (4.10) and (4.12), with a response seen in Fig. 4.10. The output word rate of the decimator is f_s/K and the input desired spectrum is limited to DC to $f_s/2K$ (as seen in Fig. 4.10).

Notice that, in the filters we've covered so far in this chapter, we've put the comb filter or differentiator before the integrators or resonators. This was to eliminate unwanted overflow problems in the latter (which are recursive filters). After reviewing Fig. 4.28 we might try putting the integrators before the comb filters. While this may work for small signals, or by using very wide registers in the integrators, it would be better to avoid overflow problems altogether. In order to move towards this goal lets equate Eq. (4.8) and (4.29)

$$H(z) = \left[\frac{1-z^{-K}}{1-z^{-1}}\right]^L = \left[\sum_{n=0}^{K-1} z^{-n}\right]^L = [1+z^{-1}+z^{-2}+\ldots+z^{1-K}]^L$$
$$= \left[(1+z^{-1})\cdot(1+z^{-2})\cdot\ldots\cdot(1+z^{-2^{\log_2 K-1}})\right]^L$$
$$= (1+z^{-1})^L\cdot(1+z^{-2})^L\cdot\ldots\cdot(1+z^{-2^{\log_2 K-1}})^L \quad (4.34)$$

or we can implement decimation using only non-recursive averagers (see Fig. 1.16 and the associated discussion). The word size increases by 1-bit through each averager. Note that we are assuming the decimation is a factor of 2 here. Figure 4.32 shows the implementation of the decimator. The registers are used to re-sample the data. The higher factors of delay are implemented by clocking the registers with the slower clock (clocking an averager, $1+z^{-1}$, with a clock of f_s/K implements $1+z^{-K}$). This topology is called, by some authors, a CIC filter (see comment at the bottom of previous page) even though there aren't any integrators present in the topology. Figure 4.33 shows example spectrums when using this decimator. Note that the sidelobes alias into the desired spectrum. By increasing the order of the filter L, we can reduce the effects of aliasing.

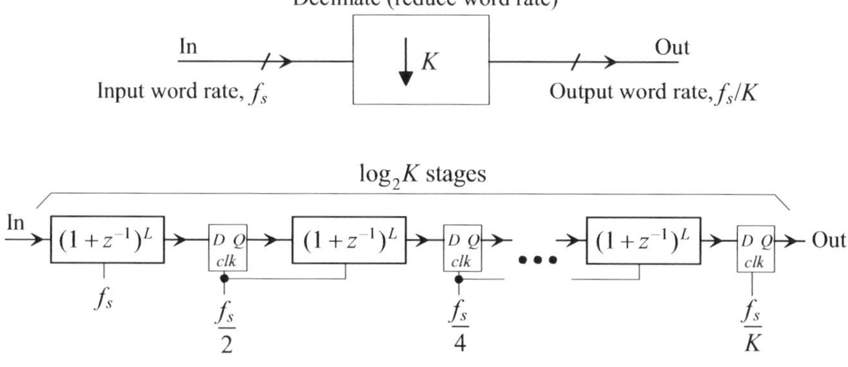

Figure 4.32 Decimation using Sinc anti-aliasing filters.

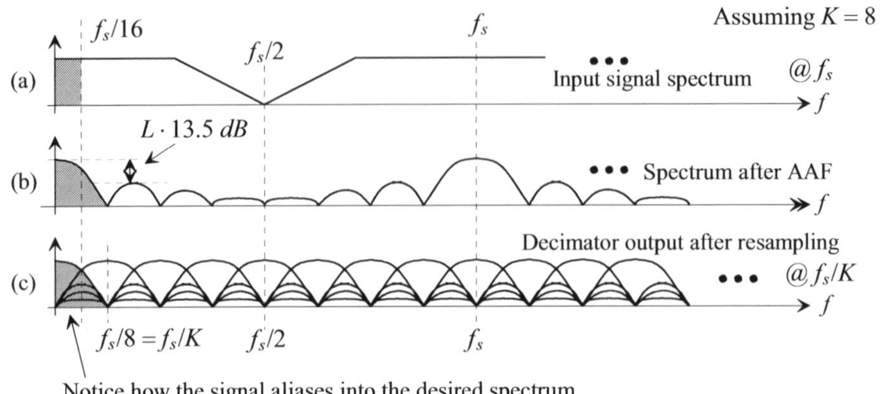

Figure 4.33 Spectrums when decimating and a Sinc anti-aliasing filter used.

Example 4.8b

Using an 8-bit word, an input clock rate of 100 MHz, an input frequency of 1.1 MHz, and $K = 8$ determine the output word size, magnitude, and output clock rate for Sinc-decimators with $L = 1$ and 2. Verify your results with SPICE.

The transfer function of the anti-aliasing filter inherent in the decimator is given by Eq. (4.34) with $K = 8$ and $L = 1$ or 2. The output word rate is 12.5 MHz. The simulation results are seen in Fig. 4.34. Note that the decimated output with $L = 1$ looks essentially the same as the output with $L = 2$ (except for more delay when $L = 2$). This (the shape of the outputs being more or less the same) wasn't the case for the interpolator outputs seen in Fig. 4.29. For the interpolator, where aliasing isn't a concern, increasing L results in a reduction in the aliased signals present on the output of the interpolator. Increasing L allows the interpolator output to move closer to the original input (in the extreme, $L \to \infty$, and we get the original analog waveform). By increasing the order, L, in the decimation filter we reduce the aliased signals present in our desired spectrum, Fig. 4.34. ■

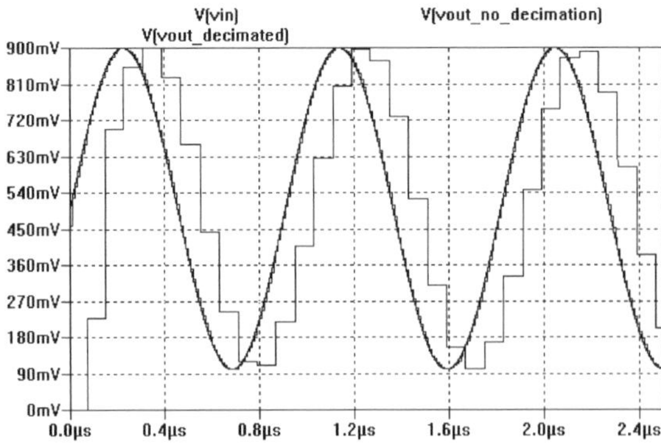

Figure 4.34 Decimating by 8, an example.

4.3 Filtering Topologies

The Sinc filters discussed in the last section are very useful for general mixed-signal circuit design, especially when interpolating or decimating, since they don't employ complicated multiplications. Unfortunately, they also don't have sharp filtering characteristics and there isn't a lot of flexibility when selecting the filter's frequency response. In this section we'll present some additional filtering topologies mostly based on the integrator. Most of these topologies are directly related to topologies discussed in the last chapter covering analog filters.

4.3.1 FIR Filters

Towards understanding the reason for this approach, consider the non-recursive FIR filter topology seen in Fig. 4.35. The transfer function of the filter is

$$H(z) = A_0 + A_1 z^{-1} + A_2 z^{-2} + A_3 z^{-3} \tag{4.35}$$

If, for example, we set all of the filter's coefficients to 1 then we can write, as seen in Eqs. (4.9) and (4.10),

$$H(z) = \frac{1 - z^{-4}}{1 - z^{-1}} \tag{4.36}$$

or, again, a Sinc-shaped lowpass comb filter. When compared to the comb filters used earlier, Figure 1.24, this topology has more adders.

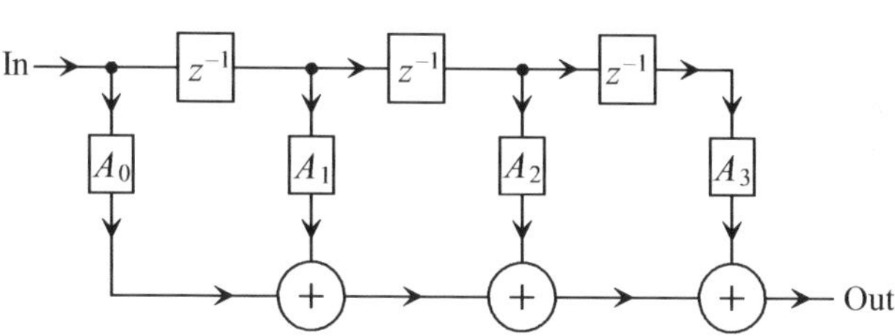

Figure 4.35 A four-stage FIR filter.

As another example consider setting all of the coefficients to 0.25. This is done by shifting the word left two bits or by extending the sign-bit by two bits. Extending the sign bit increases the word size and ensures we don't lose resolution (but results in more hardware). Figure 4.36 shows the resulting filter's step response. Note the similarity to a first-order RC circuits step response. Also note that the impulse response for this filter simply reveals the filter's coefficients (here all 0.25). Based on a desired step response a heuristic approach can be used to design the filter.

The benefits of using FIR filters are that they are inherently stable (they are non-recursive so no feedback is used) and they can have linear phase (constant delay). The drawback is that they, for a given number of delays, aren't as good at filtering as the recursive structures we'll talk about in the rest of this section. Unfortunately, recursive structures are subject to instability. In addition, topologies using integrators are subject to overflow. Let's discuss these two issues before going any further.

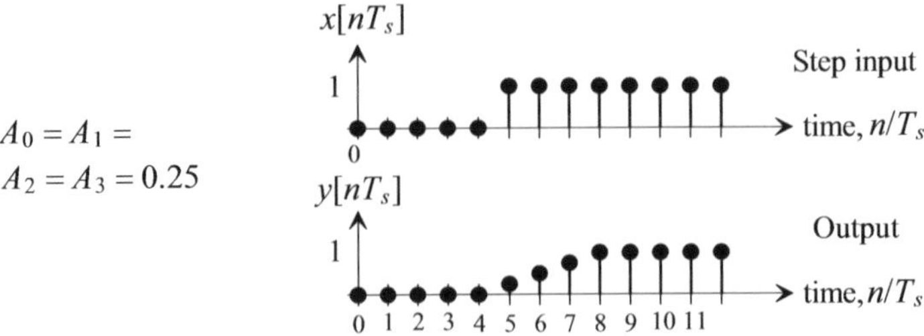

Figure 4.36 Step response of a 4-stage FIR filter with all coefficients set to 0.25.

4.3.2 Stability and Overflow

A weighted integrating filter is seen in Fig. 4.37. The output of the circuit is fed back to the input after it is multiplied by a. The output of the circuit in the time-domain may be written as

$$y[nT_s] = x[(n-1)T_s] + a \cdot y[(n-1)T_s] \tag{4.37}$$

or

$$y[nT_s] = x[(n-1)T_s] + a \cdot x[(n-2)T_s] + a^2 \cdot x[(n-3)T_s] + a^3 \cdot x[(n-4)T_s] + \ldots, \tag{4.38}$$

which will obviously blow up if $a > 1$.

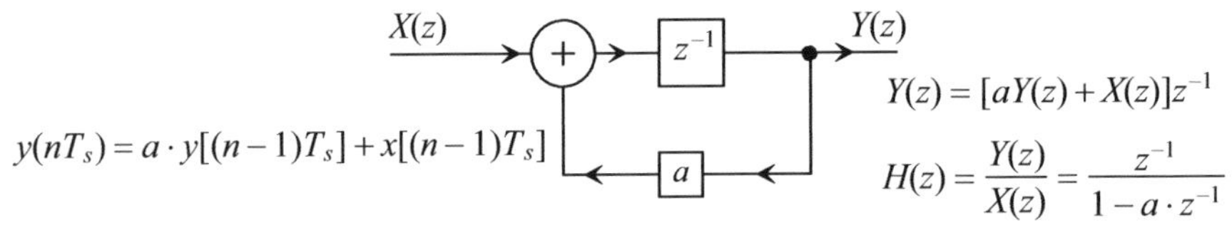

$X(z)$ + z^{-1} $Y(z)$

a

$y(nT_s) = a \cdot y[(n-1)T_s] + x[(n-1)T_s]$

$Y(z) = [aY(z) + X(z)]z^{-1}$

$H(z) = \dfrac{Y(z)}{X(z)} = \dfrac{z^{-1}}{1 - a \cdot z^{-1}}$

Figure 4.37 A weighted integrating filter.

The z-domain representation of Eq. (4.37) is

$$H(z) = \frac{1}{z-a} \tag{4.39}$$

Figure 4.38 shows the z-plane and magnitude plots for this equation. If $a > 1$ then $H(z)$ becomes unstable. So *for a stable system* we must require our poles to reside within the unit circle. (There are no restrictions on the location of zeroes.) This sounds simple enough; however, notice that we have, in most of the previously discussed digital filters, placed poles right on the unit circle. If there is rounding in our digital numbers, we could be faced with an unstable digital filter. This would be a very common occurrence in a digital filter implemented using software, if care was not taken to avoid rounding errors. Since we use integer numbers in our hardware implementations, instability shouldn't be a problem unless we start to try to round numbers to decrease hardware complexity (performing divisions or multiplications) without being careful.

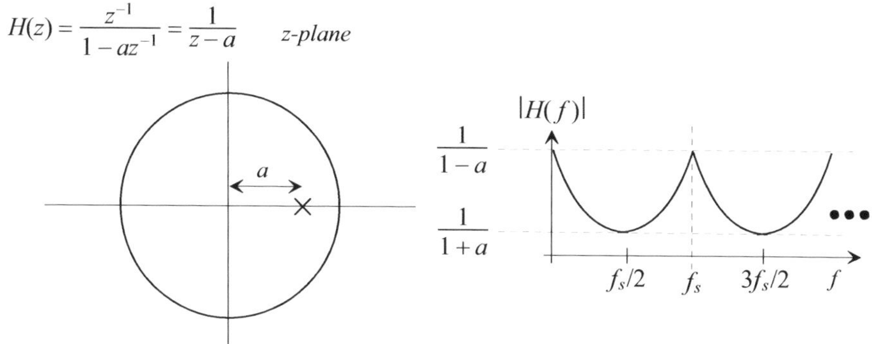

Figure 4.38 The z-plane representation and magnitude response for a weighted integrating filter.

Overflow

Recall that when we used integrators with comb filters earlier in the chapter we increased the input/output word size of the integrators by $\log_2 K - 1$ while the comb filter outputs increased by 1-bit for an overall increase of $\log_2 K$. We calculated these values by determining the register size required to sum K words. In a recursive filter the output is fed back and summed at various points in the filter, so determining the exact register size required can be challenging. Figure 4.39a shows how an integrator overflows causing the output to wrap around (go from the most positive value to the most negative value or vice-versa). In Fig. 4.39b we show the more desirable situation where the output saturates. This keeps the filter from becoming unstable; however, it will introduce nonlinearities in the filter's response. Nonlinearity at some extreme is usually better than instability. The question is, "How do we determine overflow?" We know that we can't look at the carry out bit because, as we saw in Figs. 4.7 and 4.8, the adder overflowing is a normal occurrence when adding two's complement numbers. What we do know is that if the input sign bits are both 0 then the output sign bit must be 0 (if both inputs are positive then the output must be positive). The same can be said for adding negative numbers. Figure 4.40 shows how we can modify an integrator to avoid overflow.

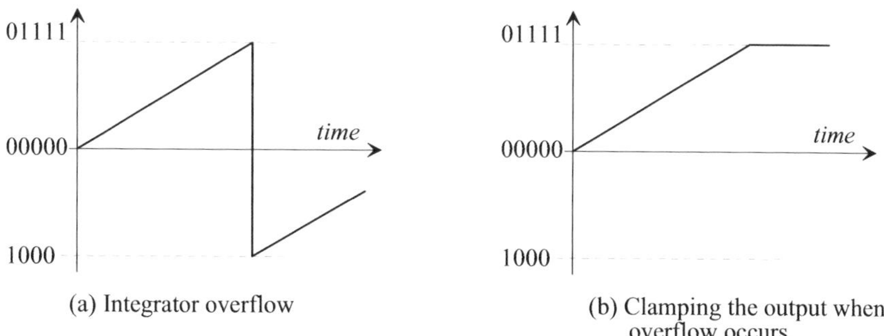

Figure 4.39 Integrator overflow and clamping the integrator's output.

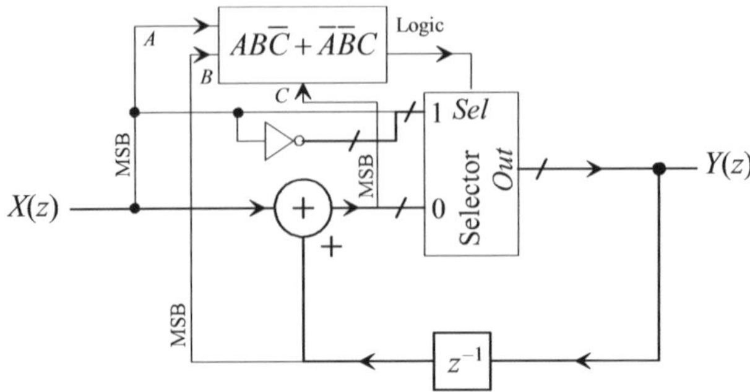

Figure 4.40 Modifying the integrator to avoid overflow.

4.3.3 The Bilinear Transfer Function

The bilinear transfer function and its implementation using an integrator and differentiator were presented back in Sec. 3.2.1. Figure 4.41 shows the digital implementation of the bilinear filter seen in Fig. 3.29. It's important, at this point, to see how the continuous-time implementation in Fig. 3.29 is directly implemented in Fig. 4.41. In particular we note that

$$G_1 = G_{1D} \cdot f_s \text{ and } G_3 = \frac{G_{3D}}{f_s} \tag{4.40}$$

and so

$$\frac{v_{out}(f)}{v_{in}(f)} = \frac{1}{G_2} \cdot \frac{1 + \frac{s}{1/G_3}}{1 + \frac{s}{G_1 G_2}} = \frac{1}{G_2} \cdot \frac{1 + j \cdot \frac{f}{f_s/(2\pi G_{3D})}}{1 + j \cdot \frac{f}{f_s G_{1D} G_2 / 2\pi}} \tag{4.41}$$

The location of the pole is given by

$$f_{3dB,pole} = \frac{f_s G_{1D} G_2}{2\pi} \tag{4.42}$$

while the location of the zero is at

$$f_{3dB,zero} = \frac{f_s}{2\pi G_{3D}} \tag{4.43}$$

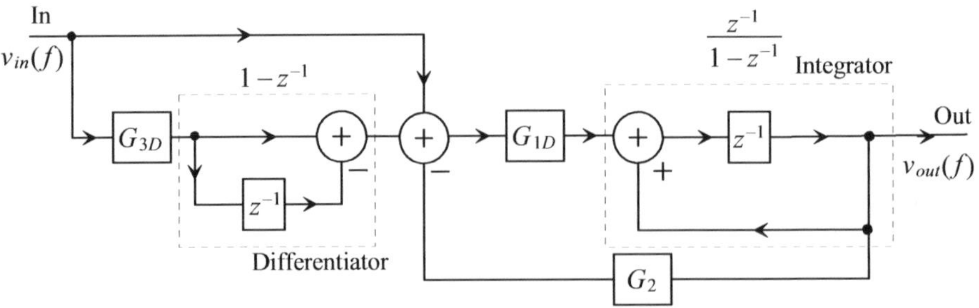

Figure 4.41 Digital implementation of the bilinear transfer function.

Note that in order for our filter to be useful the frequencies of interest must be much lower than the filter's clocking frequency, f_s. In other words, the frequencies where the pole and zero are located must be much less than f_s. This means, assuming $G_2 = 1$, that $G_{1D} << 1$ and $G_{3D} >> 1$.

Let's attempt to simplify this filter. If we look at the transfer function, Eq. (4.41), we see that the feedback gain, G_2, simply scales the amplitude of the transfer function and can be used to further adjust the location of the filter's pole. Because we can scale the amplitude of the signal either before or after the filter, and independent of the filter's operation, and we can precisely set the pole of the filter using G_{1D}, we can, without loss of functionality, set G_2 to 1. We can then rearrange the summing, delaying, and multiplying blocks, as seen in Fig. 4.43. Using these results we can write the z-domain representation of the transfer function, Eq. (4.41), as

$$\frac{v_{out}(z)}{v_{in}(z)} = \frac{G_{1D}(1+G_{3D})}{z-(1-G_{1D})} \cdot \frac{z - G_{3D}/(1+G_{3D})}{z} \tag{4.44}$$

Before attempting to simplify the filter implementation seen in Fig. 4.43 further, let's show that, indeed, Eq. (4.44) is equivalent to Eq. (4.41) when $f << f_s$. It will be helpful to remember that

$$z \approx 1 + \frac{s}{f_s} \text{ and } z^{-1} \approx 1 - \frac{s}{f_s} \text{ if } f << f_s \text{ or } z \approx 1 + \frac{s}{f_s} \approx \frac{1}{1-\frac{s}{f_s}} \approx \frac{1}{z^{-1}} \text{ if } \frac{s^2}{f_s^2} \approx 0 \tag{4.45}$$

where $s = j\omega = j2\pi f$. Rewriting Eq. (4.44) in the s-domain results in

$$\frac{v_{out}(f)}{v_{in}(f)} = \frac{G_{1D}(1+G_{3D})}{1+\frac{s}{f_s}-1+G_{1D}} \cdot \left[1 - \left(1 - \frac{s}{f_s}\right) G_{3D}/(1+G_{3D})\right] \tag{4.46}$$

$$= \frac{1+\frac{s}{f_s/G_{3D}}}{1+\frac{s}{G_{1D} \cdot f_s}} = \frac{1+j \cdot \frac{f}{f_s/(2\pi G_{3D})}}{1+j \cdot \frac{f}{G_{1D} \cdot f_s/2\pi}} \tag{4.47}$$

which is clearly the same as Eq. (4.41) when $G_2 = 1$.

Example 4.9

Sketch, and determine the transfer function for, the digital filter equivalent of the following RC circuit. Assume the digital filter is clocked at 100 MHz.

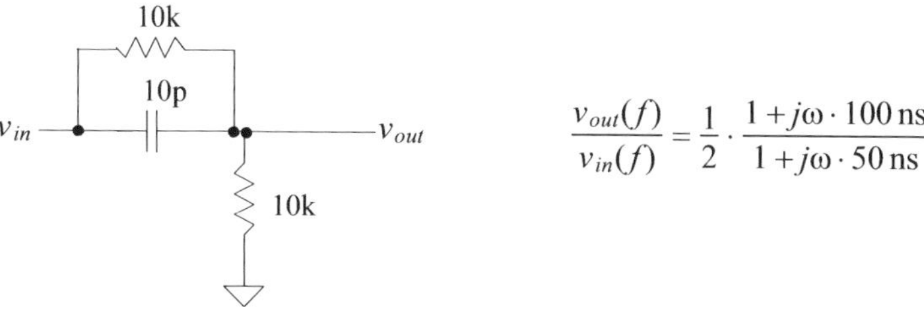

Figure 4.42 A simple first-order RC circuit.

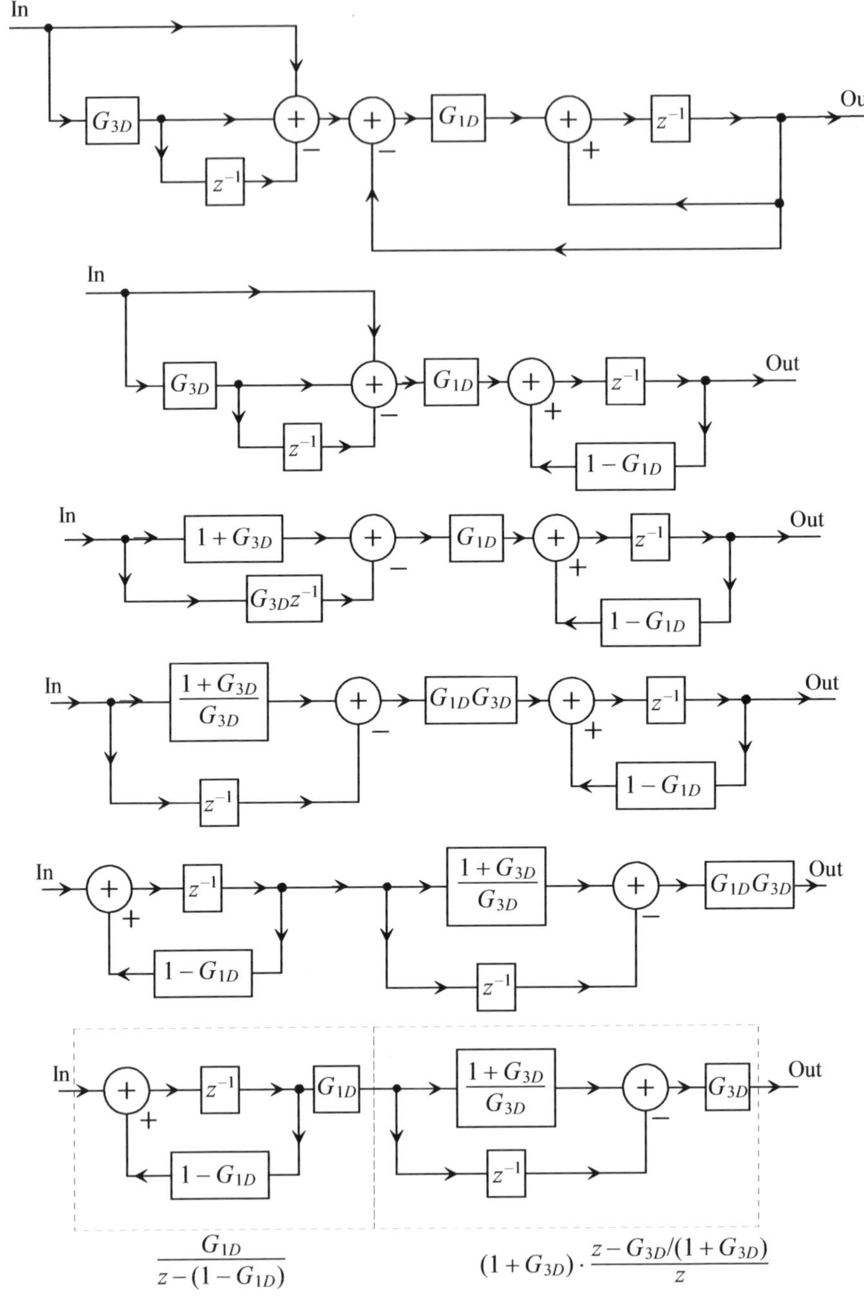

Figure 4.43 Simplifying the digital implementation of the bilinear filter.

Comparing the transfer function in Fig. 4.42 to Eq. (4.47), we see that if

$$\frac{G_{3D}}{f_s} = 100\ ns \rightarrow G_{3D} = 10 \text{ and } \frac{1}{G_{1D} \cdot f_s} = 50\ ns \rightarrow G_{1D} = 0.2$$

then we can use any of the filters in Fig. 4.43. The sketch of the digital filter is seen in Fig. 4.44. The multiplication of the transfer function by 1/2 is nulled by the multiplication by $G_{1D}G_{3D}$ (= 2). To verify that the filter in Fig. 4.44 functions as desired at DC, we see that the output of the first stage is 0.5 when the input is 0.1, and the output of the second stage is 0.05 (with a 0.5 on its input). ■

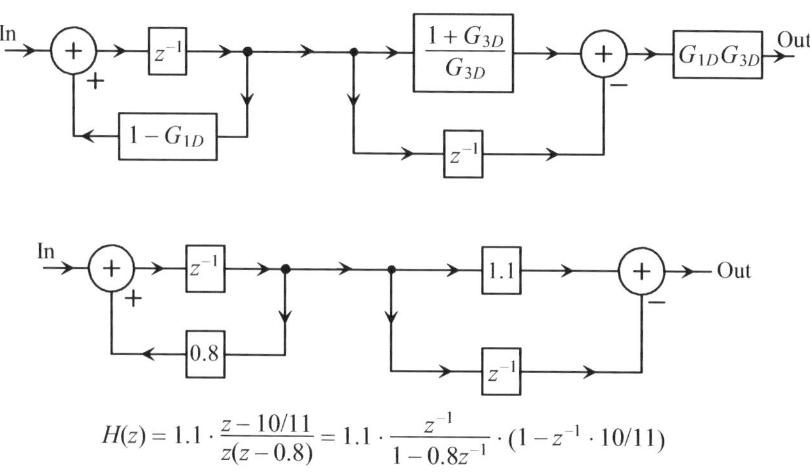

$$H(z) = 1.1 \cdot \frac{z - 10/11}{z(z-0.8)} = 1.1 \cdot \frac{z^{-1}}{1-0.8z^{-1}} \cdot (1 - z^{-1} \cdot 10/11)$$

Figure 4.44 Digital filter from Ex. 4.9.

The Canonic Form (or Standard Form) of a Digital Filter

Studying Fig. 4.44, we might wonder if we can further reduce the size of the digital filter. We see in this figure that it may be possible to eliminate the second delay element (which, of course, is a register) and use only a single delay. The result of this modification is seen in Fig. 4.45. Intuitively we would think that the phase response of the filter will change because, now, there is less delay in series with input of the second adder. We can write the output as

$$\frac{v_{out}(z)}{v_{in}(z)} = G_{1D}G_{3D}\left[\frac{1+G_{3D}}{G_{3D}} \cdot \frac{1}{1-z^{-1}(1-G_{1D})} - \frac{z^{-1}}{1-z^{-1}(1-G_{1D})}\right] \quad (4.48)$$

or

$$\frac{v_{out}(z)}{v_{in}(z)} = G_{1D}(1+G_{3D}) \cdot \frac{z - G_{3D}/(1+G_{3D})}{z-(1-G_{1D})} \quad (4.49)$$

which is clearly the same response as derived in Fig. 4.43 or Eq. (4.44) except that the output is one clock cycle, z, earlier. This reduced delay has no effect on the magnitude response of the filter and little effect, assuming $f << f_s$, on the phase response of the filter.

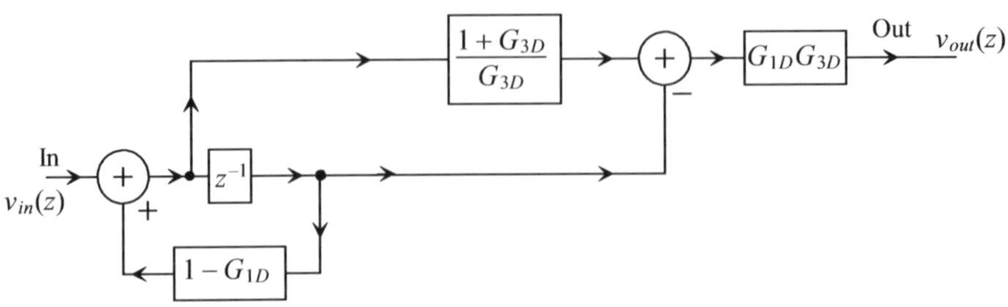

Figure 4.45 Canonic form of a first-order digital filter.

The general form of this first-order canonic (or standard form) filter is seen in Fig. 4.46. The filter is termed *canonic* because the minimum number of delays are used. One delay is used for each pole. Remember that in order for a digital filter to be realizable in hardware there must be fewer or an equal number of zeroes than poles in a filter's transfer function.

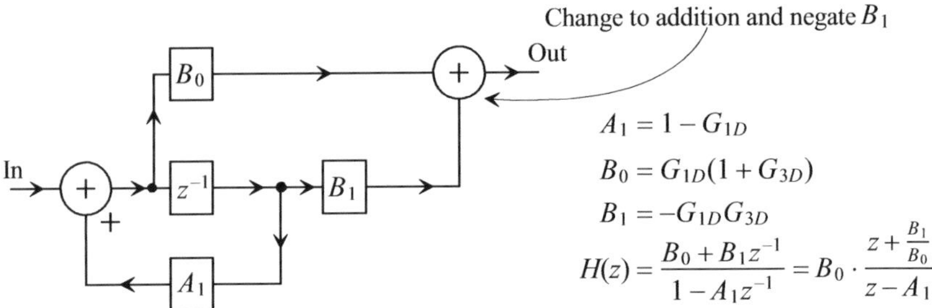

Figure 4.46 General canonic form of a first-order digital filter.

We can, again, derive the transfer function for the first-order bilinear digital filter. This time, however, let's use the variables in Fig. 4.46. Again, assuming $f << f_s$, and using Eq. (4.45) results in

$$H(f) = \frac{B_0 + B_1}{1 - A_1} \cdot \frac{1 + \frac{s}{f_s\left(1 + \frac{B_1}{B_0}\right)}}{1 + \frac{s}{f_s(1 - A_1)}} \tag{4.50}$$

where

$$f_{3dB,pole} = \frac{f_s(1 - A_1)}{2\pi} \tag{4.51}$$

and

$$f_{3dB,zero} = \frac{f_s}{2\pi}\left(1 + \frac{B_1}{B_0}\right) \tag{4.52}$$

and the gain at DC is

$$A_{DC} = \frac{B_0 + B_1}{1 - A_1} \tag{4.53}$$

Example 4.10
Using the canonic form of the first-order digital filter, repeat Ex. 4.9.

Comparing Eq. (4.50) with the transfer function in Fig. 4.42, we can write

$$2\pi f_{3dB,pole} = \frac{1}{50 \text{ ns}} = f_s \cdot (1 - A_1) = 100\,\text{MHz}(1 - A_1) \rightarrow A_1 = 0.8$$

$$A_{DC} = \frac{1}{2} = \frac{B_0 + B_1}{1 - A_1} = \frac{B_0 + B_1}{1 - 0.8} \rightarrow B_0 + B_1 = 0.1$$

$$2\pi f_{3dB,zero} = \frac{1}{100 \text{ ns}} = f_s \cdot \left(1 + \frac{B_1}{B_0}\right) = 100\,\text{MHz}\left(\frac{0.1}{B_0}\right) \rightarrow B_0 = 1$$

and thus $B_1 = -0.9$. The filter's sketch is seen in Fig. 4.47. We will discuss how to implement the multipliers later.

Let's do a quick check to see if the filter functions as desired at DC. If we apply 0.1 to the input of the filter then, according to the transfer function in Fig. 4.42, the output of the filter should be 0.05 or one-half the input. Because the input to the filter is a DC signal, both sides of the delay will have the same value. According to Fig. 4.38, this value will be 0.5 (the output of the weighted integrator is 1/[1 – 0.8] times the input signal, here 0.1, at DC). The output will then be 0.5 – 0.45 or 0.05 (as we would expect). ■

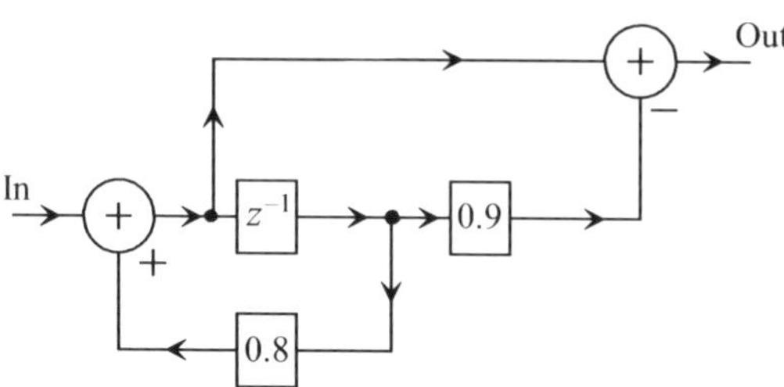

Figure 4.47 Canonic form of the first-order digital filter in Ex. 4.10.

Example 4.11
Sketch the digital filter implementation of the lowpass filter in Ex. 3.1 from the last chapter that has a DC gain of 1 and a 3 dB frequency of 1.59 MHz. Assume the filter is clocked at 100 MHz.

The filter's continuous-time frequency response is given by

$$H(f) = \frac{1}{1 + j \cdot \frac{f}{1.59\text{ MHz}}}$$

Using Eqs. (4.50) to (4.53), we begin by calculating A_1

$$f_{3dB,pole} = 1.59 \text{ MHz} = \frac{f_s(1-A_1)}{2\pi} = \frac{100 \text{ MHz}}{2\pi}(1-A_1) \rightarrow A_1 = 0.9$$

and then

$$A_{DC} = \frac{B_0 + B_1}{1-A_1} = 1 = \frac{B_0 + B_1}{1-0.9} \rightarrow B_0 + B_1 = 0.1$$

Let's put the zero at infinity so it doesn't affect the transfer function

$$f_{3dB,zero} = \infty = \frac{f_s}{2\pi}\left(1 + \frac{B_1}{B_0}\right) \rightarrow B_0 = 0 \text{ and } B_1 = 0.1$$

A sketch of the filter is seen in Fig. 4.48. ■

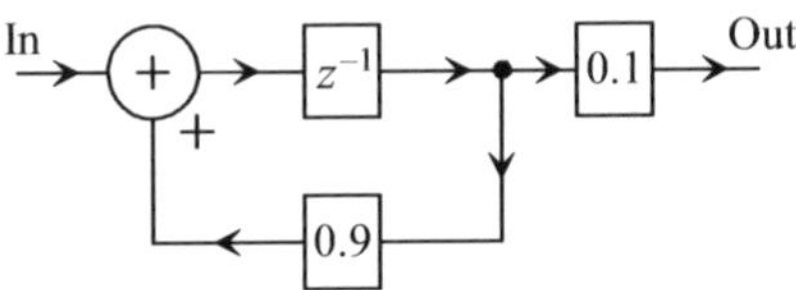

Figure 4.48 First-order digital filter in Ex. 4.11.

General Canonic Form of a Recursive Filter

Before leaving this section, let's show in Fig. 4.49 the general form of an nth-order canonic digital filter (where n indicates the number of poles in the transfer function). The z-domain transfer function of the filter is given by

$$H(z) = \frac{\sum_{i=0}^{m} B_i z^{-i}}{1 - \sum_{i=1}^{n} A_i z^{-i}} = \frac{\sum_{i=0}^{m} B_i z^{n-i}}{z^n - \sum_{i=1}^{n} A_i z^{n-i}} \tag{4.54}$$

If we want to write the frequency domain transfer function, we write, again assuming $f \ll f_s$ and using Eq. (4.45),

$$H(f) \approx \frac{\sum_{i=0}^{n-1} B_i \left(1 - j\frac{2\pi f}{f_s}\right)^i}{1 - \sum_{i=1}^{n} A_i \left(1 - j\frac{2\pi f}{f_s}\right)^i} \tag{4.55}$$

or

$$H(f) \approx \frac{\sum_{i=0}^{n-1} B_i \left(1 + j\frac{2\pi f}{f_s}\right)^{n-i}}{\left(1 + j\frac{2\pi f}{f_s}\right)^n - \sum_{i=1}^{n} A_i \left(1 + j\frac{2\pi f}{f_s}\right)^{n-i}} \tag{4.56}$$

While we can design higher-order digital filters using the topology of Fig. 4.49, we will restrict our analysis to first- and second-order filters where hand calculations are relatively easy to perform. Note that we can increase the attenuation of a filter by using several of these sections in cascade.

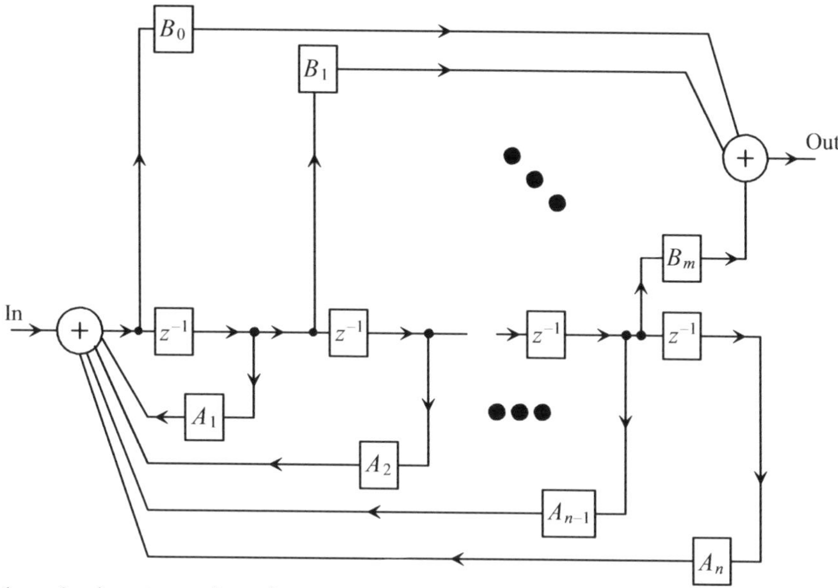

Number of poles $n \geq$ number of zeroes.

Figure 4.49 General canonic form of a digital filter.

4.3.4 The Biquadratic Transfer Function

The digital biquad filter based on the canonic form seen in Fig. 4.49 is shown in Fig. 4.50. The transfer function of this filter is

$$H(z) = \frac{Y(z)}{X(z)} = \frac{B_0 + B_1 z^{-1} + B_2 z^{-2}}{1 - A_1 z^{-1} - A_2 z^{-2}} = \frac{B_0 z^2 + B_1 z + B_2}{z^2 - A_1 z - A_2} \tag{4.57}$$

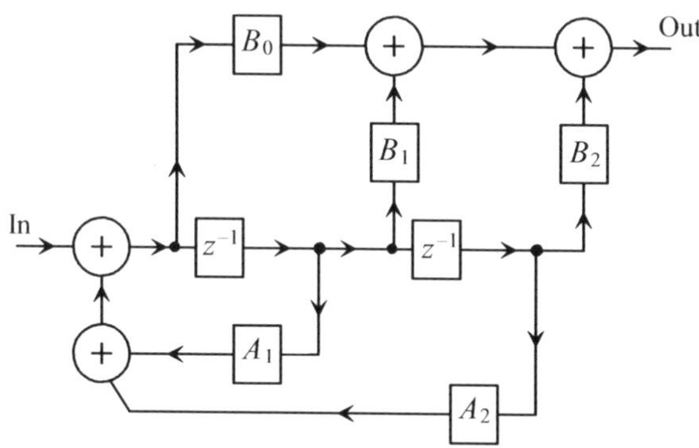

Figure 4.50 The digital biquad filter (see Fig. 4.49).

In order to translate this transfer function into the frequency domain, we use Eq. (4.45) and assume our frequencies of interest are much less than the sampling frequency

$$H(f) = \frac{B_0\left(1+\frac{s}{f_s}\right)^2 + B_1\left(1+\frac{s}{f_s}\right) + B_2}{\left(1+\frac{s}{f_s}\right)^2 - A_1\left(1+\frac{s}{f_s}\right) - A_2} \tag{4.58}$$

After some algebraic manipulation we can put this equation in the form seen in Eq. (3.62)

$$H(f) = \frac{B_0 \cdot s^2 + f_s(2B_0 + B_1)\cdot s + f_s^2(B_0 + B_1 + B_2)}{s^2 + f_s(2 - A_1)\cdot s + f_s^2(1 - A_1 - A_2)} \tag{4.59}$$

where

$$a_2 = B_0 \tag{4.60}$$

$$a_1 = f_s(2B_0 + B_1) \tag{4.61}$$

$$a_0 = f_s^2(B_0 + B_1 + B_2) \tag{4.62}$$

$$\frac{2\pi f_0}{Q} = f_s(2 - A_1) \tag{4.63}$$

and finally,

$$f_0 = \frac{f_s}{2\pi}\cdot\sqrt{(1 - A_1 - A_2)} \tag{4.64}$$

Example 4.12

Repeat Ex. 3.8 using the digital biquad clocked at 100 MHz.

In this example a lowpass filter is designed with $f_0 = 1.59$ MHz and $Q = 0.707$. Reviewing Fig. 3.35, we see that for a lowpass filter a_2 and a_1 are zero. This means, in Eq. (4.59), B_0 and B_1 are zero. Further,

$$Q = \frac{2\pi \cdot 1.59 \times 10^6}{100 \times 10^6(2 - A_1)} = 0.707 \rightarrow A_1 = 1.859$$

and

$$1 - A_1 - A_2 = \left(\frac{2\pi f_0}{f_s}\right)^2 \rightarrow A_2 = -0.869$$

and finally, because the gain at DC is 1,

$$B_2 = 1 - A_1 - A_2 = 0.01$$

Note that if a scaling in the amplitude is allowable, we can remove this multiplication or approximate it with shifts in the digital word.

The simulation results are seen in Fig. 4.51. In order to implement this simulation in SPICE, we used transmission lines for the delay elements and voltage- controlled voltage sources for both the multiplications and the adders. This method allows us to simulate the filter's frequency response. Note that the frequency response is periodic with the filter's clocking frequency. ■

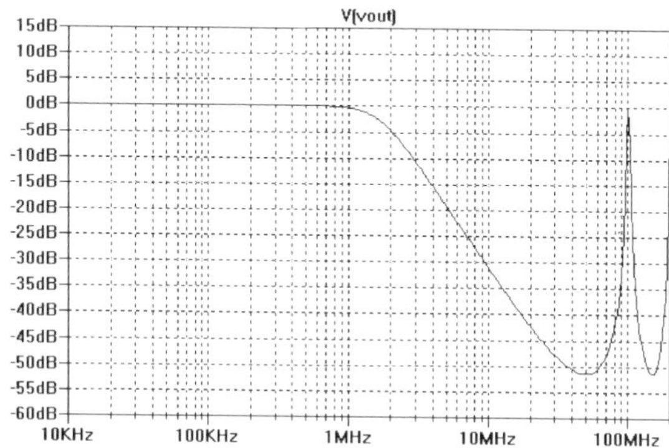

Figure 4.51 Simulating the digital filter in Ex. 4.12

Comparing Biquads to Sinc-Shaped Filters

Consider the frequency response of the Sinc-shaped lowpass filter shown in Fig. 4.52. This filter uses a clocking frequency of 100 MHz and a K of 16 or

$$|H(z)| = \left|\frac{1-z^{-16}}{1-z^{-1}}\right|^3 \tag{4.65}$$

Note the significant droop in the filter's response. It's desirable to design a filter that doesn't have this droop or, even more desirable, contains a small amount of peaking to compensate for a Sinc-shaped attenuation from a S/H, decimation filter, etc. Using Eqs. (4.59) - (4.64) we can set, for a digital biquad equivalent of this filter, $B_2 = 0.03125$, $A_1 = 1.75$, $A_2 = -0.78125$ (there are several other solutions, depending on the desired complexity or performance of the filter). The simulation results comparing the Sinc and biquad filters are seen in Fig. 4.53. The transfer function of the biquad filter is

$$H(z) = \frac{0.03125}{z^2 - 1.75z + 0.78125} \tag{4.66}$$

The block diagram of the filter is seen in Fig. 4.54.

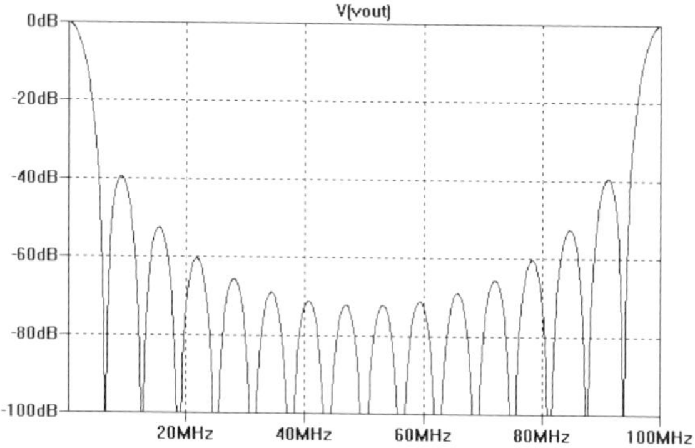

Figure 4.52 Frequency response of a third-order Sinc filter with K = 16.

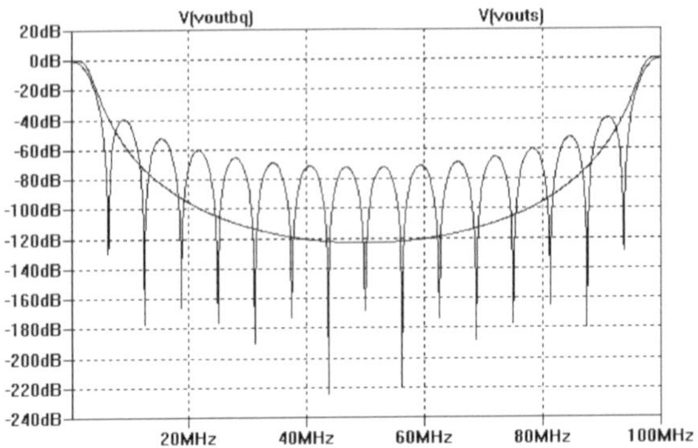

Figure 4.53 Comparing a third-order lowpass Sinc filter to a third-order biquad.

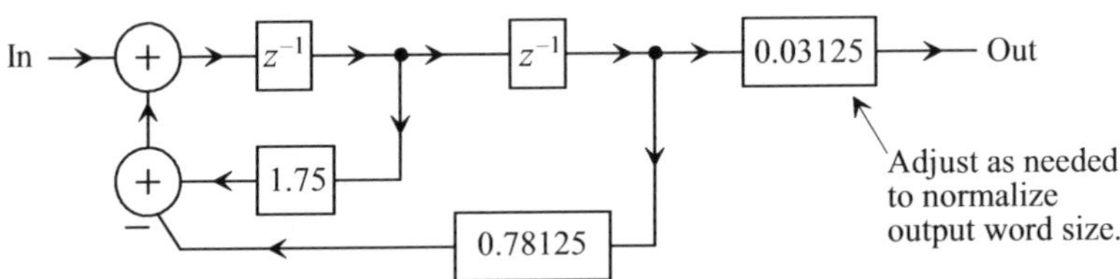

Figure 4.54 The digital biquad filter described by Eq. (4.66).

Example 4.13

Redesign the filter in Fig. 4.54 so that the response has a small amount of peaking at 3.125 MHz. Compare, with simulations, the biquad's response to the Sinc filter's response seen in Fig. 4.52.

Reviewing Eq. (4.63), we see that to increase the Q we need to increase A_1. Keeping in mind that we want to have simple multiplications relying heavily on shifts, let's try increasing A_1 to 1.78125 and A_2 to 0.8125. The simulation results are seen in Fig. 4.55. In this figure we compare the modified filter response to the response of the Sinc filter. Note that we have a couple of dB peaking in the cascaded biquad filter's output. ■

A Comment Concerning Multiplications

While discussing the implementation of digital multipliers is outside the scope of this book a comment is in order about simple multiplier implementations. While, in some filtering applications, simple shifts can prove very useful, we can implement more useful multipliers using adders. Figure 4.56 shows one possible implementation using a single adder along with the associated multiplication factors. We could implement the coefficients in Ex. 4.13 using a similar scheme. For example, $A_1 = 1.78125 = 2 - 0.25 + 0.03125$ and $A_2 = 0.8125 = 1 - 0.25 + 0.0625$ (both requiring two adders). Other creative ways can be used to implement multipliers. For example, a multiplication of 0.5625 by cascading two simple multipliers with multiplication factors 0.75.

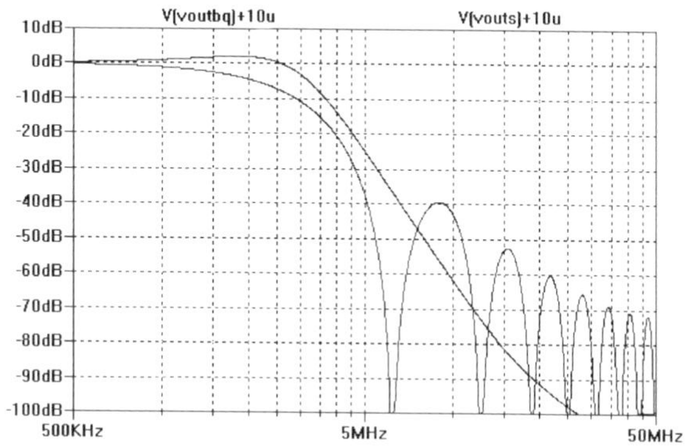

Figure 4.55 Designing a biquad filter with peaking, see Ex. 4.13.

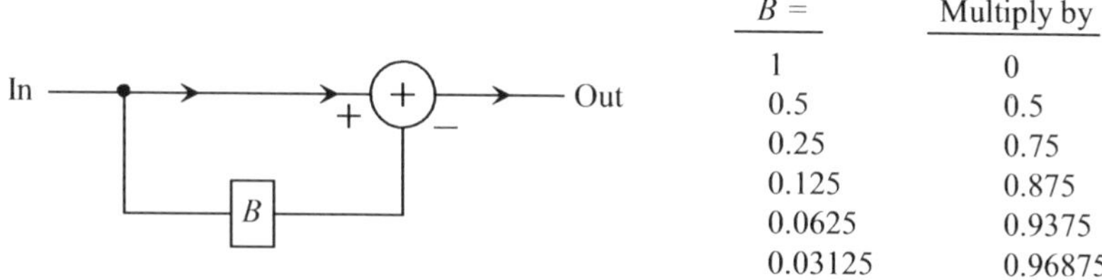

B =	Multiply by
1	0
0.5	0.5
0.25	0.75
0.125	0.875
0.0625	0.9375
0.03125	0.96875

Figure 4.56 A simple multiplier using a single adder.

ADDITIONAL READING

[1] M. Weeks, *Digital Signal Processing Using MATLAB and Wavelets*, Infinity Science Press, 2007. ISBN 978-0977858200

[2] T. B. Welch, C. H. G. Wright, M. G. Morrow, *Real-Time Digital Signal Processing from Matlab to C with the TMS320C6x DSK*, CRC, 2006. ISBN 978-0849373824

[3] S. Haykin and M. Moher, *An Introduction to Analog and Digital Communications*, Second Edition, John Wiley and Sons, 2006. ISBN 978-0471432227

[4] R. G. Lyons, *Understanding Digital Signal Processing*, Second Edition, Prentice-Hall, 2004. ISBN 978-0131089891

[5] P. A. Lynn and W. Fuerst, *Introductory Digital Signal Processing*, Second Edition, John Wiley and Sons, 1998. ISBN 978-0471976318

[6] L. W. Couch, *Modern Communication Systems: Principles and Applications*, Prentice-Hall, 1995. ISBN 978-0023252860

[7] E. P. Cunningham, *Digital Filtering: An Introduction*, John Wiley and Sons, 1995. ISBN 978-0471124757

QUESTIONS

4.1 If $V_{REF+} = 1.0$ V and $V_{REF-} = 0$ regenerate Fig. 4.1 using SPICE. (Design a 3-bit ideal DAC model in SPICE.) The y-axis will be voltages in decimal form.

4.2 If, again, $V_{REF+} = 1.0$ V and $V_{REF-} = 0$, sketch Fig. 4.1 for a 1-bit DAC. Note that the digital input code will either be a 0 or a 1 and the analog voltage out of the DAC will be either 0 or 1.0 V. Using Eq. (4.1) what is the voltage value of 1 LSB? How does this compare to the value of 1 LSB we get from the sketch? Is Eq. (4.1) valid for a 1-bit DAC? Why? The 1-bit DAC will be a ubiquitous component in our noise-shaping modulators later in the book (see Fig. 7.15).

4.3 Why do the transfer curves of Fig. 4.3 show a shift of 1/2 LSB to the left? How do we implement this shift in SPICE?

4.4 Use SPICE to implement 4-bit ADC and DAC. If the converters are clocked at 100 MHz (and the outputs of the ADC are connected to the inputs of the DAC), apply an input sinewave (to the ADC) that has an amplitude of 500 mV peak centered around 500 mV DC with a frequency of 5 MHz. Again, use $V_{REF+} = 1.0$ V and $V_{REF-} = 0$. Show the DAC's analog output.

4.5 Using SPICE generate the spectrums of the input and output signals in question 4.4.

4.6 Suppose we think of the 1-bit input, 0 or 1, in Fig. 4.9 as +1 or −1 (two's complement numbers). What is the output of the digital filter when the input is always 0? Is the magnitude response seen in Fig. 4.10 correct? Why?

4.7 Suppose the 1-bit input signal seen in Fig. 4.9 is an alternating sequence of 101010... In terms of two's complement numbers, what is the output of the digital filter (what is the output of the counter)? What is the frequency of the input signal? Is the frequency response seen in Fig. 4.10 correct?

4.8 Repeat Ex. 4.3 for a filter with a transfer function of

$$\frac{1-z^{-7}}{1-z^{-1}}$$

Also, plot the location of the filter's poles and zeroes in the z-plane.

4.9 Repeat Ex. 4.3 for a filter with a transfer function of

$$\frac{1-z^{-9}}{1-z^{-1}}$$

4.10 Repeat Ex. 4.5 if L is increased to 3.

4.11 Sketch the impulse response of the filter seen in Fig. 4.19.

4.12 What are the transfer functions of the bandpass filters, indicated in Fig. 4.22, with center frequencies of $f_s/6$ and $f_s/3$? Sketch the frequency responses and the location of their poles and zeroes in the z-plane.

4.13 Simulate, using an ideal 8-bit ADC on the input, and an ideal DAC on the output (calculate the size of the DAC), the operation of the digital resonator seen in Fig. 4.23.

4.14 Qualitatively explain why the desired spectrum of an input signal can't be increased by passing data through an interpolator. Using the simulations given in Ex. 4.8, verify that this is indeed the case.

4.15 In Fig. 4.32, which blocks serve as the AAF and which serve as the S/H?

4.16 For the FIR filter seen in Fig. 4.35 with all coefficients set to 0.25, sketch the filter's frequency response.

4.17 For the filter seen in Fig. 4.57 determine the range of values for *a* and *b* where the filter will be stable. What is the filter's transfer function? Sketch the location of the filter's poles and zeroes.

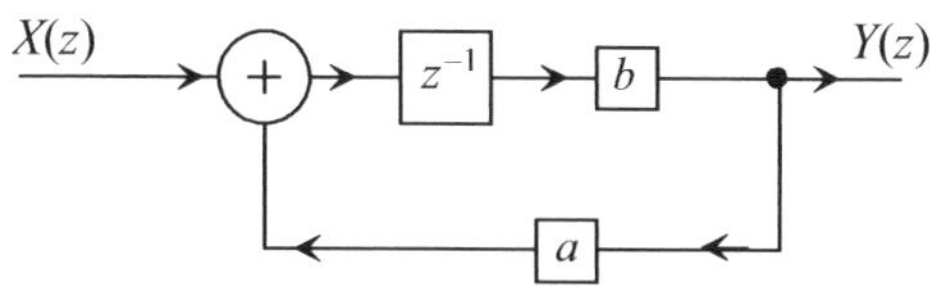

Figure 4.57 A weighted integrating filter.

4.18 Repeat Ex. 4.9 for a filter with a transfer function of

$$\frac{v_{out}}{v_{in}} = \frac{1 + j \cdot \frac{f}{4\ MHz}}{1 + j \cdot \frac{f}{800\ kHz}}$$

4.19 Repeat Question 4.18 using the canonic form of the first-order digital filter.

4.20 Repeat Ex. 4.12 if the Q is increased to 1.

4.21 Show that the filter shown in Fig. 4.58 can be implemented using a single multiplier.

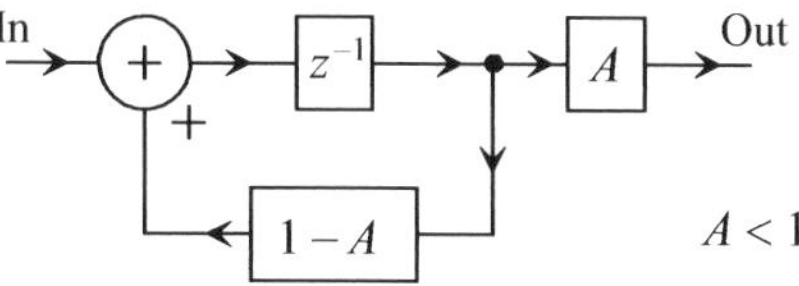

Figure 4.58 Filter used for question 4.21.

4.22 Show that if the values of A and B are restricted to 1, 0.5, 0.25, 0.125, etc. that the circuit of Fig. 4.59 can be used to implement multiplication by coefficients that aren't directly powers of two. How would a multiply by 0.75 be implemented? a multiply-by-0.9375? a multiply-by-0.5625?

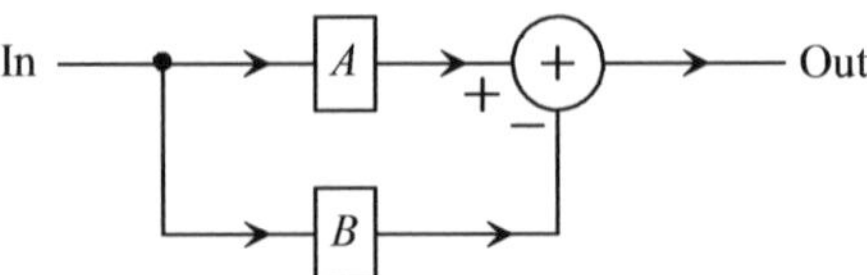

Figure 4.59 A simple multiplier where A and B simply shift the data.

Chapter

5

Data Converter SNR

After studying Chs. 1-4 we should understand the sampling process, including analog sampling, decimation, and interpolation, the operation of the ideal ADC and DAC, and the basics of filtering. In this chapter we turn our attention towards quantization noise. Quantization noise is the effective noise added to a signal after it passes through an ADC (aka quantizer), a comparator (a 1-bit ADC or quantizer)), or, for digital signals, a circuit that reduces the word size (removes the LSBs of the word). One of the key things we'll focus on is the shape of the quantization noise spectrum and how it's added to the spectrum of an input analog or digital signal.

5.1 Quantization Noise

Examine the clocked comparator seen in Fig. 5.1. In this chapter, like the rest of the book, we'll use a VDD of 1 V and a common-mode voltage, V_{CM}, of 500 mV. When the input to the comparator, the non-inverting input, is 600 mV then, since the inverting input is held at V_{CM}, the output goes to 1 V on the rising edge of the clock signal. The difference between the input and output of the comparator is 400 mV. Note that we've gone from an analog signal, the input, to a digital signal, the output. However, while doing this we added noise to our input signal. It's useful to think of this analog-to-digital (quantization) process, from a block diagram point of view, as simply adding noise to our signal. This model is shown in Fig. 5.2. Notice that we've assumed that the noise we added to our

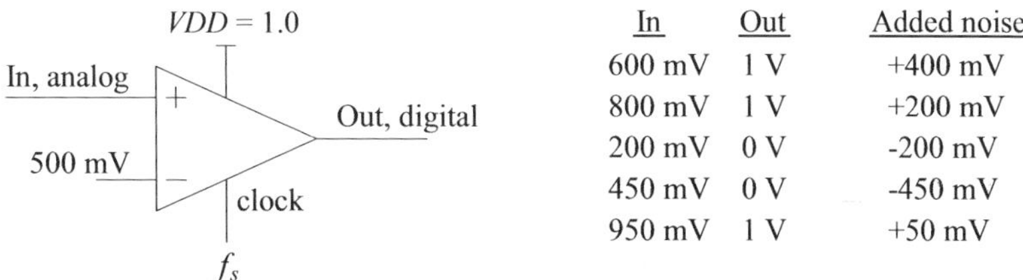

In	Out	Added noise
600 mV	1 V	+400 mV
800 mV	1 V	+200 mV
200 mV	0 V	-200 mV
450 mV	0 V	-450 mV
950 mV	1 V	+50 mV

Figure 5.1 How a comparator adds noise to an input signal.

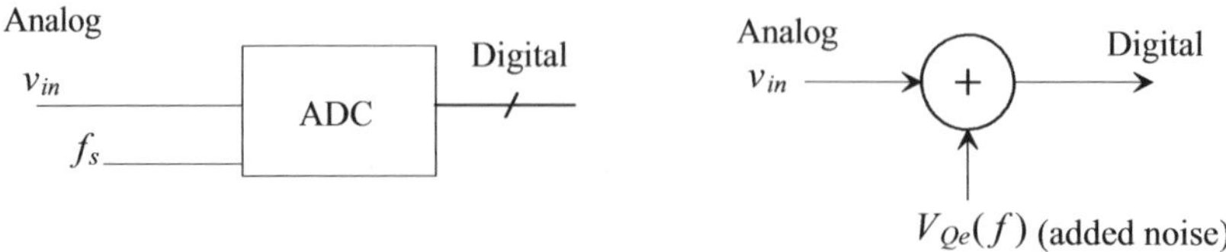

Figure 5.2 Modeling ADC quantization noise.

input signal has a spectrum of $V_{Qe}(f)$. Determining the shape and range of this noise spectrum is one of the goals of this chapter. Finally, while a good portion of our studies will focus on the quantization noise introduced during the analog-to-digital process we can also apply the same results when truncating a digital word's size, Fig. 5.3.

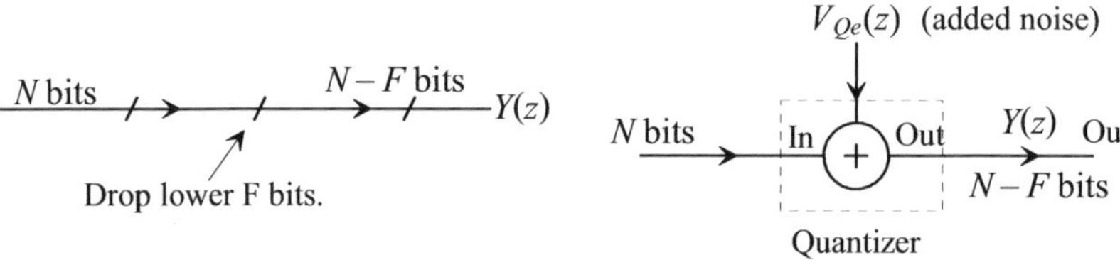

Figure 5.3 Quantizing a digital word and modeling the added noise.

5.1.1 Viewing the Quantization Noise Spectrum Using Simulations

Consider the simple connection of an ideal 8-bit ADC to an ideal 8-bit DAC seen in Fig. 5.4. If we apply a 7 MHz sinewave to the ADC with an amplitude of 0.4 V and an offset of 0.5 V (so the sinewave swings from 100 mV to 900 mV]) and clock the ADC at 100 MHz we get the signals seen in Fig. 5.5. Note that the output of the DAC looks very similar to the output of an ideal S/H (see Fig. 2.14). Now, however, the amplitude of the DAC output signal is quantized, that is, within 1 LSB (= 1.0/256 or 3.906 mV, see Eq. [4.1], for the present simulation) of the ADC input. This quantization is not obvious after looking at Fig. 5.5 (the time domain response). However, looking at the spectrums of the

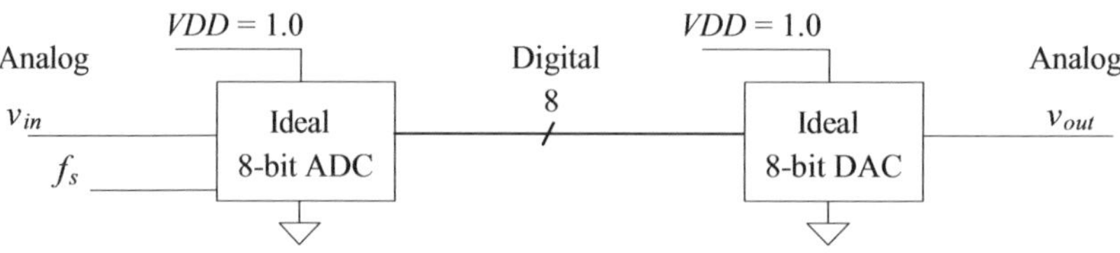

Figure 5.4 Passing a signal through an ADC and then through a DAC.

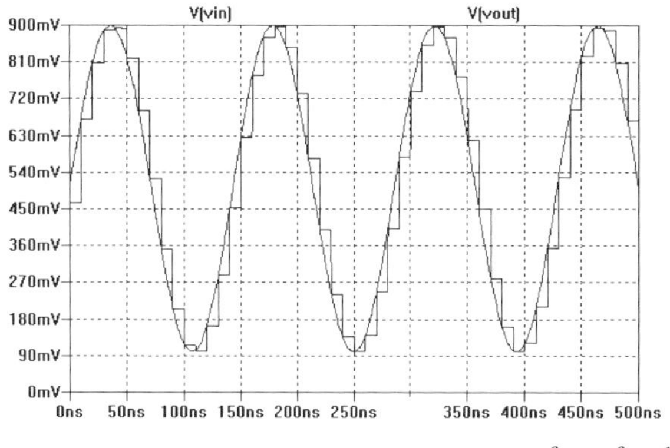

Figure 5.5 Seven MHz ADC input and the corresponding DAC output.

ADC input and the DAC output reveals the difference in the noise floor between the two, Fig. 5.6. The inherent noise floor in the simulation that is associated with the input signal is approximately –80 dB (0.1 mV, RMS.) The noise floor associated with the DAC's output (the signal + quantization noise) is approximately –70 dB (0.316 mV, RMS). It is desirable to determine what sets this value and its spectral content. Again, note that the ADC quantizes the signal which results in the quantization noise (an ideal S/H and DAC don't introduce quantization noise).

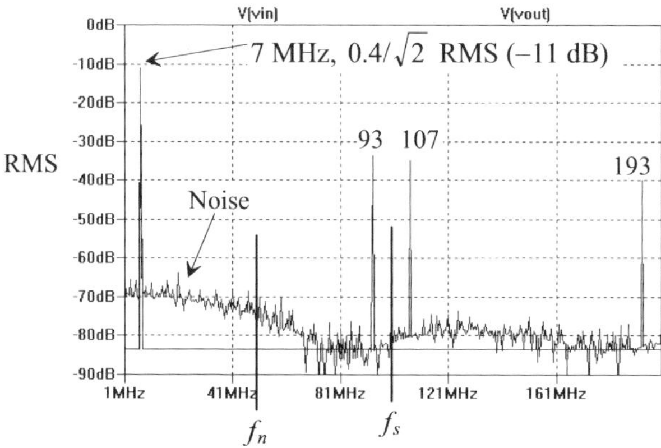

Figure 5.6 ADC input and output spectrums for Fig. 5.4 with signals seen in Fig. 5.5.

Bennett's Criteria

In order to characterize the spectral characteristics of the quantization noise let's make the following assumptions (Bennett's criteria) concerning the signal we are converting:

1. The input (to the ADC) signal's amplitude variation falls between V_{REF+} and V_{REF-} so that no saturation of the digital output code occurs. Exceeding the normal operating range of the ADC affects the quantization noise spectrum by adding spurs or spikes to the output spectrum.

2. The ADC's LSB is much smaller than the input signal amplitude. When this isn't the case, the output of the ADC can appear squarewave-like (when converted back into an analog waveform) and result in a spectrum, once again, that contains spikes or spurs. We'll see later in the book that adding or subtracting a fed-back signal (from the output based on the expected or past quantization noise) to the input modifies this requirement.

3. The input signal is busy (not DC or a low frequency input). We define busy, for the moment, as meaning that no two consecutive outputs of the ADC have the same digital code. For the ideal ADC in Fig. 5.4 1 LSB = 3.906 mV and T_s = 10 ns so that the input must change at least 3.906 mV every 10 ns. We'll see that adding a *high-frequency* dither or pseudorandom noise signal to the input, which can be filtered out later (either using a digital filter or when we pass the output through the reconstruction filter), can make the requirement on the input of being busy practical in an actual circuit. We use these assumptions (Bennett's criteria) in the following discussion unless otherwise indicated.

An Important Note

It's important to note that simply sampling an input waveform, using a S/H, does not result in quantization noise. The amplitude into the ideal S/H, at the sampling instant, is exactly the same as the amplitude out of the ideal S/H. In order to understand why this is important, consider the test setup shown in Fig. 5.7. If we apply a 3 MHz sinewave centered around the common-mode voltage with a 400 mV amplitude we get the outputs seen in Fig. 5.8. Clearly there *is* a difference between the S/H's input and its output. However, this difference has nothing to do with noise, an unwanted signal, since passing the output of the S/H, v_{outsh}, through the ideal reconstruction filter of Fig. 2.19 results in an exact replica of the S/H input v_{in}.

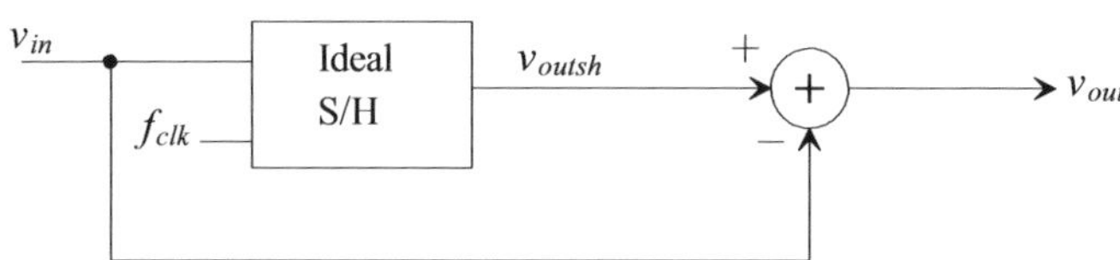

Figure 5.7 Taking the difference between the S/H input and output.

RMS Quantization Noise Voltage

If we were to set up a test configuration similar to that shown in Fig. 5.7 (see Fig. 5.9), where the input to the ADC is subtracted from the DAC output, the resulting output waveform would have little to do, in every case, with the quantization noise. This is true when the input to the ADC contains a broad frequency spectrum extending from DC to the Nyquist frequency, $f_n = f_s/2$. However, if we apply a slow linear ramp to this test setup (to limit the input frequency spectrum) we can (1) see the resulting quantization noise over a wide frequency spectrum and (2) observe the transfer curve, in the time domain. Note that this input violates Bennett's criteria (which, as we'll see, means the noise power spectral density is flat from DC to the Nyquist frequency).

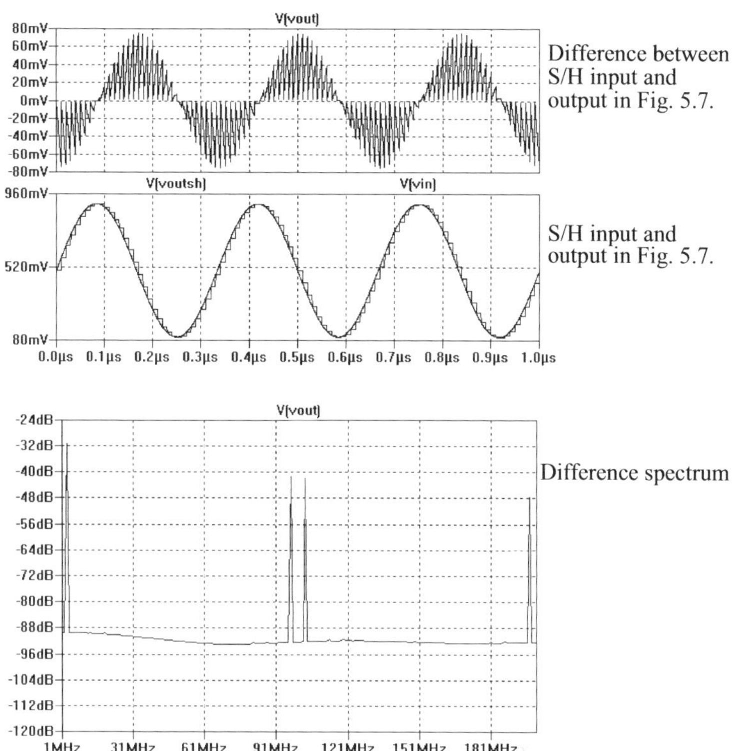

Figure 5.8 Time-domain difference between S/H input and output along with the spectrum.

A section of the input and output, using the test setup of Fig. 5.9, is shown in Fig. 5.10a. It's important to understand the input/output relationship between the ideal ADC and DAC shown in this figure. (Note that clocking the ADC too slow or putting in a ramp that rises too quickly will distort this waveform.) As an example, when the ADC input is slightly above 482 mV, in this figure, the ADC output code (input to the DAC) changes. The ADC output code can be calculated as 482 mV/1 LSB (1 LSB = 1/256 = 3.906 mV for the present simulation) or changing from 123 to 124. Looking at the transfer curves in this figure it appears as though the output changes when the ADC code is 123 or 480.4

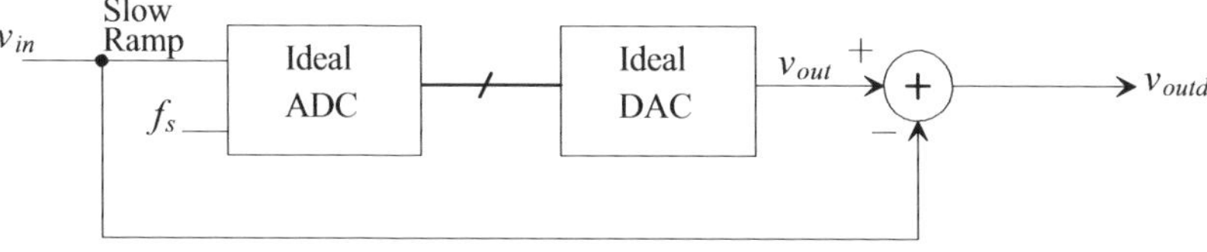

Figure 5.9 Taking the difference between an ADC input and the DAC output.

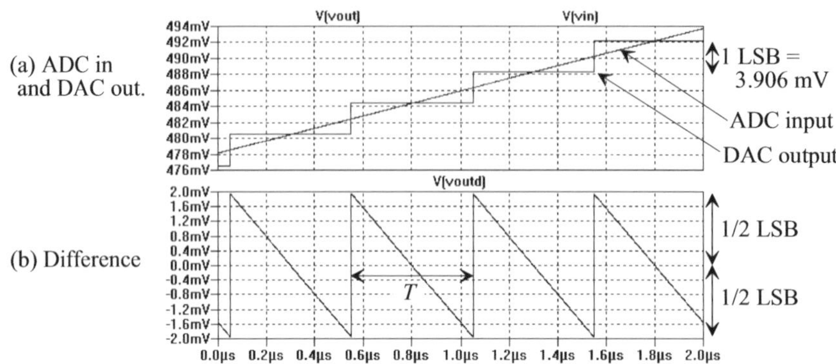

Figure 5.10 Difference between an ADC input and DAC output.

mV/1 LSB. This, as seen in Fig. 5.10b and discussed below, results in centering the quantization error around the input. This is the reason we shifted the ADC transfer curves by 1/2 LSB when we developed our ideal ADC model.

The difference output, between the two signals of Fig. 5.10a, is shown in Fig. 5.10b. Some points to note about this sawtooth waveform are that 1) its average value is zero, 2) the waveform contains an abrupt transition (and so we expect a wideband output spectrum similar to that which occurs after sampling a waveform), and 3) its peak-to-peak amplitude is 1 LSB. Like a sinewave, which also has zero average value, we can characterize this quantization error waveform by looking at its root-mean-square (RMS) value. This value can be calculated using

$$V_{Qe,RMS} = \sqrt{\frac{1}{T}\int_0^T \left(0.5\text{ LSB} - \frac{1\text{ LSB}}{T}\cdot t\right)^2 dt} = \frac{1\text{ LSB}}{\sqrt{12}} = \frac{V_{LSB}}{\sqrt{12}} \quad (5.1)$$

This value is the RMS quantization noise voltage for a specific data converter. Note that the value of the period for this sawtooth waveform, T, doesn't appear in the evaluated result of this equation. Also note that the sampling frequency, f_s, isn't present in this equation. For our present discussion where 1 LSB is 3.906 mV, $V_{Qe,RMS}$ = 1.13 mV or –59 dB (RMS).

Treating Quantization Noise as a Random Variable

If Bennett's criteria hold, then the quantization noise voltage can be thought of as a random variable falling in the range of ±0.5 LSB, as seen in Fig. 5.11. The probability that the quantization error is –0.2 LSB is the same as the probability that the error is 0.4 LSB. In other words, there is no reason why the quantization error should have one value more often than another value.

The quantization error noise power is the variance of the probability density function. The RMS quantization error voltage is the square root of the quantization noise power. The variance of the probability density function (the quantization noise power, P_{Qe}) is given, knowing the average of the quantization error, $\overline{Qe}$, is zero, by

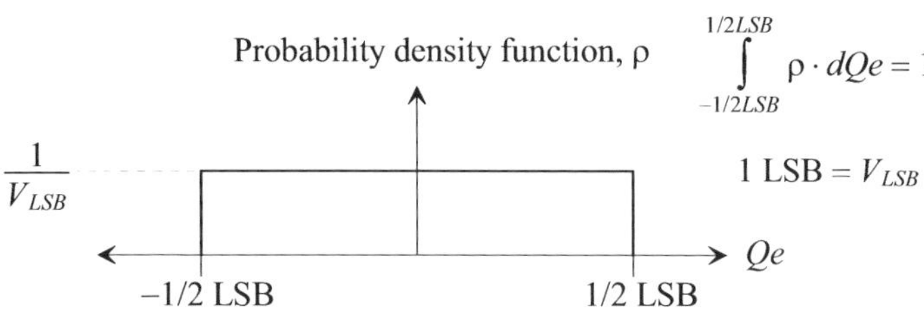

Figure 5.11 Probability density function for the quantization error in an ADC assuming Bennett's criteria hold.

$$P_{Qe} = \int_{-1/2LSB}^{1/2LSB} \rho \cdot (Qe)^2 \cdot dQe = \frac{V_{LSB}^2}{12} \tag{5.2}$$

so that, once again, the RMS quantization noise voltage is

$$V_{Qe,RMS} = \frac{V_{LSB}}{\sqrt{12}} \tag{5.3}$$

Again, if our LSB voltage is 3.096 mV, then, once again, $V_{Qe,RMS}$ = 1.13 mV or −59 dB (RMS). If we look at Fig. 5.6, we see that the RMS noise voltage varies essentially over the entire spectrum (white noise) and has a value ranging from around −70 dB down to less than −80 dB. Note that although the entire spectrum contains quantization noise it is not because of the sampling process used in the ADC (and so quantization noise doesn't experience aliasing). Quantization noise is added to the signal after the sampling process during the analog-to-digital conversion process. In order to qualitatively understand why the quantization error spectrum is white, in Fig. 5.6, we remember that there are abrupt transitions in the DAC output, and if the quantization error is truly random, the times between the changes have varying periods. We might speculate that by simulating a longer time or using a multiple frequency input so as to "exercise" the ADC, the resulting quantization errors are further randomized and the resulting error spectrum will be flatter than what is seen in Fig. 5.6.

5.1.2 Quantization Noise Voltage Spectral Density

If the quantization noise voltage spectrum is truly flat (Bennett's criteria hold) we can determine the noise power spectral density of $V_{Qe,RMS}$, $V_{Qe}^2(f)$ with units of V^2/Hz, or the noise voltage spectral density, $V_{Qe}(f)$ with units of $V/\sqrt{Hz}$ by solving

$$\frac{V_{LSB}^2}{12} = 2\int_0^{f_s/2} V_{Qe}^2(f) \cdot df \tag{5.4}$$

where the factor of 2 accounts for the power in the negative frequencies of the spectrum. Notice that we assumed a spectrum from DC to the Nyquist frequency, $f_s/2$. Assuming all of the quantization noise falls below the Nyquist frequency is the **worse-case** situation. Solving this equation yields

$$V_{Qe}(f) = \frac{V_{LSB}}{\sqrt{12 f_s}} = \frac{V_{REF+} - V_{REF-}}{2^N \sqrt{12 f_s}} \tag{5.5}$$

with units of $V/\sqrt{Hz}$. Note that the quantization noise spectral density is inversely proportional to the sampling frequency. Figure 5.2 shows how we can model the ADC as a summation of the input signal and the quantization noise.

After looking at Eq. 5.5 we might think that by simply increasing the sampling frequency we can reduce the amount of quantization noise an ADC introduces into an analog input signal. While increasing the sampling frequency spreads the quantization noise spectral density out over a wider range of frequencies (see Fig. 5.12) with a corresponding reduction in amplitude, the sampling frequency doesn't affect the total RMS quantization noise voltage. However, bandlimiting the spectrum using a filter reduces the amount of quantization noise introduced into an input signal (**this is important** and the reason mixed-signal design is so powerful). In the simplest case a lowpass filter is used on the output of the ADC to reduce the amount of quantization noise introduced into the signal. We can write the amount of noise introduced into an input signal over a range of frequencies using

$$V_{Qe,RMS}^2 = 2\int_{f_L}^{f_H} V_{Qe}^2(f)\cdot df \text{ where } f_L < f_H \le f_s/2 \tag{5.6}$$

Again, the factor of 2 is used to account for the contributions to $V_{Qe,RMS}$ in the negative frequency spectrum. Let's show that the sampling frequency doesn't affect the quantization noise, assuming Bennett's criteria are valid.

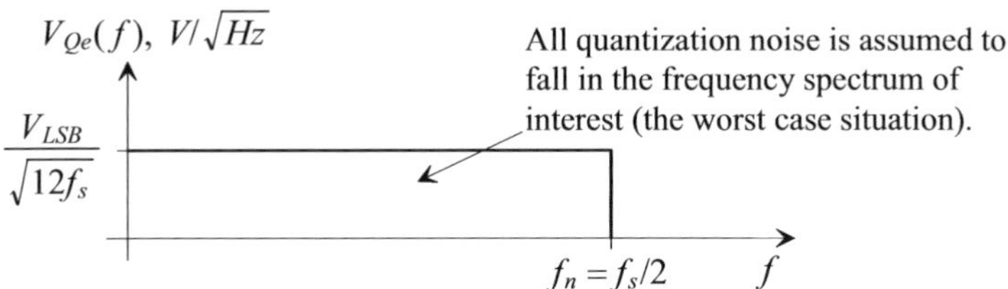

Figure 5.12 Quantization noise spectral density.

Example 5.1
Using the SPICE simulation that was used to generate Fig. 5.10b with sampling frequencies of 100 and 200 MHz calculate and simulate the amount of quantization noise introduced into the input signal.

Doubling the sampling frequency has little effect on the output quantization noise. Our calculated value was 1.13 mV using Eqs. (5.1) and (5.3) while the simulated values are 1.127 mV for both 100 and 200 MHz sampling frequencies. ■

Example 5.2
Suppose a 7 MHz sinewave with a peak amplitude of 400 mV is applied to the topology seen in Fig. 5.13. Calculate and compare to simulations the amount of quantization noise introduced into the sinewave when sampling frequencies of 100 and 200 MHz are used.

Again, the RMS value of the quantization noise added to the input signal is 1.13 mV. The simulated values are 1.12 mV and 1.15 mV for 100 and 200 MHz sampling frequencies. Figure 5.13 shows the simulation results. Again, note that we assume Bennett's criteria are valid (input busy, large compared to an LSB, etc.)

In order to see the quantization noise introduced by the ADC we have to remove tones (input signal and its aliases) from the DAC's output signal. As seen in Fig. 5.13 the way we do this is by subtracting the S/H input signal (remembering simply sampling a waveform doesn't introduce noise) from the DAC's output. Using ideal components the delay through the system is nearly zero. In most systems we have to adjust the delay to match the delay through the mixed-signal system. This only works if the phase shift through the mixed-signal system is linear (constant delay). A more useful technique is to look at the spectrum of the output signal and remove, using for example Matlab™, the wanted signal and its aliases. ■

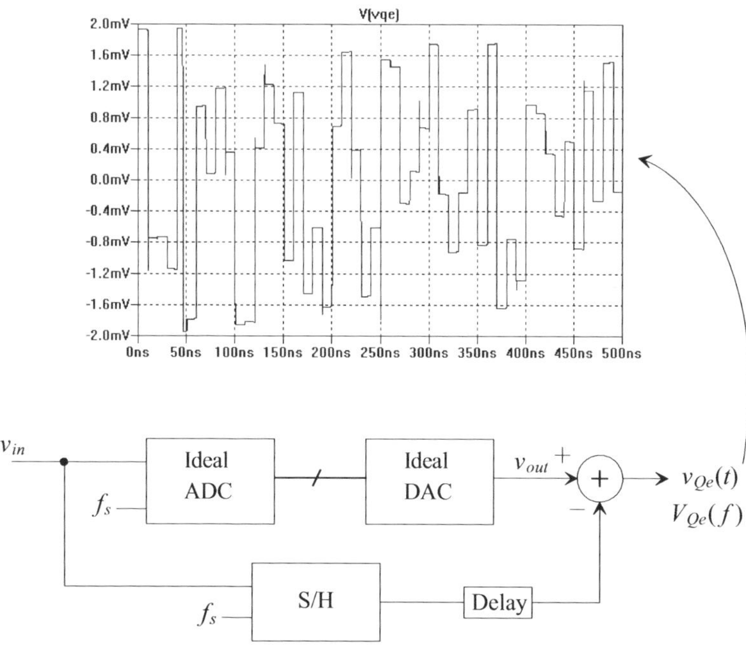

Figure 5.13 Determining quantization noise in a mixed-signal system.

Calculating Quantization Noise from a SPICE Spectrum

The spectrum of a signal in a SPICE simulation is generated using a Fast-Fourier Transform (FFT). We can estimate the RMS value of this spectrum by summing the individual components using

$$V_{Qe,RMS} = \sqrt{\sum_{k=0}^{M-1} V_{FFT}^2(k \cdot f_{res})} \tag{5.7}$$

where M is the number of points used in the FFT and f_{res} is the resolution of the FFT (the distance between adjacent points). Note that here we are assuming that $V_{FFT}(f)$ is an *RMS* value. If it's a peak value then the right-side of Eq. (5.7) should be divided by $\sqrt{2}$.

Figure 5.14 shows a portion of the quantization noise spectrum for the signal seen in Fig. 5.13. We can estimate the RMS noise from this spectrum assuming, as seen in Fig. 5.12, that the noise is bandlimited. While we can look at the specifics of the FFT in the simulation to get exact numbers, let's simply count the number of points seen in the plot (Fig. 5.14) and see how close we get to the exact answer of 1.13 mV. First note that –74 dB is 200 μV RMS and, as seen in Fig. 5.14, we are assuming a constant spectrum (white noise). Looking at the plot there are roughly 32 points from DC to 50 MHz so

$$V_{Qe,RMS} = \sqrt{32 \cdot (0.2\ mV)^2} = 1.13\ mV$$

is an exact match. Note, again, that since we are assuming all of the noise falls between DC and the Nyquist frequency, our estimated value (–74 dB) for the quantization noise is a little above the actual value we see in this range.

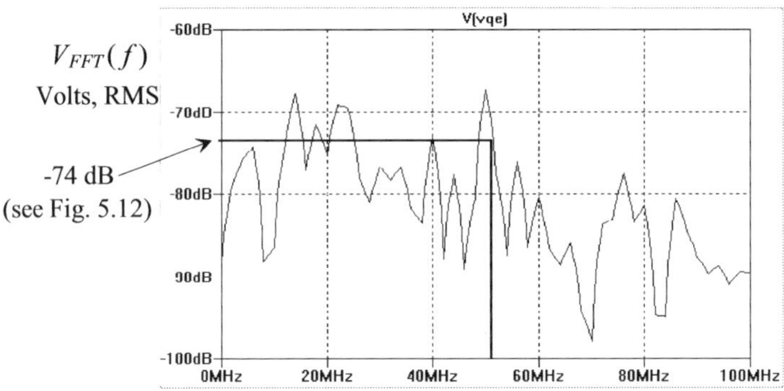

Figure 5.14 Spectrum of the quantization noise seen in Fig. 5.13.

Power Spectral Density

The variable $V_{Qe}^2(f)$ in Eq. (5.6) is a power spectral density (PSD) with units of V^2/Hz. To convert our SPICE plot above (RMS amplitudes) to a power spectral density we use

$$V_{Qe}^2(f) = \frac{V_{FFT}^2(f)}{f_{res}} \text{ with units of } V^2/Hz \tag{5.8}$$

and change the y-axis from $20 \cdot \log y$ to $10 \cdot \log y$. Note that we can then use Eq. (5.6) to calculate $V_{Qe,RMS}$ (without the factor of two since SPICE uses one-sided spectrums). To generate a SPICE plot similar to what is seen in Fig. 5.12 we use

$$V_{Qe}(f) = \frac{V_{FFT}(f)}{\sqrt{f_{res}}} \text{ with units of } V/\sqrt{Hz} \tag{5.9}$$

Again the quantization noise in the simulation would extend beyond $f_s/2$, unlike the assumed shape seen in Fig. 5.12.

5.2 Signal-to-Noise Ratio (SNR)

In the last section we developed the idea of treating an analog-to-digital converter (ADC) as a noisy circuit block where the output of the ADC is the sum of quantization noise and the input signal. Logically, the next step in our development of concepts is to characterize the mixed-signal system in terms of the signal-to-noise ratio (SNR).

If we apply a sinewave with an amplitude of V_p (and thus an RMS value of $V_p/\sqrt{2}$) to an ADC input, then, knowing the RMS quantization noise added to a busy ADC input signal is $V_{LSB}/\sqrt{12}$ (see Eqs. [5.1] and [5.3]), the resulting SNR is

$$SNR_{ideal} = 20 \cdot \log \frac{V_p/\sqrt{2}}{V_{LSB}/\sqrt{12}} \tag{5.10}$$

If we remember that

$$V_{LSB} = 1 \text{ LSB} = \frac{V_{REF+} - V_{REF-}}{2^N} \tag{5.11}$$

and we assume that the largest possible amplitude sinewave is the ADC input (to maximize the SNR), that is,

$$2V_p = V_{REF+} - V_{REF-} \tag{5.12}$$

then Eq. (5.10) can be rewritten as

$$SNR_{ideal} = 20 \cdot \log \frac{2^N\sqrt{12}}{2\sqrt{2}} = 6.02N + 1.76 \text{ (in dB)} \tag{5.13}$$

Effective Number of Bits

Equation (5.13) relates the number of bits used in a data converter to the ideal SNR when the input signal is a sinewave that ranges from V_{REF+} to V_{REF-}. In reality the measured SNR, in most cases, will be different from the ideal value calculated using this equation. When the SNR is measured, we relate it to the *effective number of bits* using

$$N_{eff} = \frac{SNR_{meas} - 1.76}{6.02} \tag{5.14}$$

where the measured SNR (SNR_{meas}) is specified in dB.

Example 5.3

Determine the effective number of bits for an ADC with , $V_{REF+} = 1.0$, $V_{REF-} = 0$ and a measured $V_{Qe,RMS}$ of 2 mV.

If we assume that the input peak amplitude, V_p, is $0.5 \cdot (V_{REF+} - V_{REF-})$ or 500 mV, then the measured SNR is given by

$$SNR_{meas} = \frac{0.5/\sqrt{2}}{2 \text{ mV}} = 177 = 45 \text{ dB}$$

The effective number of bits, N_{eff}, is, from Eq. (5.14), 7.18 bits. Note that to calculate $V_{Qe,RMS}$ we either take the output of the DAC and feed it into a spectrum analyzer or take the FFT of the digital output data of an ADC so that we get a plot, in either case, similar to what is seen in Fig. 5.14. ■

Example 5.4
Using the ideal 8-bit ADC and DAC shown in Fig. 5.4 with a sampling frequency of 100 MHz show, using SPICE, that applying a full-scale sinewave at 24 MHz to this configuration will cause the resulting SNR to approach the ideal value given by Eq. (5.10).

Let's begin by calculating SNR_{ideal}. From Eq. (5.13), SNR_{ideal} is roughly 50 dB, as the data converters have 8-bit resolution.

The time-domain input and output of the circuit and the corresponding DAC output spectrum, are shown in Fig. 5.15. The input to the ADC in Fig. 5.4 is a 24 MHz sinewave with a peak amplitude of 0.5 V centered on a DC voltage of 0.5 V (the peak-to-peak voltage of the input waveform is *VDD* or 1 V). The $V_{Qe,RMS}$ measured with SPICE, in a separate simulation using a S/H to remove the input signal and aliased signals from the DAC's output spectrum, is 1.4 mV. The simulated SNR is then $(0.5/\sqrt{2})/1.4\,\text{mV}$ or 253 (48 dB), which is close to the value, 50 dB, calculated at the beginning of the example. ■

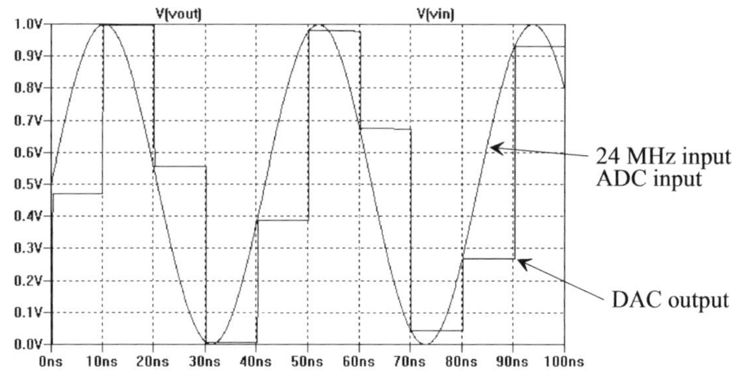

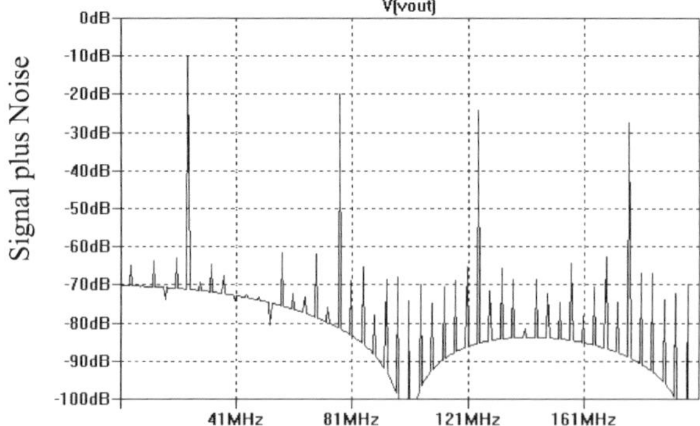

Figure 5.15 Signals from Ex. 5.4, simulating Fig. 5.4 with a 24 MHz sinewave input.

Coherent Sampling

It's important to understand that poor selection of the input frequency can result in an SNR that is different from the ideal value calculated using Eq. (5.13). For example, if our input frequency is 10 MHz while the clock frequency is 100 MHz then the sampling is *coherent*. The sampled points repeat with every cycle of the input signal. The amount of quantization error can then be, repeatedly, near $\pm 1/2$ LSB or much larger than $V_{LSB}/\sqrt{12}$.

Coherent sampling can be useful to minimize the undesired effects of the FFT, namely spectral leakage. In order to understand what is meant by "spectral leakage," consider the sinewave with infinite duration shown in Fig. 5.16a. When an FFT is performed on a time-domain waveform, the first step is to "window" the waveform. The simplest window is the rectangular window. In a simulation the duration of the sinewave is finite and set by the simulation time or transient stop time, T_{stop}. We can think of taking the infinite duration sinewave of (a) and multiplying it by the rectangular waveform of Fig. 5.16b to obtain the waveform used in the simulation, Fig. 5.16c. This multiplication means the resulting waveform is the convolution of the original sinewave spectral response (an impulse) and the frequency domain transform of the squarewave (a Sinc waveform) in the frequency domain. The result is that instead of the sinewave spectral response being an impulse function, as seen in Fig. 5.16d, it is a weighted Sinc waveform, Fig. 5.16e. Note that the FFT spectral response of the sinewave in (e) is spread out or "leaks" into the frequencies around the actual or continuous time response. The large ratio of the peak value of the Sinc pulse to its first sidelobe is usually undesirable. Rather, to minimize these sidelobes, other windowing functions are used. One commonly used

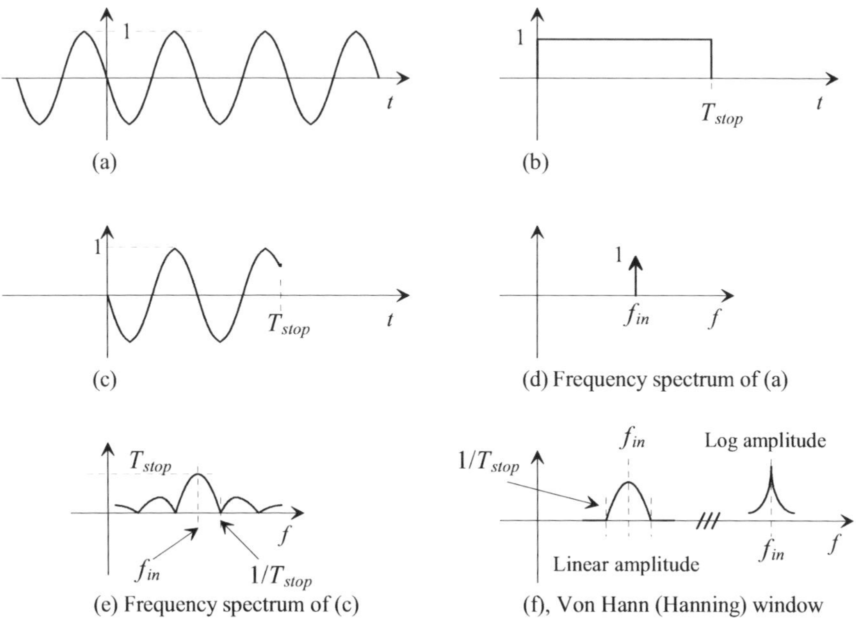

Figure 5.16 Showing how spectral leakage in an FFT affects the spectrum of a waveform.

window is the von Hann (a.k.a. Hanning or Cosine) window represented, without the sidelobes, in Fig. 5.16f. The response is shown on both linear and log amplitude scales and the width of the window is $2/T_{stop}$ at its base.

Selecting an input sinewave frequency, f_{in} , such that f_s/f_{in} is a whole number creates a condition where an integral number of input sinewave cycles fit perfectly into the simulation or measurement time (so the windowing function isn't important). This results in an output spectrum that contains isolated tones (no spectral leakage). Coherent sampling may be used to reduce the effects of spectral leakage when determining the SNR by ensuring $f_{in} << f_s$ (to randomize the quantization noise).

Signal-to-Noise Plus Distortion Ratio

In a practical data converter the output spectrum contains not only quantization noise but distortion resulting from nonlinearities and mismatch in the data converter circuitry. When we calculate the RMS quantization noise voltage using Eq. (5.7) and nonideal components, we are actually calculating the noise plus the distortion in the spectrum. Until this point we have only used ideal components, so that distortion in the output spectrums was absent. We can rewrite Eq. (5.7) to indicate that when it is used with a measured spectrum, both noise and distortion are included in the result as

$$V_{Qe+D,RMS} = \sqrt{\sum_{k=0}^{M-1} V_{FFT}^2(k \cdot f_{RES})} \tag{5.15}$$

The *signal-to-noise plus distortion ratio* is then given by

$$\text{SNDR} = 20\log\frac{V_p/\sqrt{2}}{V_{Qe+D,RMS}} \tag{5.16}$$

The effective number of bits, from Eq. (5.14), can then be calculated using

$$N_{eff} = \frac{\text{SNDR} - 1.76}{6.02} \tag{5.17}$$

Example 5.5

Suppose that the test setup shown in Fig. 5.9 is used with an input sinewave having a frequency of 7 MHz, a peak amplitude of 0.5 V, and centered around 0.5 V (so that, once again, the sinewave swings from 0 V to 1 V.) Using SPICE simulation, determine the SNDR if there is a gain error in the ideal ADC in Fig. 5.9 (it's no longer ideal) so that each stage in the pipeline algorithm used to implement the ideal SPICE ADC has a gain of 2.1 instead of the ideal 2.0.

The resulting DAC output spectrum is shown in Fig. 5.17. The RMS noise plus distortion voltage, $V_{Qe+D,RMS}$, is 22.58 mV, using SPICE and remembering to remove, with the S/H, the desired terms at DC and 7 MHz as well as the undesired images at 93 MHz, 107 MHz, and 193 MHz. The SNDR is then

$$\text{SNDR} = 20\log\frac{0.5/\sqrt{2}}{22.58\text{ mV}} \approx 24\text{ dB}$$

The effective number of bits is 3.7. In other words, a 5% gain error in the ADC amplifiers results in an effective resolution of less than half of the ideal value of 8-bits. ■

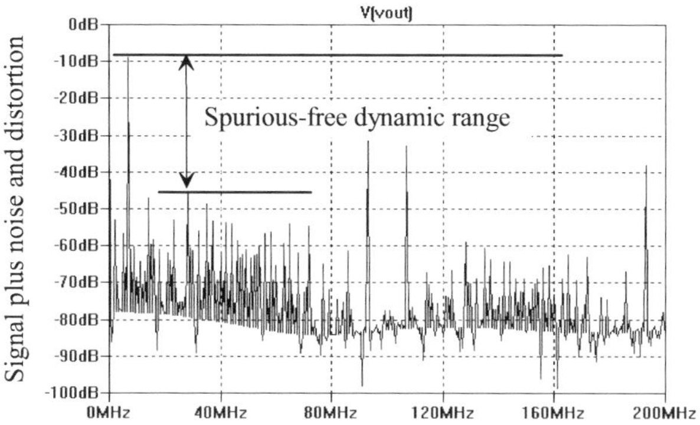

Figure 5.17 Output spectrum with ADC gain error (see Ex. 5.5).

Measuring SNDR requires a spectrum analyzer, when looking at the output of a DAC, or loading digital data into a program that can perform an FFT when looking at the output of an ADC. Trying to measure SNDR using a time domain instrument, such as an oscilloscope, is usually a waste of time because the dynamic range of the instrument is comparable to the dynamic range of the data converter under test. Spectrum analyzers utilize narrow band filtering on their input to reduce the inherent noise measured in a circuit and can have dynamic ranges in excess of 120 dB over a very wide frequency spectrum. Also note that the SNDR is sometimes abbreviated as *SINAD* (signal-to-noise and distortion.)

Spurious Free Dynamic Range

Another specification of interest is the data converter's spurious free dynamic range. This term relates the peak signal in the output spectrum (the input sinewave or carrier) to the largest spike in the output spectrum up to the Nyquist frequency. This can be written using

$$\text{SFDR(dBc)} = \text{input carrier(dB)} - \text{unwanted tone(dB)} \tag{5.18}$$

For the spectrum shown in Fig. 5.17, the input sinewave (carrier) has an amplitude of 0.5 V (–9 dB RMS), while the largest unwanted tone has an amplitude of –46 dB. The SFDR of this data converter is then 37 dBc.

Dynamic Range

The dynamic range of a data converter can be specified in several ways. One definition is as the ratio of the largest output signal change (e.g., $[V_{REF+} - 1\ LSB] - V_{REF-}$) over the smallest output signal change (1 LSB). Remembering 1 LSB = $(V_{REF+} - V_{REF-})/2^N$ the dynamic range (DR) can be written as

$$\text{DR} = 20\log\frac{V_{REF+} - (V_{REF+} - V_{REF-})/2^N - V_{REF-}}{(V_{REF+} - V_{REF-})/2^N} = 20\log 2^N = 6.02N \tag{5.19}$$

If a 1,000 to 1 dynamic range (60 dB) is required, then a data converter with at least 10 bits is needed.

Another way to specify DR is as the ratio of the RMS full-scale input sinusoid amplitude, $V_p/\sqrt{2}$, to the input sinusoid amplitude (RMS) that results in an SNDR of 0 dB. (The RMS amplitude of the input signal is equal to the RMS quantization noise plus distortion, $V_{Qe+N,RMS}$, when the SNDR is 0 dB.) This is nothing more than saying that the SNDR can be used to specify DR.

Example 5.6

Determine the DR for the ideal ADC in Ex. 5.4 using Eq. (5.19). Compare the result to the SNDR calculated in Ex. 5.5.

Using Eq. (5.19), the DR is 48.16 dB (the ideal value). The SNDR calculated in Ex. 5.5 was 24 dB. Clearly, the SNDR is a better indication of DR than is the value obtained using Eq. (5.19). ■

Specifying SNR and SNDR

The SNR and the SNDR are usually specified as a function of input sinewave amplitude at a fixed frequency, Fig. 5.18. The x-axis in Fig. 5.18 is normalized so that an input sinewave with a peak-to-peak amplitude of $V_{REF+} - V_{REF-}$ corresponds to 0 dB. We might be wondering how we differentiate between SNR and SNDR as both, up to this point, have been calculated in the same way (Eqs. [5.7] and [5.15]). We continue to calculate SNDR using a data converter output spectrum, remembering to zero out the desired tones and images, and Eq. (5.15) as was done in Ex. 5.5. When we calculate the SNR, we follow the same procedure except that now we also zero out any *spikes* or *spurs* (spurious responses) in the spectrum that are "sticking up" above the noise floor in the spectrum. These spikes come from imperfections in the data converter and result in distortion in the output waveform. Note in Fig. 5.18 how the SNR and the SNDR coincide until the input signal amplitude gets reasonably large (so the distortion tones increase in amplitude above the quantization noise).

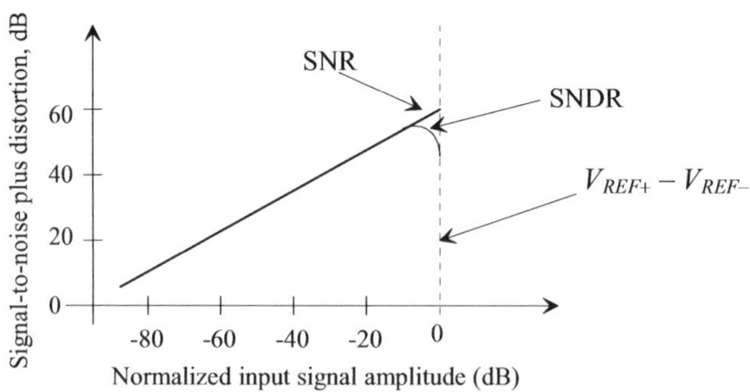

Figure 5.18 Specfying SNR and SNDR for a data converter.

5.2.1 Clock Jitter

We might assume that by using the ideal data converters in a system with "real world" input and clock signals we would get a resolution (number of bits) limited by the resolution of the ideal data converters used. However, if the clock signal isn't ideal, the resolution will be less than ideal because of a problem known as aperture jitter. In this section we relate the sampling clock's jitter to the data converters SNR and thus the effective number of bits. Clock jitter is the variation in the period of the clock signal around the ideal value.

Figure 5.19 shows the basic problem. In this figure we have assumed the input sinewave frequency is running at the Nyquist frequency f_n $(= f_s/2)$ so that the sampling point (when the sinewave crosses zero in this figure) is seeing the fastest transition in the input signal. We assume the peak amount of jitter in the clock signal is ΔT_s. For example, if the sampling clock frequency is 100 MHz (T_s = 10 ns) and the peak-to-peak clock jitter is 50 ps ($= \Delta T_s$), then the specification of the sampling clock stability is 5,000 ppm (where parts per million [ppm] = 10^{-6} and ΔT_s = (stability, ppm) $\cdot$ $(1/f_s)$).

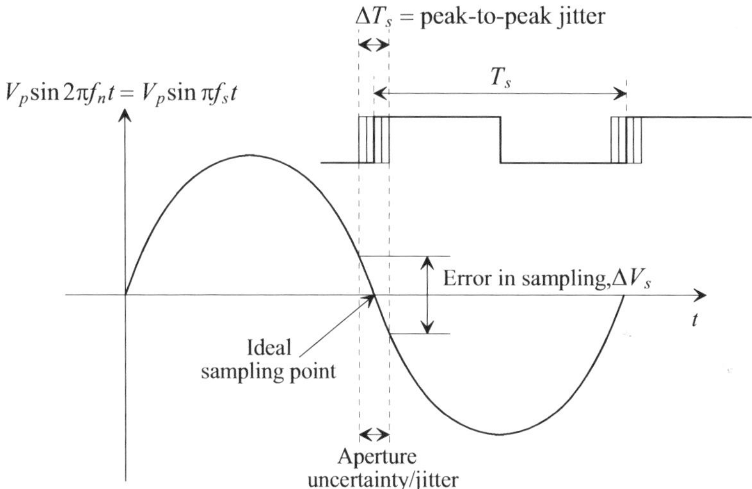

Figure 5.19 Data converter input signal and clock jitter.

The slew rate of the signal in Fig. 5.19, at the sampling point (when the clock signal transitions high), is given by

$$\frac{d}{dt}(V_p \sin \pi f_s t) = \pi f_s V_p \overbrace{\cos \pi f_s t}^{=1} = \pi f_s V_p \tag{5.20}$$

We can relate the uncertainty in the sampling instant, ΔT_s, to the uncertainty in the sampled voltage, ΔV_s, using

$$\frac{\Delta V_s}{\Delta T_s} = \pi f_s V_p, \text{ or } \Delta V_s = \Delta T_s \cdot \pi f_s V_p \tag{5.21}$$

If we require the uncertainty in the sampled voltage, ΔV_s, to be at most 0.5 LSB = $(V_{REF+} - V_{REF-})/2^{N+1}$ and we remember $V_p = (V_{REF+} - V_{REF-})/2$, then our maximum allowable peak-to-peak clock jitter can be determined for a particular data converter using

$$\Delta T_s \leq \frac{1}{2^N} \cdot \frac{1}{\pi f_s} \tag{5.22}$$

or in terms of the sampling clock stability

$$\text{Stability, ppm} = \Delta T_s \cdot f_s = \frac{\Delta T_s}{T_s} = \frac{1}{\pi \cdot 2^N} \tag{5.23}$$

Table 5.1 relates the stability requirements placed on a sampling clock for a data converter resolution, N, if less than 0.5 LSBs aperture error or sampling voltage uncertainty is required of the data converter.

Table 5.1 Maximum jitter, ΔT_s, for 0.5 LSB sampling uncertainty.

Resolution, N	Stability, ppm	ΔT_s(max), ps	ΔT_s(max), ns
		If f_s = 100 MHz	If f_s = 44.1 kHz
6	5,000	50	113.4
8	1,250	12.5	28.3
10	312.5	3.13	7.1
12	78.1	0.78	1.77
14	19.5	0.2	0.44
16	4.9	0.05	0.11

Example 5.7

Suppose a phase-locked loop (PLL) is used to generate a clock signal for a data converter. If the resolution of the data converter is 10 bits and the frequency of the sampling clock coming from the PLL is 900 MHz, then specify the maximum jitter allowed in the output of the PLL. Assume that the maximum sampling error allowed is 0.5 LSB and that the data converter is sampling a sinewave with a frequency of 100 MHz.

Because the sinewave being sampled has a frequency below the Nyquist value, Eq. (5.22) cannot be used directly. Instead, after reviewing the derivation of this equation, we can rewrite it in terms of any input signal frequency, f_{in}, as

$$\Delta T_s \leq \frac{1}{2^N} \cdot \frac{1}{2\pi \cdot f_{in}} \tag{5.24}$$

noting that when $f_{in} = f_n = f_s/2$, Eq. (5.24) reduces to Eq. (5.22). Using Eq. (5.24) with the numbers in this problem results in a peak-to-peak jitter of 1.56 ps! The reader familiar with PLL design will recognize that this is a very challenging requirement when designing a PLL (that is, to design a PLL with an output frequency of 900 MHz and an output jitter less than 1.56 ps). ■

We're now in a position to answer how, given the peak-to-peak clock jitter, the SNR of a data converter is degraded from the ideal value (given in Eq. [5.13]) when the input sampling clock isn't ideal. Rewriting Eq. (5.24) and assuming

$$\Delta T_s \geq \frac{1}{2^N} \cdot \frac{1}{2\pi \cdot f_{in}} \tag{5.25}$$

(a resolution loss ≥ 0.5 LSB), we get

$$\Delta T_s = \frac{1}{2^{N-N_{Loss}}} \cdot \frac{1}{2\pi \cdot f_{in}} \tag{5.26}$$

where f_{in} is, once again, the frequency of the input sinewave, and N_{Loss} is the number of bits lost due to the excess jitter. Assuming Eq. (5.25) is valid, then when N_{Loss} is zero, the loss in resolution is 0.5 LSB and Eq. (5.26) reduces to the equality condition in Eqs. (5.25) or (5.26). The ideal data converter's SNR, assuming the only non-ideal factor in the system is clock jitter, can be written as

$$SNR = 6.02 \cdot \overbrace{(N - N_{Loss} - 0.5)}^{\text{effective bits, } N_{eff}} + 1.76 \text{ (in dB)} \tag{5.27}$$

Example 5.8

For an ideal 8-bit ADC clocked at 100 MHz, determine the SNR of the data converter with 100 ps of peak-to-peak jitter in the input sampling clock, ΔT_s, assuming the ADC's input is a full-scale sinewave at 25 MHz.

We can write the number of bits lost by solving Eq. (5.26) as a function of peak-to-peak jitter as

$$N_{Loss} = N + 3.32 \cdot \log(2\pi \cdot f_{in} \cdot \Delta T_s) \text{ assuming } \Delta T_s \geq \frac{1}{2^N} \cdot \frac{1}{2\pi \cdot f_{in}} \tag{5.28}$$

or

$$N_{Loss} = 8 + 3.32 \cdot \log(2\pi \cdot 25\text{MEG} \cdot 100\text{ps}) = 2 \text{ bits}$$

The effective number of bits, N_{eff}, is then 5.5 and the SNR is 34.87 dB. ■

Using Oversampling to Reduce Sampling Clock Jitter Stability Requirements

Suppose we limit the maximum input frequency coming into an ADC to f_{in}, such that the sampling frequency is related to the maximum ADC input frequency by

$$\frac{f_s}{2} = K \cdot f_{in} = f_n \text{ or } f_{in} = \frac{f_s}{2K} \tag{5.29}$$

In other words, we are getting at least $2K$ samples for every cycle of the input sinewave (we are oversampling the input signal). If we were sampling at twice the Nyquist frequency (f_s), where $K = 1$, then we would get two samples for every cycle of the input signal. Notice that Eq. (5.24) gives the maximum jitter specification for a given input frequency and data converter resolution, but it doesn't specify the sampling frequency, f_s, or the sampling frequency period, T_s.

For a given maximum jitter, ΔT_s, we can reduce the requirements placed on the stability of the oscillator by increasing the sampling frequency. This can be written as

$$\text{Stability(new), ppm} = [\text{stability(old), ppm}] \cdot K \tag{5.30}$$

If we were sampling at 1 MHz and the stability required was 10 ppm, then the jitter in the sampling clock would be at most 10 ps, peak-to-peak. Increasing the sampling rate to 100 MHz, with 10 ps jitter would require an oscillator stability of 1,000 ppm. If we were to increase the sampling clock frequency to 1 GHz, then the stability of the clock would be at least 10,000 ppm (the period is 1 ns and the jitter is 10 ps or 1% [10,000 ppm]).

Example 5.9

In Table 5.1 we saw that the 16-bit data converter clocked at 44.1 kHz could have at most 111 ps peak-to-peak jitter to limit the sampling uncertainty to 0.5 LSB. We saw that the stability required of the oscillator under these circumstances was 5 ppm at 44.1 kHz. What would happen to the stability requirements of the oscillator generating the sampling clock if we increased the sampling clock frequency to $128 \cdot 44.1 \text{ kHz} = 5.645 \text{ MHz}$?

We know that the input bandwidth, prior to increasing the sampling frequency, is limited to 44.1 kHz/2 or 22.05 kHz (= B, the bandwidth of the input signal). We then assume the maximum input frequency, f_{in}, remains at or below 22.05 kHz even after we increase the sampling frequency. We can define the oversampling factor, in this example, as

$$K = \frac{f_s/2}{f_{in}} = \frac{2.822 \times 10^6}{22.05 \times 10^3} = 128$$

The jitter requirement remains 111 ps whether we use a sampling frequency of 44.1 kHz or 5.645 MHz. However, now that the clock frequency has increased to 5.645 MHz, the stability required of the oscillator has gone from approximately 5 ppm to 640 ppm. ■

It's important to note that the oversampling ratio, K, is given by

$$K = \frac{f_s/2}{B} = \frac{f_n}{B} \quad \text{for } f_{in} \leq B \tag{5.31}$$

If we desire less than 0.5 LSBs aperture error, and we are using oversampling, then we can use Eqs. (5.22) and (5.31) to write

$$\Delta T_s \leq \frac{1}{2^N} \cdot \frac{K}{\pi f_s} = \frac{1}{2^N} \cdot \frac{1}{2\pi B} \tag{5.32}$$

where, once again, B is the bandwidth of the input signal and K is the oversampling ratio. As shown by this equation and in Eq. (5.30), using oversampling reduces the requirements placed on the stability of the sampling clock.

A Practical Note

We need to point out that the effects of clock jitter are possible even if the clock is perfectly stable because of the clock's finite transition times (rise and fall times). If the rise time of the clock signal in Fig. 5.19 is finite, say 50 ps, then the same derivations and discussions concerning jitter in this section can be applied to determine how the SNR of the data converter is affected. We would assume the aperture window is a function of the transition times of the sampling clock signal. The slower the transition times, the larger the sampling uncertainty. In any practical data converter the SNR, and thus the effective number of bits, will be reduced because of the clock jitter and finite transition times as the input signal frequency increases.

5.2.2 A Tool: The Spectral Density

The observant reader may have noticed, in the last section, that we only discussed the peak-to-peak jitter, ΔT_s, and how it effects the data converter's performance. It is very useful, in many situations, to also have an idea how the spectrum or spectral characteristics of the data converter's output change as a function of the random sampling jitter or a random variable such as noise. In this section we discuss tools useful in describing the spectrum of a random signal.

The Spectral Density of Deterministic Signals: An Overview

Consider the simple sinewave signal of the form

$$v_{in}(t) = V_p \sin 2\pi f_{in} t \quad \text{(units, V)} \tag{5.33}$$

This signal is termed "deterministic" because the signal has a well-defined shape whether it is continuous or sampled. We can find the average power of this signal, as a function of time $R_{in}(t)$, using the *autocorrelation function* (ACF) for continuous signals given by

$$R_{in}(t) = \lim_{T_0 \to \infty} \frac{1}{T_0} \int_{-T_0/2}^{T_0/2} v_{in}(\tau) \cdot v_{in}(\tau + t) \cdot d\tau \text{ (units, V}^2\text{)} \tag{5.34}$$

The average value of Eq. (5.33) as a function of time is then

$$R_{in}(t) = \lim_{T_0 \to \infty} \frac{1}{T_0} \int_{-T_0/2}^{T_0/2} [V_p \sin 2\pi f_{in} \tau] \cdot [V_p \sin 2\pi f_{in}(\tau + t)] d\tau \tag{5.35}$$

or knowing

$$\sin A \cdot \sin B = \frac{1}{2}[\cos(A - B) - \cos(A + B)] \tag{5.36}$$

we can write

$$V_p^2 \cdot \sin 2\pi f_{in} \tau \cdot \sin 2\pi f_{in}(t + \tau) = \frac{V_p^2}{2}[\cos 2\pi f_{in} t - \cos 2\pi f_{in}(t + 2\tau)] \tag{5.37}$$

When we integrate this result, the term $\cos[2\pi f_{in}(t + 2\tau)]$ represents a sinusoid with a frequency of $4\pi f_{in}$ (remembering our integration variable is τ) and a phase shift of $2\pi f_{in} t$. Over a long period of time this term averages to zero. Therefore, we can write the average value of Eq. (5.33) as a function of time (the autocorrelation function)

$$R_{in}(t) = \lim_{T_0 \to \infty} \frac{1}{T_0} \int_{-T_0/2}^{T_0/2} \frac{V_p^2}{2} \cdot \cos 2\pi f_{in} t \cdot d\tau = \frac{V_p^2}{2} \cdot \cos 2\pi f_{in} t \quad \text{(units, V}^2\text{)} \tag{5.38}$$

The spectrum of the average value of a function can be found by taking the Fourier transform of the autocorrelation function. The result is called the *power spectral density* function (PSD) and is given by

$$P_{in}(f) = \int_{-\infty}^{\infty} R_{in}(t) \cdot e^{-j \cdot 2\pi f \cdot t} \cdot dt \left(\text{units, V}^2\text{/Hz or V}^2 \cdot \text{s}\right) \tag{5.39}$$

The power spectral density function of Eq. (5.33) is then, with the help of Eq. (5.38),

$$P_{in}(f) = \frac{V_p^2}{4} \cdot [\delta(f + f_{in}) + \delta(f - f_{in})] \quad \text{(units, V}^2\text{/Hz)} \tag{5.40}$$

This is simply two impulses in the frequency spectrum located at $\pm f_{in}$ with an amplitude of $V_p^2/4$ (V^2/Hz). The *total average power* of this signal is given by

$$P_{AVG} = \int_{-\infty}^{\infty} P_{in}(f) \cdot df = 2 \cdot \int_{0}^{\infty} P_{in}(f) \cdot df \text{ (units, } V^2/\Omega \text{ or } watts) \quad (5.41)$$

assuming a 1-Ω (normalized) load, which, for Eq. (5.33), is $V_p^2/2$ (V^2).

The *voltage spectral density*, with units of V/$\sqrt{\text{Hz}}$, is simply the square root of Eq. (5.39) (that is, the square root of the PSD [= $\sqrt{P_{in}(f)}$]). The *root mean square* (RMS) voltage of a signal is given by

$$V_{RMS} = \sqrt{P_{AVG}} = \sqrt{2\int_{0}^{\infty} P_{in}(f) \cdot df} = \sqrt{2\int_{0}^{\infty} (voltage\ spectral\ density)^2 \cdot df} \quad (5.42)$$

The RMS value of Eq. (5.33) is simply, as one would expect for a sinewave, $V_p/\sqrt{2}$. Note the similarity between Eq. (5.42) and Eq. (5.7).

Example 5.10

Determine the ACF, PSD, average power, and RMS value of a signal $V(t)$ made up of three sine waves with peak amplitudes of V_1, V_2, and V_3 with frequencies of f_1, f_2, and f_3.

Using Eqs. (5.34) and (5.38), the ACF is

$$R(t) = \frac{V_1^2}{2}\cos 2\pi f_1 t + \frac{V_2^2}{2}\cos 2\pi f_2 t + \frac{V_3^2}{2}\cos 2\pi f_3 t \quad \text{(units, } V^2)$$

The PSD (positive frequencies) is determined using Eqs. (5.39) and (5.40)

$$P(f) = \frac{V_1^2}{4}\cdot\delta(f-f_1) + \frac{V_2^2}{4}\cdot\delta(f-f_2) + \frac{V_3^2}{4}\cdot\delta(f-f_3) \text{ (units, } V^2/\text{Hz})$$

The average power, using Eq. (5.41), is

$$P_{AVG} = \frac{V_1^2 + V_2^2 + V_3^2}{2} \quad \text{(units, } watts)$$

Finally, the RMS value of the signal is given by

$$V_{RMS} = \sqrt{\frac{V_1^2 + V_2^2 + V_3^2}{2}} \quad \text{(units, V)}$$

Note that if we added phase shifts to our signals the results would be the same; the phase shift doesn't change the signal's average value, so we get the same results whether sines or cosines are used in our original spectrum. ■

Next, suppose that the sinewave specified by Eq. (5.33) is sampled at a rate of f_s

$$v_{in}(nT_s) = V_p \sin(2\pi f_{in} \cdot nT_s) \quad (5.43)$$

The ACF for a sampled signal can be written as

$$R_{in}(nT_s) = \lim_{N\to\infty} \frac{1}{(2N+1)} \sum_{k=-N}^{N} v_{in}(kT_s) \cdot v_{in}(kT_s + nT_s) \quad (5.44)$$

which results in

$$R_{in}(nT_s) = \frac{V_p^2}{2} \cos 2\pi f_{in} \cdot nT_s \text{ (units, V}^2\text{)} \tag{5.45}$$

The PSD is the Fourier transform of this equation,

$$P_{in}(f) = \frac{V_p^2}{4T_s} \sum_{k=-\infty}^{\infty} [\delta(f - f_{in} + kf_s) + \delta(f + f_{in} + kf_s)] \tag{5.46}$$

The RMS value of the sampled sinewave, Eq. (5.43), assuming we have passed the signal through an ideal reconstruction filter (RCF) with a bandwidth of $f_s/2$, is simply, once again, $V_p/\sqrt{2}$. The PSD of the signal, after passing through the RCF, has an amplitude of $V_p^2/4$ at frequencies of $\pm f_{in}$.

The Spectral Density of Random Signals: An Overview

Let's use our jitter discussion of the last section to illustrate how to look at the spectrum of a random signal. We'll do this in two parts: (1) we'll begin by assuming the jitter is a random variable that falls between two limits and has equal probability of lying anywhere in the region (just as was assumed for the quantization error probability density function when calculating the RMS quantization noise voltage in the last chapter), and (2) then assume the jitter has a Gaussian distribution around some average value (the more practical and realistic situation) and determine how the output of the ADC is affected.

Consider the representations of clock jitter shown in Fig. 5.20. Trace 1 in this figure shows the ideal position of the rising edge of a clock signal. This point is represented on the probability density function (PDF), $\rho(t)$, at time zero. On the next rising edge of the clock, trace 2, the edge is a little too early and is represented on the PDF as shown. We are assuming, probably incorrectly for most practical situations, that the rising edge of the clock is falling within the peak-to-peak boundaries with the equal probability of being in the correct position (as shown in trace 1) or at the edge of a boundary (as shown in trace 4). We also know that the area under the PDF curve in Fig.

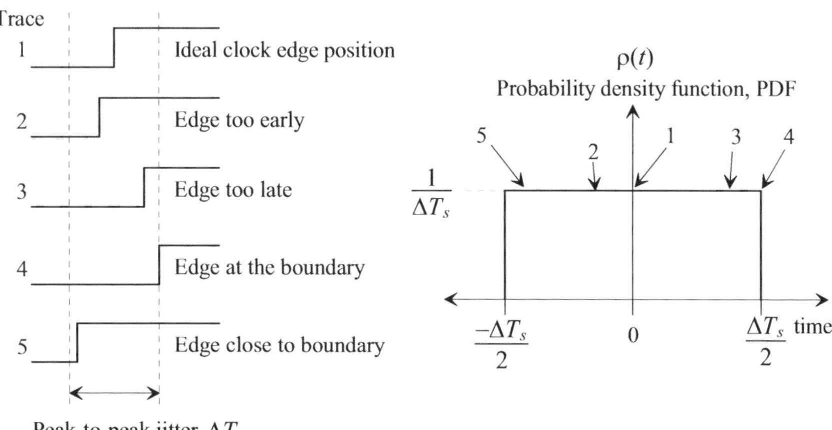

Figure 5.20 Clock jitter assuming the edge falls with the same probability anywhere within the peak-to-peak limits.

5.20 must equal unity, and the average value (also known as the mean or the expected value and denoted by $<y>$ or $\bar{y}$) of a PDF is given by

$$\text{Average value, } \bar{y}, = \int_{-\infty}^{\infty} t \cdot \rho(t) \cdot dt \tag{5.47}$$

Example 5.11
Determine the average value of the jitter with the PDF shown in Fig. 5.20.

We can use Eq. (5.47) to determine the average value of any PDF. Applying this equation to the PDF shown in Fig. 5.20 results in

$$\text{Average value, } \bar{y}, = \int_{-\Delta T_s 2}^{\Delta T_s/2} t \cdot \frac{1}{\Delta T_s} \cdot dt = 0$$

This somewhat obvious result means that the average position of the clock rising edge is the ideal position indicated by trace 1 in Fig. 5.20. Any PDF that is symmetrical about some center point will have an average equal to the center point. ■

The variance of the PDF is defined as the average of the square of the signal's departure from its average value. For a random signal this can be written as

$$\sigma^2 = \overline{(y-\bar{y})^2} = \int_{-\infty}^{\infty} (y-\bar{y})^2 \cdot \rho(y) \cdot dy \tag{5.48}$$

where σ is the standard deviation of the PDF (the square root of Eq. [5.48]). For our purposes, in this book, we can think of variance as the average power of a random (voltage) signal and the standard deviation as the RMS value of the signal (see Eqs. [5.41] and [5.42]). Example random signals include the time difference between the actual edge of a clock and the ideal edge location (jitter), the voltage difference between the input of an ADC and the ADC's reconstructed output (quantization noise), and the random fluctuations of electrons due to thermal motion in a resistor (thermal noise).

Example 5.12
Determine the RMS value of the jitter when the jitter has a probability density function, PDF, as shown in Fig. 5.20.

Using Eq. (5.48) the variance of the jitter PDF is

$$\sigma^2 = \int_{-\Delta T_s/2}^{\Delta T_s/2} t^2 \cdot \frac{1}{\Delta T_s} \cdot dt = \frac{1}{3 \cdot \Delta T_s} \cdot t^3 \Big|_{-\Delta T_s/2}^{\Delta T_s/2} = \frac{(\Delta T_s)^2}{12} \text{ (seconds}^2\text{)}$$

and thus the RMS jitter is

$$\sigma = \frac{\Delta T_s}{\sqrt{12}} \text{ RMS jitter, (seconds)}$$

where ΔT_s is the peak-to-peak jitter in the sampling clock rising edge. Note the similarity to the derivation of $V_{Qe,RMS}$ in Sec. 5.1.1. ■

A more useful discussion of jitter can be constructed if we assume the jitter has a Gaussian PDF, as shown in Fig. 5.21, and attempt to describe how the jitter in the

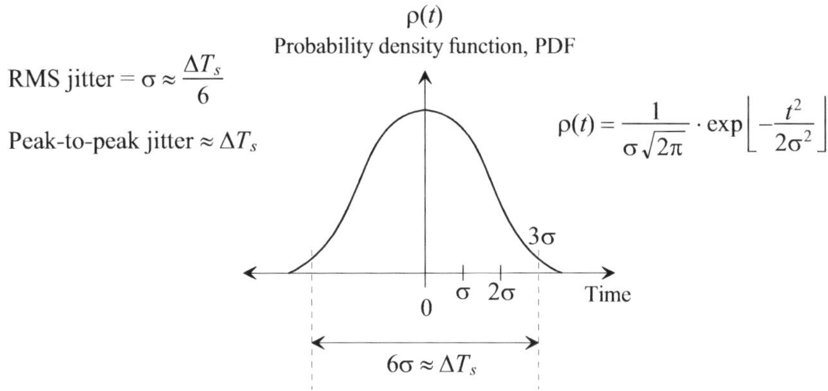

Figure 5.21 Sampling jitter with a Gaussian probability distribution.

sampling clock affects an ADC output spectrum with a single-tone input. Using Eqs. (5.20), (5.21), and (5.22), we can write the sampling error voltage (review Fig. 5.19), at a given time, as

$$\Delta V_s(t) = \delta T_s(t) \cdot V_p \cdot 2\pi f_{in} \cdot \cos 2\pi f_{in} t \tag{5.49}$$

where $\delta T_s(t)$ is a random variable indicating the jitter in the sampling clock at a given time. (The variable $\delta T_s(t)$ is the time difference between the actual clock transition time and the expected transition times that are spaced by T_s [see Fig. 5.20].) The peak-to-peak value of $\delta T_s(t)$ is ΔT_s, while its average value is zero. Again, we assume that the jitter probability distribution function is Gaussian, as seen in Fig. 5.21.

Rewriting Eq. (5.49) using a discrete time step nT_s, the sampling error can be written as

$$\Delta V_s(nT_s) = \overbrace{\delta T_s(nT_s) \cdot V_p \cdot 2\pi f_{in}}^{\text{Sampling error amplitude}} \cdot \overbrace{\cos 2\pi f_{in} n T_s}^{\text{Carrier term}} \tag{5.50}$$

We're interested in the spectrum of this error signal as it will add to our RMS quantization noise plus distortion voltage, effectively lowering the data converter's SNDR. Notice that the spectrum of Eq. (5.50) will have aliased components (and so will the sampled signal) so we need to filter out these components above $f_s/2$ (with the reconstruction filter.) Also note that multiplying the sampling error by the cosine term in Eq. (5.50) simply shifts the error spectrum to a frequency f_{in}. The cosine terms act like a carrier in an amplitude-modulated signal. This is illustrated in Fig. 5.22.

Example 5.13

Repeat Ex. 5.8 assuming the clock jitter has a Gaussian PDF.

In this example the peak amplitude of the input signal, V_p, is 0.5 V, the input frequency, f_{in}, is 25 MHz, and the peak-to-peak jitter is 100 ps. The average power in the sampling error amplitude spectrum is

$$P_{AVG,jitter} = \sigma^2 \cdot \frac{(V_p \cdot 2\pi f_{in})^2}{2} = \left(\frac{\Delta T_s}{6}\right)^2 \cdot \frac{(V_p \cdot 2\pi f_{in})^2}{2} \tag{5.51}$$

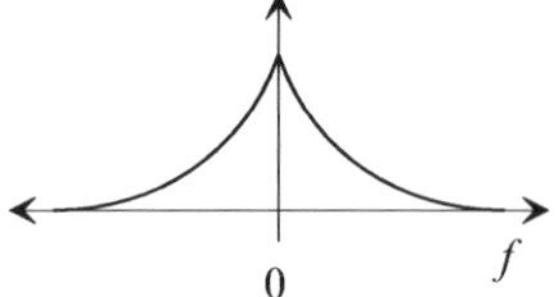

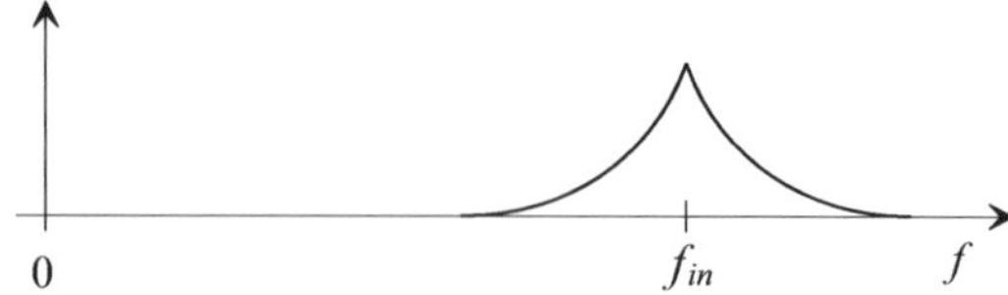

Figure 5.22 Modulating sampling error with an input sinewave frequency.

or

$$P_{AVG,jitter} = \left[\frac{100\ ps}{6}\right]^2 \cdot \frac{(0.5 \cdot 2\pi \cdot 25\ \text{MHz})^2}{2} = 0.858 \times 10^{-6}\ \text{V}^2$$

while the RMS voltage associated with this error is 0.926 mV. The quantization noise associated with this 8-bit data converter is

$$V_{Qe,RMS} = \frac{V_{LSB}}{\sqrt{12}} = \frac{V_{REF+} - V_{REF-}}{2^N\sqrt{12}} = 1.3\ \text{mV}$$

The RMS noise voltage due to clock jitter and quantization effects is then given by

$$\sqrt{0.858 + 1.3^2}\ \ \text{mV} = 1.6\ \text{mV}$$

We can calculate the SNR using

$$\text{SNR} = 20 \cdot \log \frac{0.5/\sqrt{2}}{1.6\ \text{mV}} = 46.9\ \text{dB}$$

giving an effective number of bits, from Eq. (5.14), equal to 7.53. Note that this is a *significant* improvement over what was calculated in Ex. 5.8, where the jitter variation was always the peak-to-peak value. ■

The PSD of the sampling error amplitude, described by Eq. (5.50), can be determined with the help of Eq. (5.41)

$$\sigma^2 \cdot \frac{(V_p \cdot 2\pi f_{in})^2}{2} = 2\int_0^\infty P_{jitter}(f) \cdot df \tag{5.52}$$

If the spectrum of the phase noise due to jitter is narrow, as seen in Fig. 5.22, then the spectral density of the sampling error, $P_{jitter}(f)$, is concentrated around the frequency of the input sinusoid. However, if we assume the phase noise spectrum is white and evenly distributed throughout the base spectrum (so that we integrate Eq. [5.52] from DC to $f_s/2$), we can write

$$P_{jitter}(f) = \frac{\sigma^2}{f_s} \cdot \frac{(V_p \cdot 2\pi f_{in})^2}{2} \tag{5.53}$$

The power spectral density of the sampling error voltage, assuming even distribution of the noise throughout the base spectrum, is shown in Fig. 5.23.

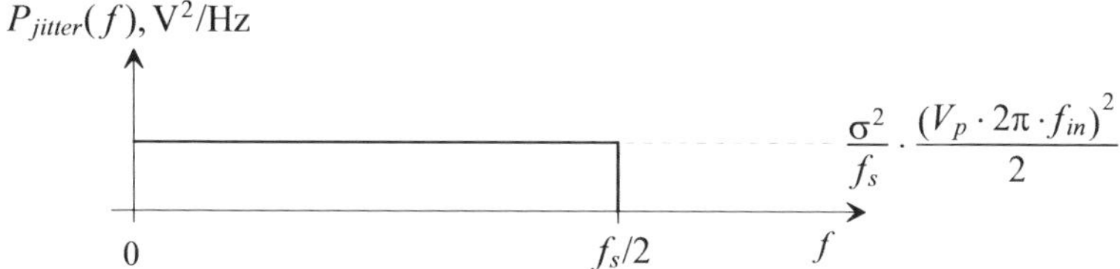

Figure 5.23 Sampling amplitude error PSD assuming sampling error spectrum is white.

Specifying Phase Noise from Measured Data

It's important to note that we have been discussing clock signals that are square waves (that is, have odd order harmonics) and so discussing jitter (a time-domain term) is, generally, more appropriate than discussing phase noise (a frequency domain term). However, because the terms are both widely used to indicate the same, basic, effect (a variation in the period of a periodic waveform), we will briefly discuss phase noise specification from measured oscillator data.

Consider the representation of a measured oscillator spectrum (power spectral density) shown in Fig. 5.24. In general, oscillator noise is specified in terms of the carrier voltage (or power) with units of dBc (decibels with respect to the carrier). The ratio of the power of the fundamental (called the carrier or sampling clock) at f_s is taken to the noise power in a bandwidth at some offset from the fundamental

$$\text{Phase noise, dBc/Hz} = \overbrace{10 \cdot \log\left[\int_{f_{L1}}^{f_{H1}} P_{osc}(f) \cdot df\right]}^{10 \cdot \log\left(\mathrm{V}^2\right)} - \overbrace{10 \cdot \log P(f_s)}^{10 \cdot \log\left(\mathrm{V}^2/\mathrm{Hz}\right)} \tag{5.54}$$

where the first term is the noise power at an offset from f_s.

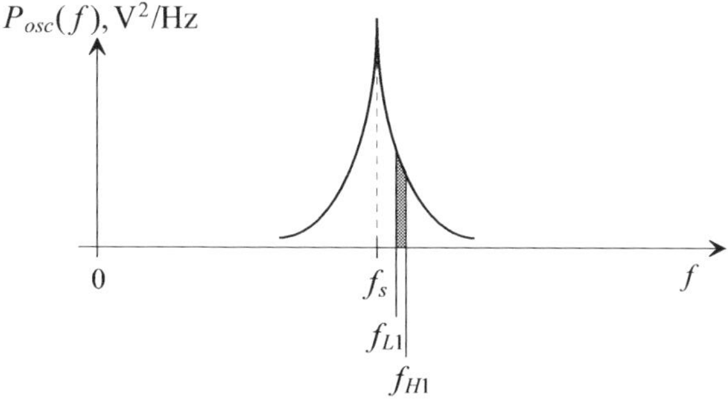

Figure 5.24 Measured oscillator spectrum.

5.3 Improving SNR using Averaging

Examine the two-path combination of ADCs and DACs shown in Fig. 5.25. The top ADC and DAC are the path we had back in Fig. 5.4 clocked at 100 MHz. The bottom path is a mirror image of the top except that its clock signal is inverted. The two resistors are used to sum (or more correctly average) the outputs of the DACs into a single output. This configuration effectively samples the input at 200 MHz (200 Msamples/s $[2 \cdot f_s]$).

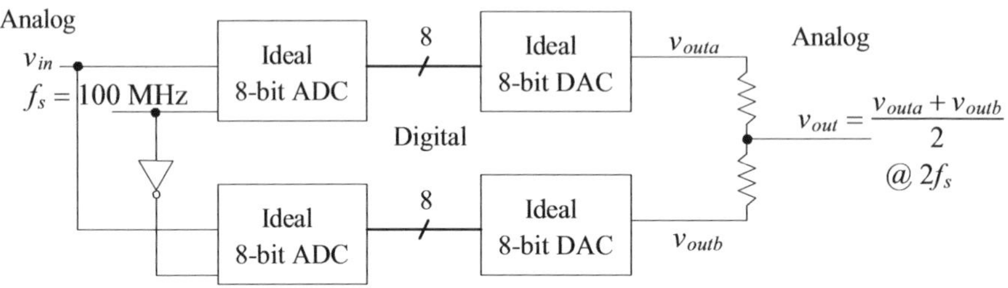

Figure 5.25 Using two paths to average the quantization noise.

Figure 5.26 shows the quantization noise PSD spectrums when the single path ADC/DAC is clocked at f_s and then at $2f_s$. Notice how the quantization noise is spread out over a wider frequency range when a higher sampling frequency is used. As seen in Ex. 5.1 the sampling frequency doesn't affect the RMS value of quantization noise. The area of the spectrums in Fig. 5.26 is equal to $V_{LSB}^2/12$ (negative frequencies are not shown) for both sampling frequencies.

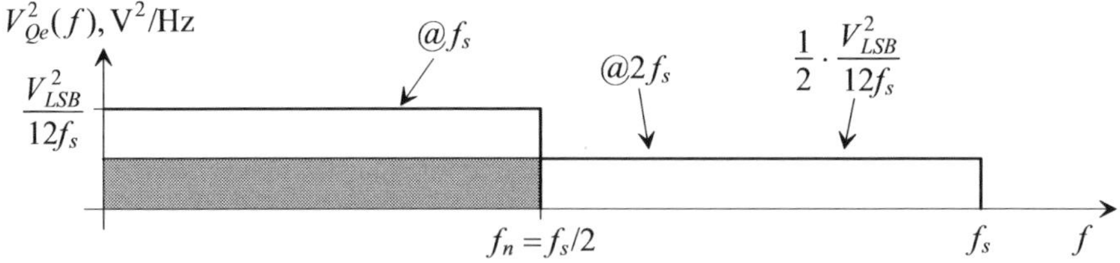

Figure 5.26 Quantization noise power spectral density for two sampling frequencies.

Now if we were to compare the quantization noise added to an input signal using a single path clocked at $2f_s$ to a system using two paths clocked at f_s we would find that the two path system added less noise. In order to understand why, notice that when we take two random uncorrelated variables, the quantization noise added to the input signal from each path in Fig. 5.25, and average (sum the power from each) them we get

$$V_{Qe,RMS,2-path}^2 = \frac{V_{Qe,RMS1}^2 + V_{Qe,RMS2}^2}{2} \tag{5.55}$$

or, generalizing this to K-paths

$$V_{Qe,RMS} = \frac{1}{\sqrt{K}} \cdot \frac{V_{LSB}}{\sqrt{12}} \tag{5.56}$$

Further note that if we were to add a lowpass filter to the output of our DAC, we could filter out part of the quantization noise and further decrease $V_{Qe,RMS}$. We use these ideas in this section to improve SNR; however, let's do an example before going any farther.

Example 5.14
Repeat Ex. 5.1 but use, to increase the sampling frequency, the two path topology seen in Fig. 5.25.

As discussed in Ex. 5.1, the RMS value of the quantization noise added to the input signal is 1.13 mV. Using two paths to increase the sampling frequency results in adding, using Eq. (5.56),

$$V_{Qe,RMS} = \frac{1}{\sqrt{2}} \cdot \frac{3.906 \text{ mV}}{\sqrt{12}} = 0.797 \text{ mV}$$

to the input signal. Figure 5.27 shows the simulation results (the quantization noise added to the input signal when two paths are used). The simulated $V_{Qe,RMS}$ is 0.787 mV. ■

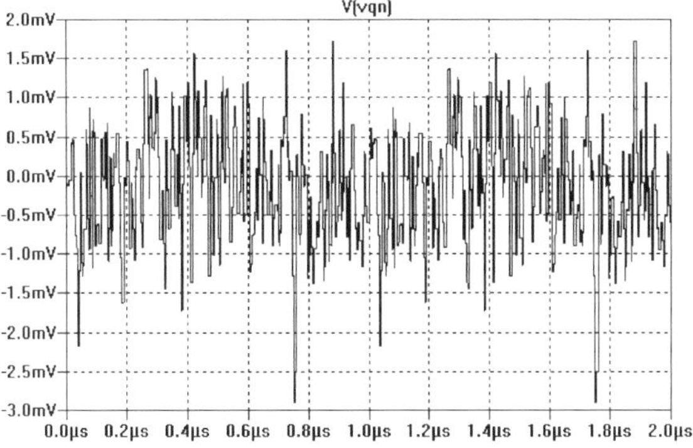

Figure 5.27 Simulated quantization noise using two-paths, see Ex. 5.14.

An Important Note

For averaging to effectively reduce the RMS quantization noise, the ADC and DAC must be linear to the final output resolution. In order to understand this in more detail examine Fig. 5.28. In the ideal situation, two adjacent codes are averaged to give an output code that falls exactly in between the outputs of the data converter. In the case where the data converter has a nonlinearity, the averaged point doesn't necessarily provide an output that is much different from the data converter outputs themselves. If the data converter contains a missing code (an input difference between two inputs at consecutive sampling times of 1-LSB results in the same output), then the averaging does nothing. If the data converter is nonmonotonic (an increase in the data converter's input doesn't result in an

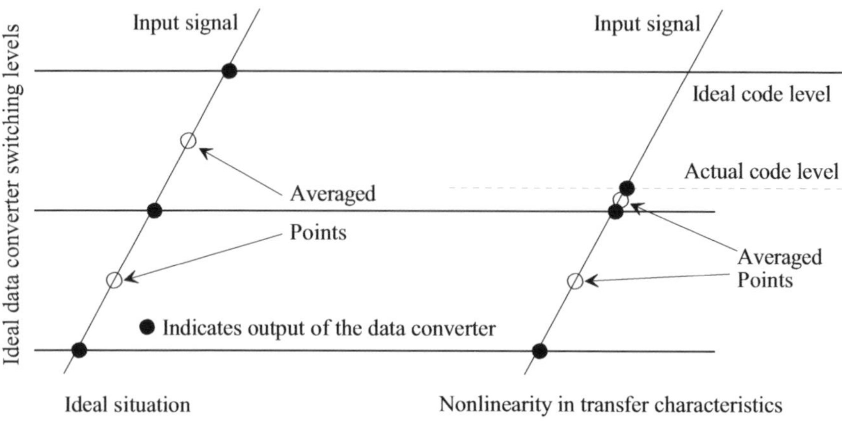

Figure 5.28 How ADC or DAC linearity affects averaging.

increase in its output) then the averaged value is meaningless. Finally, note that an input DC value (a digital code that isn't changing for the DAC, or an analog voltage that isn't changing for the ADC) or a value that isn't "busy" (not changing by at least 1 LSB in between sampling instances) will not benefit from averaging.

5.3.1 Using Averaging to Improve SNR

The averaging topology shown in Fig. 5.25 is not practical in most situations (one notable exception is the parallel combination of noise-shaping converters discussed later in the book). The silicon area required to implement the extra ADC and DAC generally costs more than is gained by the reduction in quantization noise. Figure 5.29 shows how we can add a digital averaging filter (in this figure averaging two ADC's outputs, or $K = 2$) to the output of the ADC to reduce quantization noise. By lowpass filtering the quantization noise PSD seen in Fig. 5.26, we reduce the amount of noise added to our signal. Unfortunately, this lowpass filtering also limits the range of allowable (wanted) signal frequencies. Notice that both the ADC and digital averaging filter are clocked at a rate of f_s. The averaging filter seen in Fig. 5.29 is described using Figs. 1.16 and 1.17.

We might, at this point, assume that we can use a low-resolution ADC, say 6-bits, with a significant amount of averaging ($K >> 2$, see the lowpass Sinc-shaped filters seen in Ch. 4) to attain large resolutions (again the ADC must be linear). Assuming the input to the ADC is busy and we place restrictions on the bandwidth of the signals coming into the ADC then we can increase the resolution by averaging. We have to place restrictions

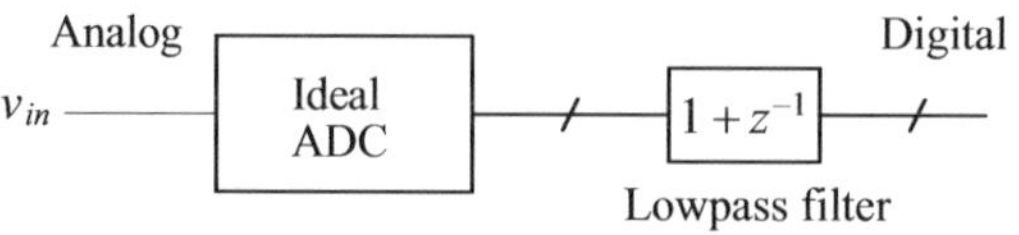

Figure 5.29 Using a digital averaging filter to reduce quantization noise.

on the bandwidth of the signal coming into the ADC because, unlike Fig. 5.25, we haven't increased the sampling rate of the signals. Therefore, the amplitude of the power spectral density seen in Fig. 5.26 remains unchanged. For an averaging of two, we would have to limit our desired input signal bandwidth to $f_s/4$, Fig. 1.17. If this wasn't the case, then an input sinewave at $f_s/2$ would average to zero.

Example 5.15

For the topology seen in Fig. 5.29, sketch the output quantization noise PSD. From this spectrum determine the RMS value of the quantization noise added to input signal. Compare this value to the one obtained using Eq. (5.56).

To begin let's write (see Eq. [1.40])

$$1+z^{-1}=2\cdot\left|\cos\pi\frac{f}{f_s}\right|$$

so the PSD of the quantization noise can be written as

$$V_{Qe}^2(f)=\frac{V_{LSB}^2}{12f_s}\cdot 4\cdot\left|\cos^2\pi\frac{f}{f_s}\right| \text{ for } 0\leq f\leq\frac{f_s}{2}$$

This PSD is plotted in Fig. 5.30.

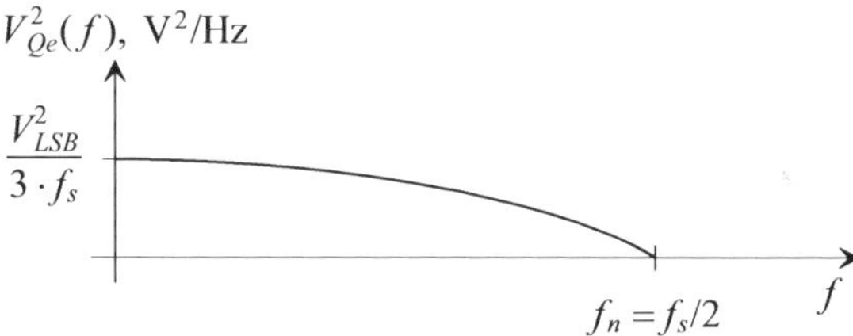

Figure 5.30 Quantization noise power spectral density after averaging two samples.

To determine the RMS quantization noise (see Eq. [1.29])

$$V_{Qe,RMS}^2=2\int_0^{f_s/2}\frac{V_{LSB}^2}{12f_s}\cdot 2\left(1+\cos\left[2\pi\frac{f}{f_s}\right]\right)\cdot df$$

or

$$V_{Qe,RMS}^2=\frac{V_{LSB}^2}{6}+\frac{V_{LSB}^2}{6\pi}\left[\sin 2\pi\frac{f}{f_s}\right]_{f=0}^{f=f_s/2}=\frac{V_{LSB}^2}{6}\neq\frac{1}{K}\cdot\frac{V_{LSB}^2}{12}$$

It would appear that our RMS quantization noise has actually increased! However, after we review the transfer characteristics of our filter seen in Fig. 1.17 we see the gain of the filter at low frequencies (where our desired signal resides) is two. Knowing this we can write the SNR as

$$SNR=20\log\frac{2V_p/\sqrt{2}}{V_{LSB}/\sqrt{6}}=20\log\frac{V_p/\sqrt{2}}{V_{LSB}/\sqrt{24}}$$

or, once again,

$$V_{Qe,RMS} = \frac{1}{\sqrt{K}} \cdot \frac{V_{LSB}}{\sqrt{12}} \tag{5.57}$$

■

Ideal Signal-to-Noise Ratio

Reviewing the derivation of Eq. (5.13) and using Eq. (5.57) we can write the ideal SNR for a data converter employing a digital (averaging or Sinc-shaped) lowpass filter as

$$SNR_{ideal} = 6.02N + 1.76 + 10\log K \tag{5.58}$$

where N is the number of bits (the resolution) of the data converter whose output is being averaged. Using no averaging, that is $K = 1$, results in this equation simplifying to Eq. (5.13). Averaging two samples causes the SNR_{ideal} to increase by 3 dB or the effective resolution of the data converter to increase by 0.5 bits. The increase in resolution due to averaging can be written as

$$\text{Increased resolution}, N_{Inc} = \frac{10\log K}{6.02} \tag{5.59}$$

Figure 5.31 shows how averaging the output of a data converter changes the effective resolution of the data converter. Again, note that the increase in resolution is based on the following assumptions: a busy input signal, the input signal is bandlimited, and the data converter is linear to the final resolution (data converter resolution, N, + improvement in resolution, N_{Inc}) coming out of the averaging circuit.

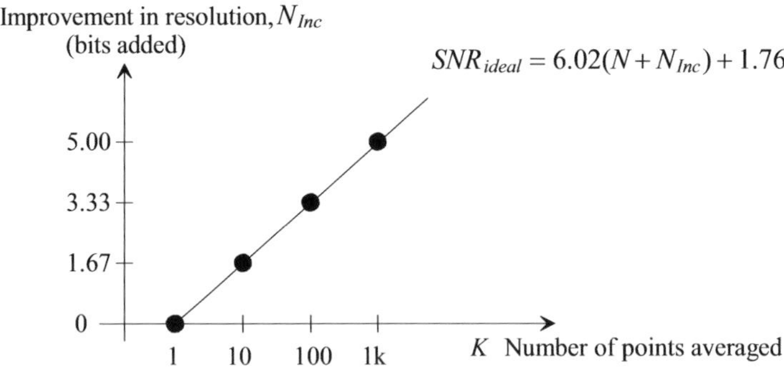

Figure 5.31 Using averaging to improve data converter resolution.

5.3.2 Linearity Requirements

Examine the cases for averaging ADC outputs shown in Fig. 5.32. In part (a) we show the ideal situation where the black dots indicate two consecutive outputs spaced by one LSB (time is not shown in this figure). The ADC outputs in part (a) are located on the ideal levels, while the averaged output falls exactly in the middle of these levels (and hence our increased resolution). Part (b) of this figure shows the situation where the ADC outputs are shifted downwards by 0.5 LSBs from their ideal levels. Following this offset, the averaged point shifts downwards as well. In part (c) the top output of the ADC (the top

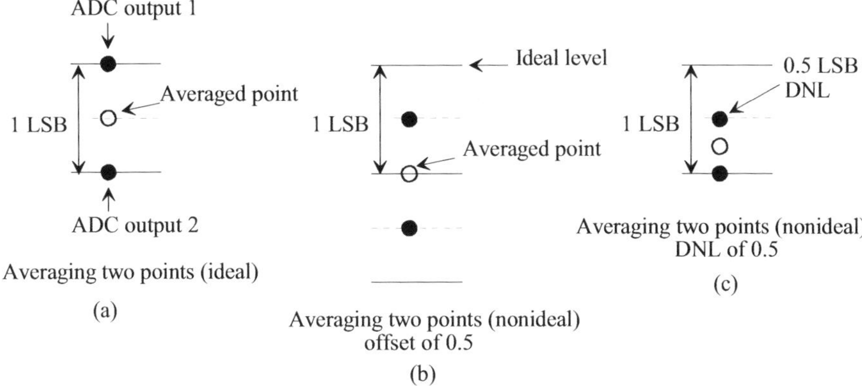

Figure 5.32 Linearity requirements when averaging.

black dot) is shifted downwards by 0.5 LSBs and so the averaged point shows a 0.25 LSB offset from its ideal position. While we used a single LSB difference to show averaging, we could use any number of LSBs to show that the ADC accuracy must be equal to or better than the desired final digital filter output accuracy.

The number of bits in the ADC (its resolution) N, and the number of bits improvement in resolution after filtering, N_{Inc}, are used with the final, total number of bits (the number of bits coming out of the digital filter) to give

$$N_{Final} = N + N_{Inc} \tag{5.60}$$

The ADC output should ideally change in increments of the exact LSB voltage. In reality, the changes will be different from the ideal output levels (as just discussed). In order to achieve an increase in the number of final bits, the output of the ADC must be accurate (its actual levels must be spaced from the ideal levels) to within

$$\pm \frac{V_{REF+} - V_{REF-}}{2^{N_{Final}+1}} = \pm\,(0.5\,LSB) \cdot \frac{1}{2^{N_{Inc}}} \tag{5.61}$$

where no averaging ($N_{Inc} = 0$ and $K = 1$) means the ADC is at least 0.5 LSBs accurate. This is a *significant limitation* when using averaging to increase the resolution of an ADC. This is especially true when a resolution greater than 10 bits is desired with INL and DNL less than ±0.5 LSBs. In the next section, and in the next chapter, we will look at feedback topologies that may relax the accuracy requirements placed on the ADC and allow averaging to more effectively remove quantization noise.

5.3.3 Adding a Noise Dither

Our assumption, when discussing the benefits of averaging or calculating the spectral density of the quantization noise, falls apart for DC or slow-moving signals (the ADC input is not "busy"). In order to help with this problem consider adding a noise signal to the ADC input that has a frequency content that falls within the range

$$\frac{f_s}{2K} \leq f < \frac{f_s}{2} \tag{5.62}$$

so that it can be filtered out with the averaging filter (see Fig. 4.10). This noise is often called *dither* (a state of indecision or agitation) because it helps to randomize the spectral content of the quantization noise, making it white.

Figure 5.33 shows the basic idea. In part (a) a DC signal is applied to the ADC that falls halfway between two ADC transition codes spaced apart by 1 LSB. The output code of the ADC remains unchanged with time. In part (b) a noise signal is added to the DC input which has two benefits: (1) the quantization noise (the difference between the input signal and the reconstructed ADC output code) changes with time, and (2) the output of the ADC has some variation which makes it possible to determine the DC voltage after averaging.

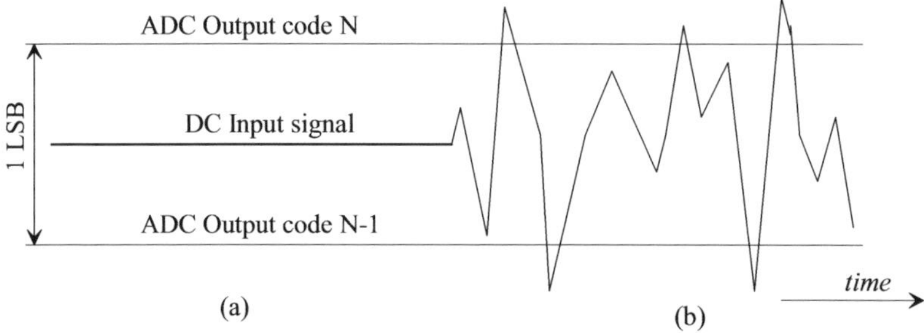

Figure 5.33 (a) DC input signal and (b) DC input signal with dither added.

We can add the noise signal to our desired input signal with a circuit similar to that shown in Fig. 5.34. Simple resistors add and reduce the noise signal applied to the ADC input. The noise signal source is, most easily, derived from some sort of asynchronous logic circuit and has a peak amplitude (before reduction) of *VDD* (= 1 V in this book). In this figure note that we have indicated that the dither signal amplitude should be approximately 0.5 LSB RMS (remembering the signal is, ideally, random and bandlimited as specified by Eq. [5.62]). This number, 0.5 LSB RMS, is subjective, and no exact rules as to its selection can be given other than the desire that the peak-to-peak

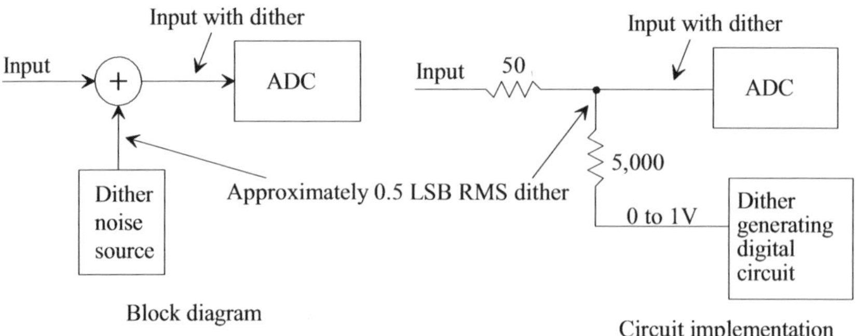

Figure 5.34 Adding dither to an ADC input signal.

amplitude be greater than 1 LSB. One *disadvantage* of adding the dither is that the allowable range of input signals shrinks. A DC signal at *VDD* – 1 LSB will not benefit from dithering since the ADC will be at its full-scale output.

Before we discuss the implementation of a dither source, consider one possibility (a Gaussian PDF) for the desired probability density function (PDF) of the dither signal and DC input shown in Fig. 5.35 (the input to the ADC). If we average this signal over a long time, we get the average or DC input signal since the dither averages to zero. This would also mean that we can have some dither spectral content below $f_s/2K$ as long as we average enough ADC output samples to make its contribution to the SNDR small. It is generally a good idea to use Eq. (5.62) as a guide for allowable dither spectral content. Finally, it's important that any dither signal we generate has a symmetrical PDF (the dither signal must average to *VDD*/2 before amplitude reduction). If not, an unknown DC offset (the known DC offset is the *VDD*/2 attenuated by the resistive divider in Fig. 5.34) in the data converter's (actually the filter's) output will result.

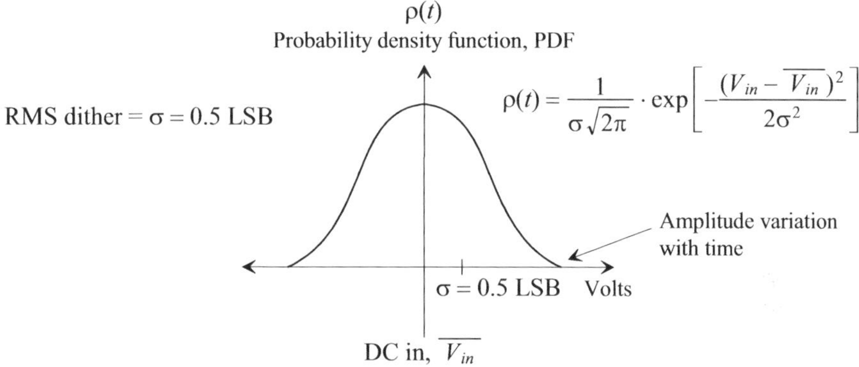

Figure 5.35 Input to the ADC, dither and DC, with a Gaussian probability distribution.

An example of an implementation of a dither noise source is shown in Fig. 5.36. The outputs of the rows of inverters, which are tied together, will occur asynchronously and fight against each other causing the amplitude of the dither signal to occupy levels

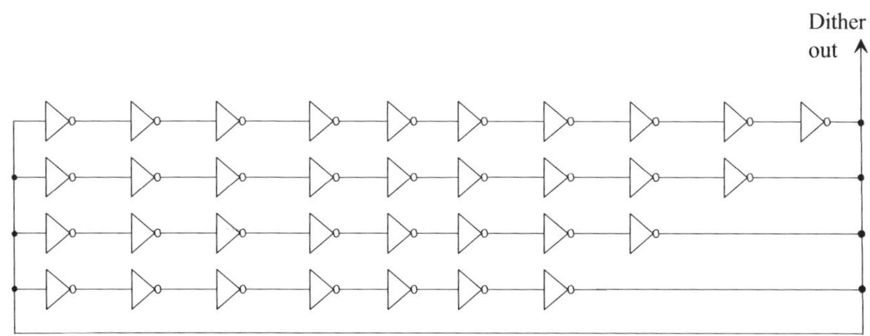

Figure 5.36 One possible implementation of a dither circuit.

other than the normal logic levels of VDD and ground for significant amounts of time. The dither signal can be made more random by adding more rows of inverters. The challenge to this design is setting the number of inverters used in each row so that the spectral content falls within the desired range (which may require a large number of inverters) and keeping the output of the dither circuit uncorrelated with the sampling clock. Other techniques for generating random noise, such as using linear serial feedback registers, can be found in most books covering communication systems.

5.3.4 Jitter

We can apply the averaging discussion just developed directly to the jitter discussion presented earlier in the chapter and answer the question, "How does averaging effect the sampling amplitude error power (resulting from jitter) in a data conversion system?" If we assume that the jitter has a Gaussian PDF, then the average power in the sampling error amplitude, from Ex. 5.13, is

$$P_{AVG,jitter} = \left[\sigma \cdot \frac{V_p}{\sqrt{2}} \cdot 2\pi f_{in}\right]^2 \tag{5.63}$$

where σ is the standard deviation of the jitter (see Fig. 5.21). It may be helpful to rewrite Eq. (5.57) in terms of the quantization error power as

$$P_{Qe,AVG} = (V_{Qe,RMS})^2 = \frac{1}{K} \cdot \frac{V_{LSB}^2}{12} \tag{5.64}$$

and apply the same derivation to Eq. (5.63) to give

$$P_{AVG,jitter} = \frac{1}{K} \cdot \left[\sigma \cdot \frac{V_p}{\sqrt{2}} \cdot 2\pi f_{in}\right]^2 \tag{5.65}$$

This equation shows that the sampling error amplitude power, $P_{AVG,jitter}$, introduced into the data converter's output spectrum decreases with averaging. Averaging two samples causes the sampling error amplitude power to decrease by 3 dB. This effectively reduces the jitter requirements placed on the sampling clock. While this may not appear to be very significant at first glance, consider what happens if, for example, 256 samples are averaged ($K = 256$). The sampling error power decreases by 48 dB, making clock jitter, when using a reasonably stable oscillator, almost not an issue. Also note that a doubling in the jitter's standard deviation, σ, results in a 6 dB increase in sampling error amplitude power.

5.3.5 Anti-Aliasing Filter

The use of averaging will also lead to relaxed requirements of the anti-aliasing filter (AAF). Figure 5.37a shows the requirements placed on the AAF without averaging. As we saw in Ch. 2, ideally, the transition from the 3 dB frequency to the "stop frequency" or Nyquist frequency should be infinitely sharp (the filter should abruptly change from a gain of unity to a gain of zero [something small]). When using averaging, Fig. 5.37b, we have to limit our desired input signal bandwidth to B. The rolloff of the filter in part (b) of the figure can be much more gradual and in many cases a simple, single pole, RC filter is all that's needed for an AAF. Also, our averaging filter will attenuate the ADC output spectrum, as seen in Fig. 1.17, and help to remove input signal power above $f_s/2K$. The

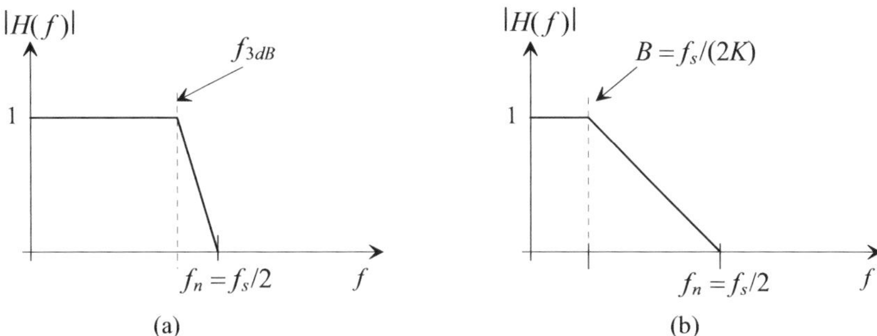

Figure 5.37 (a) AAF requirements without averaging, and (b) AAF requirements with averaging.

significance of this will be easier to see as the number of points averaged increases and our averaging filter's response gets sharper with more attenuation (see Fig. 4.10). Of course, the penalty for the relaxed requirements of the AAF is reduced signal bandwidth for a fixed sampling frequency.

5.4 Using Feedback to Improve SNR

By averaging the outputs of an ADC, or interpolating between inputs of a DAC, the effective data converter resolution can be increased. As specified by Eq. (5.58), every doubling in (octave increase in) K (where K is the number of points averaged) results in a 0.5-bit increase in effective resolution. An effective ADC resolution increase of 6-bits requires averaging 4,096 samples. If a 1 MHz signal bandwidth is of interest, our sampling clock frequency, f_s, will have to be 8.192 GHz!

In this section we briefly introduce the idea that feedback can be used with data converters (ADCs and DACs) to improve overall data conversion system performance (lower the amount of averaging or oversampling needed to attain a given resolution over a certain bandwidth). A topology of this nature is called a *modulator* or *coder* (for analog-to-digital conversion) or a *demodulator* or *decoder* (for digital-to-analog conversion). The complete analog-to-digital interface (a circuit block that functions as an ADC) would be made up of a modulator and a lowpass (decimating) filter, while the digital-to-analog interface (a circuit block that functions as a DAC) would consist of an interpolating filter and a demodulator. This can be confusing since, for example, a modulator will contain a low-resolution ADC in a feedback configuration which, together with the decimating filter, behaves like a high-resolution ADC.

The basic topology of a feedback modulator or coder is shown in Fig. 5.38. Depending on the circuit blocks used for $A(f)$ and $B(f)$ feedback modulators can be separated into two categories: *predictive modulators* and *noise-shaping modulators*.

Predictive modulators (a.k.a. predictive coders), such as delta-modulation, attempt to feed back an analog signal with the same value as the input signal. This drives the output of the summer to zero, reducing the required input range of the ADC and, possibly, the quantization error introduced by the ADC. *Predictive modulators effectively output the change in the input signal over time.* Noise-shaping modulators, an example being delta-sigma-modulation, also known as sigma-delta-modulation, on the other hand,

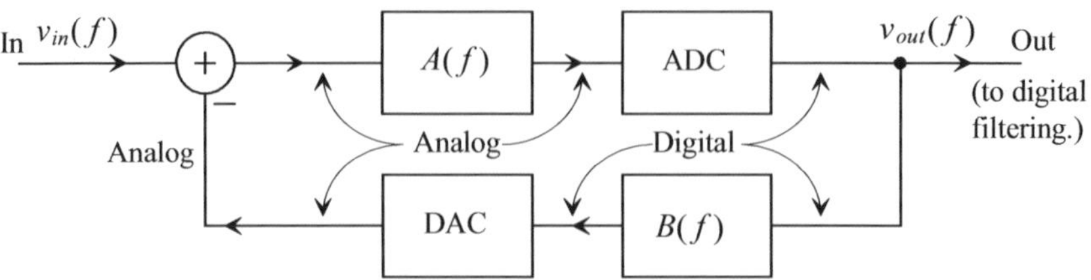

Figure 5.38 Block diagram of a feedback modulator.

feed back, and output, the average value of the input signal. This signal can be filtered (averaged) to reduce the accuracy required of the analog circuit components. *Noise-shaping modulators effectively output the average of the input signal over time.* In a noise-shaping data converter the averaging and decimating filter, as discussed earlier, is connected to the output of the modulator. Because of the averaging used in noise-shaping modulators, the analog components, in the forward path of Fig. 5.38, require less accuracy. However, the DAC's output, in the feedback path (which is subtracted from the input), doesn't experience the averaging so, once again, the DAC must be linear to the final desired resolution of the data converter. DAC linearity concerns have led to the use of a single-bit DAC (an inverter), in many noise-shaping data converter applications. The one-bit DAC is inherently linear. (Two output points determine a line!) Because of the relaxed requirements placed on the analog circuit components, *we will concentrate the next chapters, in detail, on noise-shaping topologies for both ADCs and DACs.* Notice that both predictive and noise-shaping modulators utilize oversampling.

In order to understand these statements in more detail, let's use the additive quantization noise model for the ADC developed in this chapter, Fig. 5.2. Figure 5.39 shows Fig. 5.38 redrawn using this model where the quantization noise is represented in the frequency domain by $V_{Qe}(f)$. We can relate the inputs (the wanted input signal and the unwanted quantization noise) to the output of the feedback modulator by

$$v_{out}(f) = \overbrace{\frac{A(f)}{1 + A(f) \cdot B(f)}}^{\text{Signal transfer function, } STF(f)} \cdot v_{in}(f) + \overbrace{\frac{1}{1 + A(f) \cdot B(f)}}^{\text{Noise transfer function, } NTF(f)} \cdot V_{Qe}(f) \qquad (5.66)$$

In a predictive modulator the feedback filter, $B(f)$, has a large gain so that, ideally, the fed back signal equals the input signal. If $A(f) = 1$ (a wire), then both the *STF* (signal transfer function) and the *NTF* (noise transfer function) have a value of, approximately, $1/B(f)$. Recovering the input signal requires passing the output of the predictive modulator through an analog filter with a transfer function of precisely $B(f)$ (noting that $B[f]$ is a digital filter in the modulator of Fig. 5.39). The required precision of the analog filter (the matching between the filter in the modulator and the filter in the demodulator) limits the attainable resolution when using predictive modulators. Notice that both the input signal and the quantization noise experience the same spectral shaping (spectral discrimination is absent in a predictive modulator). Also note that the name "predictive" comes from the modulator attempting to predict the input signal in order to drive the output of the summer to zero. If the prediction is perfect, the signal that is fed back exactly matches the input signal.

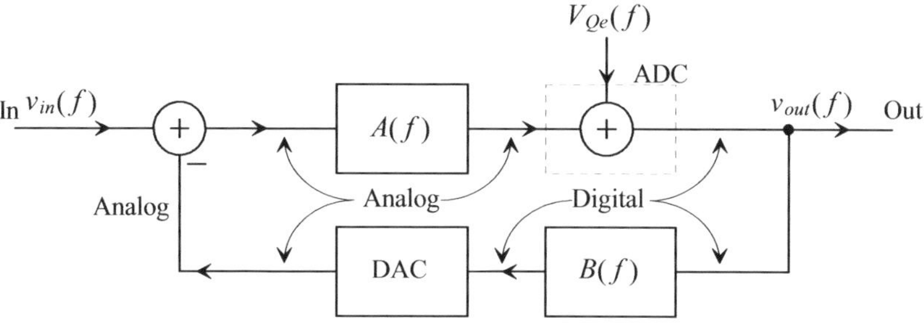

Figure 5.39 Block diagram of a feedback modulator.

In a noise-shaping modulator the gain of the forward path, $A(f)$, is large in the signal bandwidth so that the *STF* is approximately unity (assuming $B[f] = 1$). The *NTF*, on the other hand, will approach zero, ideally, in the bandwidth of interest. Note that the signal spectrum passes through the modulator essentially unchanged, while the quantization noise spectrum is shaped (and thus the name *noise-shaping*). No precision filter or analog components are required, as discussed earlier, except, perhaps, for the DAC in the feedback path of the modulator. We'll see in the next chapter that if $A(f)$ is an integrator, the quantization noise is pushed to higher frequencies so that it can be removed with the averaging filter. This is a very important concept, as a noise-shaping modulator *does not* reduce the quantization noise to attain higher resolutions, but rather pushes the noise to frequencies outside of the signal bandwidth of interest.

ADDITIONAL READING

[1] R. J. van de Plassche, *CMOS Integrated Analog-to-Digital and Digital-to-Analog Converters*, Springer, 2007. ISBN 978-1402075001

[2] Engineering Staff Analog Devices Inc., *Data Conversion Handbook (Analog Devices)*, Newnes, 2005. ISBN 978-0750678414

[3] M. Gustavsson, J. J. Wikner, and N. Tan, *CMOS Data Converters for Communications*, Springer, 2000. ISBN 978-0792377801

[4] S. R. Norsworthy, R. Schreier, and G. C. Temes (eds.), *Delta-Sigma Data Converters: Theory, Design, and Simulation*, Wiley-IEEE Press, 1997. ISBN 978-0780310452

[5] J. C. Candy and G. C. Temes (eds.), *Oversampling Delta-Sigma Data Converters*, Wiley-IEEE Press, 1992. ISBN 978-0879422851

[6] S. K. Tewksbury and R. W. Hallock, *Oversampled, Linear Predictive and Noise-Shaping Coders of Order N>1*, IEEE Trans. Circuits and Sys., Vol. CAS-25, pp. 436-447, July 1978.

[7] W. R. Bennett, "Spectra of Quantized Signals," *Bell System Technical Journal*, Vol. 27, pp. 446-472, July 1948.

QUESTIONS

5.1 Develop an expression for the effective number of bits in terms of the measured signal-to-noise ratio if the input sinewave has a peak amplitude of 50% of ($V_{REF+} - V_{REF-}$).

5.2 When using Eq. (5.14) what is the assumed ADC input signal? Put your answer in terms of the ADC reference voltages.

5.3 Describe, in your own words, the difference between specifying SNR and SNDR.

5.4 Using SPICE simulations with an ideal ADC and DAC, show how coherent sampling can result in an RMS value of quantization noise larger than what is specified by Eq. (5.3). Comment on the shape of the quantization noise's spectrum.

5.5 Suppose a perfectly stable clock is available (ΔT_s is zero in Eq. [5.21]). Would we still have a finite aperture window if the clock has a finite rise time? Describe why or why not?

5.6 How do the number of bits lost because of aperture jitter change with the frequency of an ADC input sinewave? If the ADC input is a DC signal, is aperture jitter a concern? Why?

5.7 Why must Bennett's criteria be valid for the averaging filter in Fig. 5.29 to reduce the quantization noise in the digital output signal? Give an example input signal where averaging will not reduce quantization noise.

5.8 Assuming Eq. (5.57) is valid, rederive Eq. (5.13) including the effects of averaging K ADC output samples. Is Eq. (5.13) or the equation derived here valid for a slow or DC input signal? Comment on why or why not.

5.9 If Bennett's criteria are valid, does averaging ADC outputs (or DAC inputs) put any restrictions on the bandwidth of the input signal? Why? Give an example.

5.10 How accurate does an 8-bit ADC have to be in order to use a digital filter to average 16 output samples for a final output resolution of 10-bits (see Eq. [5.59])? Assume the ideal LSB of the 8-bit converter is 10 mV. Your answer should be given in both mV and % of the full-scale.

5.11 Show the detailed derivation of Eq. (5.66).

5.12 Summarize, and compare, the advantages and disadvantages of predictive and noise-shaping data converters.

Chapter

6

Data Converter Design Basics

Mixed-signal design is powerful because it combines digital-signal processing (DSP) with analog circuit design. Using DSP, as we'll show in this chapter, reduces the required precision of the analog circuits. Perhaps a different, yet appropriate, name for this chapter is "Analog Design without using Analog Components!"

In the past five chapters, we covered topics we'll frequently use in the design of mixed-signal circuits. For example, Fig. 6.1 shows how we'll design an analog-to-digital converter (ADC) using mixed-signal circuit design techniques. Our analog input is passed through an anti-aliasing filter (AAF) to a noise-shaping (NS) modulator. The noise shaping modulator, as discussed in Sec. 5.4, uses a low-resolution quantizer (a fancy name for an ADC with 1 to a few bits resolution, like the comparator seen in Fig. 5.1) in a feedback loop to get a running average of the analog input signal. The desired digital output is extracted using a moving average filter, the lowpass filters in Sec. 4.2, and decimated. In simple words, the modulator's digital output signal is averaged to get a representation of the analog input signal. Figure 6.2 shows example modulator outputs for various DC input signals.

Note that this type of data converter is often called an *oversampled* ADC since the sampling frequency, f_s, must be much larger than the input frequencies of interest. We'll also see that these types of data converters are called *noise-shaping* ADCs or, depending on the topology used, *delta-sigma* (or *sigma-delta*) ADCs (these names are discussed in greater detail later).

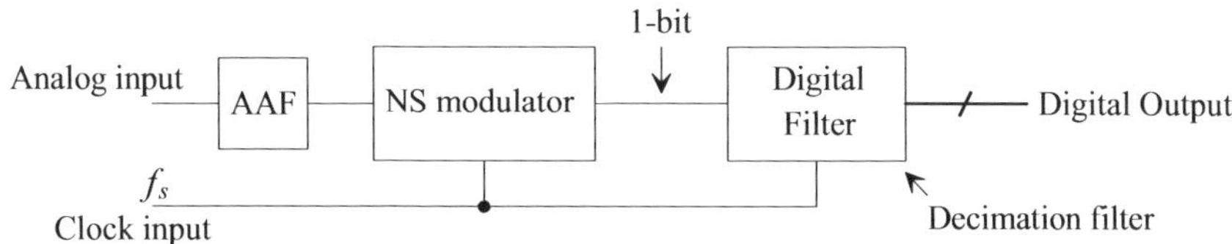

Figure 6.1 An ADC using a NS modulator and digital filter.

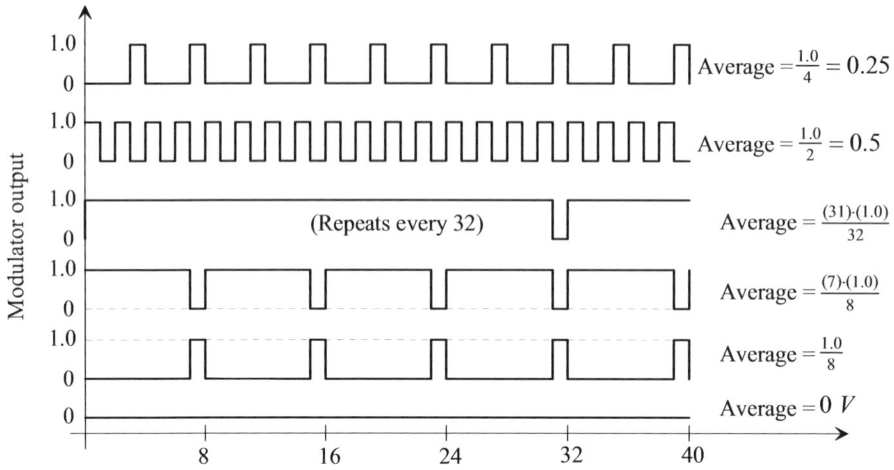

Figure 6.2 Modulator outputs and their corresponding DC averages.

The One-Bit ADC and DAC

In this chapter we'll use, exclusively, a one-bit ADC and DAC. To understand why, let's review Fig. 5.28 and 5.32. In all cases, unless the spacing of the codes is perfect in these data converters, averaging the codes will result in non-linearity. However, by using an ADC or DAC with only two output codes in our NS modulator we are guaranteed linearity (two points determine a line!). If one of the reference voltages in the ADC or DAC is offset from its ideal value then we'll get a gain error but no nonlinearity error.

In Chs. 4 and 5 we treated the clocked comparator as a 1-bit quantizer (ADC) with an LSB given by

$$1\text{ LSB} = V_{REF+} - V_{REF-} = V_{LSB} \tag{6.1}$$

as an added noise source, Fig. 5.2. The PSD of the added noise, see Eq. (5.5), is

$$V_{Qe}^2(f) = \frac{V_{LSB}^2}{12 f_s} \tag{6.2}$$

In this chapter we'll set $V_{REF+} = VDD = 1$ V and $V_{REF-} = 0$ V so that $V_{CM} = VDD/2$ or 500 mV. We can think of the output of the comparator (the output of the NS modulator) as a binary offset number with a value of 1 or 0. After a simple conversion to two's complement numbers, the output can be changed to +1 (01) or −1 (11). Since our input signals are referenced, or swing around, V_{CM} it's useful to think in terms of two's complement numbers. For example, in Fig. 6.2, if we apply 500 mV to the input of our NS modulator we get 1010101... (second sketch) or an average of 500 mV (= V_{CM}). Since all signals are referenced to V_{CM} it may be more useful to say that we aren't applying a signal (meaning just the common-mode voltage is present at the input) and thus we get a sequence of +1, −1, +1, −1, etc. that averages to zero (V_{CM}). Figure 6.3 shows how we think about this in more detail. If we pass a signal through an inverter we are multiplying it by −1 when we think in terms of two's complement numbers.

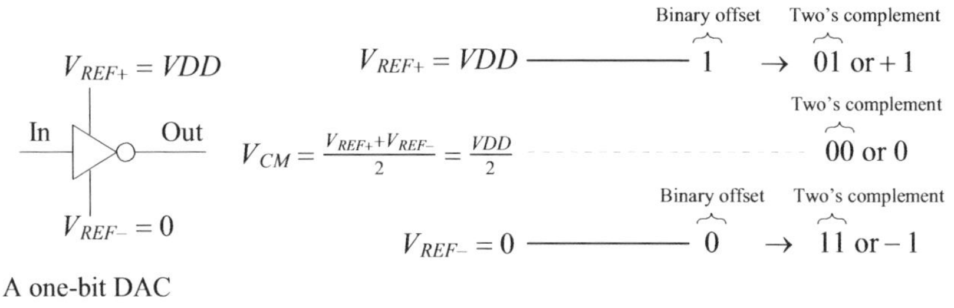

$V_{REF+} = VDD$

In Out

$V_{REF-} = 0$

A one-bit DAC

$V_{REF+} = VDD$ —— Binary offset 1 → Two's complement 01 or +1

$V_{CM} = \frac{V_{REF+}+V_{REF-}}{2} = \frac{VDD}{2}$ - - - Two's complement 00 or 0

$V_{REF-} = 0$ —— Binary offset 0 → Two's complement 11 or −1

Figure 6.3 Thinking about the inverter in terms of the common-mode voltage.

6.1 Passive Noise-Shaping

Figure 6.4 shows a schematic and block diagram of a passive NS modulator. Reviewing Fig. 5.38, we see that this is a modulator topology with $B(f) = 1$. To calculate $A(f)$ let's write the output of the summing block (the resistors), or the input to the $A(f)$ block, as

$$\frac{v_{in} - v_{int}}{R} + \frac{-v_{out} - v_{int}}{R} = i_{int} \tag{6.3}$$

Note that, in Fig. 6.4b, we are assuming that the S/H response used in the model for an ADC, Fig. 5.2, is 1 ($f \ll f_s$). The output of the $A(f)$ block is

$$v_{int} = \frac{i_{int}}{j\omega C} \tag{6.4}$$

so,

$$v_{in} - v_{out} = v_{int} \cdot (j\omega CR + 2) \tag{6.5}$$

and therefore

$$A(f) = \frac{1}{2 + j\omega RC} \tag{6.6}$$

This is nearly the transfer function of an RC circuit, see Eq. (3.1). We can also think of this response as integrating for frequencies above $1/2\pi RC$. We'll come back to this in a moment. Using Eq. (5.66) we can write

$$v_{out}(f) = \overbrace{\frac{1}{3 + j\omega RC}}^{STF(f)} \cdot v_{in}(f) + \overbrace{\frac{2 + j\omega RC}{3 + j\omega RC}}^{NTF(f)} \cdot V_{Qe}(f) \tag{6.7}$$

Notice that the AAF is built-in to this topology via the *STF*. At DC the noise transfer function appears to be 2/3. However, we know that at DC, as seen in Fig. 6.2, we can average the modulator's outputs and recover, exactly, the analog input signal (see dead zone issues for a passive modulator on page 215). This means that $NTF(0) = 0$. Let's try to get a better representation for the output of the modulator. In order to begin notice that

$$V_{Qe}(f) + v_{int}(f) = v_{out}(f) \tag{6.8}$$

Next, let's write

$$\left(\frac{v_{in} - v_{int}}{R} + \frac{-v_{out} - v_{int}}{R}\right) \cdot \frac{1}{j\omega C} + V_{Qe}(f) = v_{out} \tag{6.9}$$

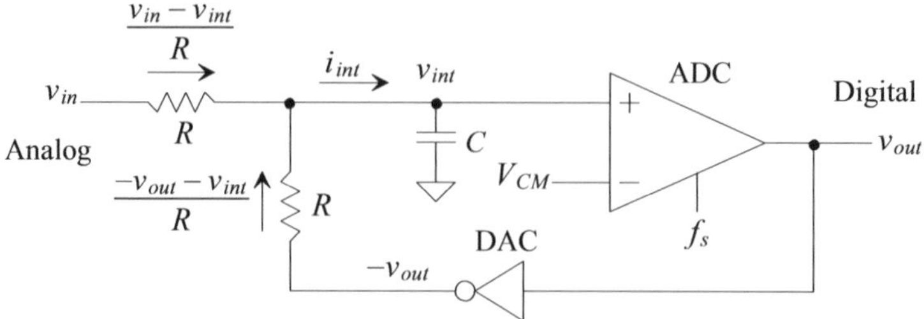

(a) Circuit implementation of a passive NS modulator

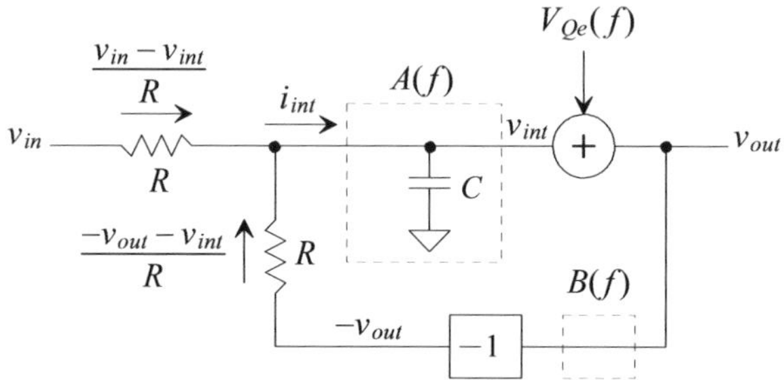

(b) Block diagram model of a passive NS modulator.

Figure 6.4 A passive-integrator NS modulator.

and

$$v_{in} - v_{out} - 2v_{int} + V_{Qe} \cdot j\omega RC = j\omega RC \cdot v_{out} \tag{6.10}$$

$$v_{in} - 2v_{int} + V_{Qe} \cdot j\omega RC = v_{out} \cdot (1 + j\omega RC) \tag{6.11}$$

so finally

$$v_{out} = \overbrace{\frac{1}{1+j\omega RC}}^{STF(f)} \cdot v_{in} + \overbrace{\frac{j\omega RC}{1+j\omega RC}}^{NTF(f)} \cdot V_{Qe} + \underbrace{\frac{-2 \cdot v_{int}}{1+j\omega RC}}_{\text{Extra noise/distortion}} \tag{6.12}$$

The key things to note are: 1) that if we can keep v_{int} from varying we eliminate this extra distortion term (this is why we use an active integrator in later chapters), 2) the signal sees a lowpass response (so we have, as already mentioned, a built-in AAF that limits input spectral content), and 3) the noise is shaped towards the higher frequencies (the noise is high passed filtered). If our quantization noise is flat, as seen in Fig. 5.12, then high-pass filtering the noise gives the PSD seen in Fig. 6.5. Quantization noise that has been shaped in this way is called *modulation noise*. Note that the faster we clock the comparator (the larger f_s) the less variation we'll get in v_{int}. Also note, again, that the (digital) output of the modulator is the average of the (analog) input signal.

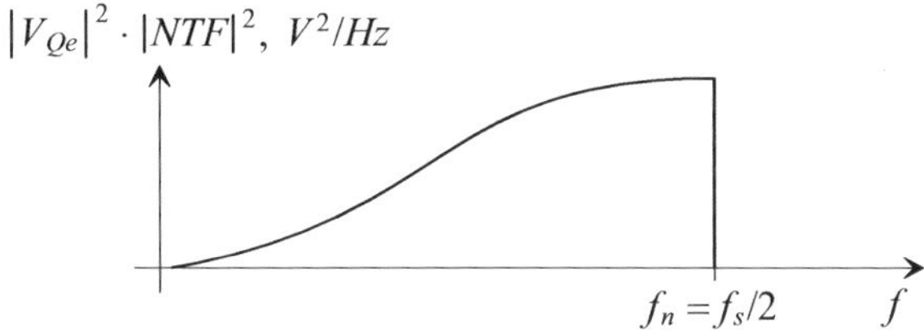

Figure 6.5 Modulation noise spectral density.

Example 6.1

Simulate the operation of the NS modulator seen in Fig. 6.4 when *R* is 10k, *C* is 10 pF, and the clocking frequency is 100 MHz. Comment on the resulting simulation results and the operation/limitations of the circuit.

Figure 6.6 shows the input (a sinewave at 500 kHz) and (digital) output of the circuit. Notice that as the input signal moves towards ground the output signal stays low more often. As the input signal moves towards *VDD* the output is high, or a logic 1, more often. While we can use a digital filter to change this one-bit code into an *N*-bit word (and we will!) let's show that a simple RC filter can be used to get our original signal back from the digital data, Fig. 6.7.

Note that for proper operation of a continuous-time (CT) NS data converter, like the one seen in Fig. 6.4, the comparator's gain and delay become very important. If the *comparator's delay varies* (and it will in a practical circuit) we get a degradation in *SNR* that appears like it's from clock jitter (see Sec. 5.2.2). Varying comparator delay is a **serious practical limitation** of CT NS converters for high-frequency conversions (it's much less of a concern for low frequency conversions) causing *amplitude modulation* in the converter's output. Adding an amplifier in the forward path of the modulator (see example in Fig. 6.24) can help ensure that the signal swing on the input of the comparator is large to minimize the comparator's delay. Adding an edge-triggered latch (a D-Flip-Flop) on the output of the comparator can be used to ensure a stable comparator output signal but adds delay (and thus makes stabilizing the system more challenging). ■

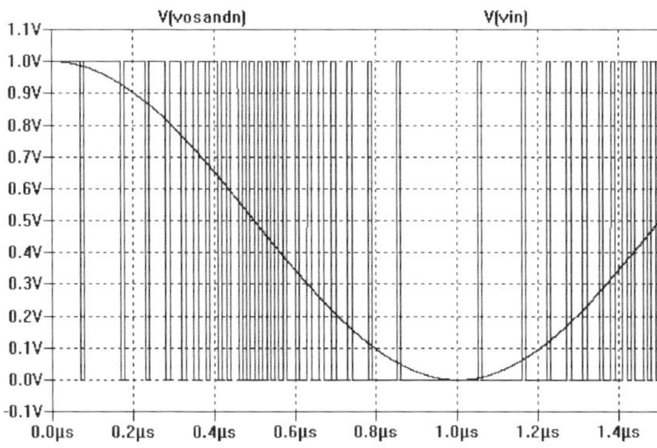

Figure 6.6 Simulating the operation of the passive NS modulator seen in Fig. 6.4.

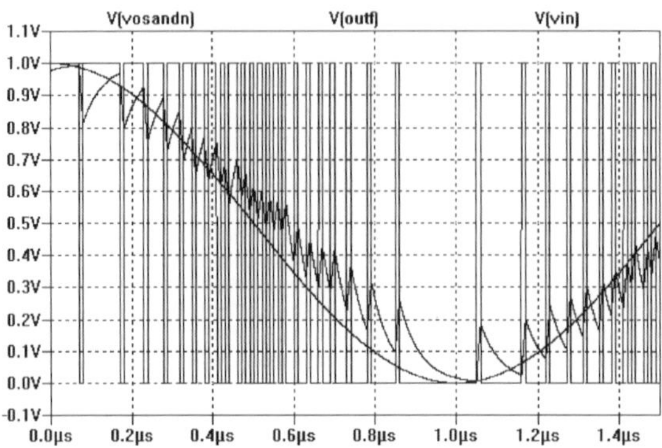

Figure 6.7 Using an RC circuit to filter the digital output in Fig. 6.6.

6.1.1 Signal-to-Noise Ratio

Let's calculate the SNR for the first-order noise-shaping modulator seen in Fig. 6.4 and characterized by Eq. (6.12). In the following we'll ignore the extra noise/distortion resulting from variations in v_{int}. Further note that the pole associated with both the STF and NTF won't affect the SNR since it's common to both. We know the output of the modulator is passed through a lowpass filter with a bandwidth B to remove the modulation noise, Fig. 6.8. The smaller B the lower the noise in the final digital output word and the larger the *SNR*. Again, the trade-off with using smaller lowpass filter bandwidth is that the frequency range of the allowable input signals shrinks. Let's calculate the RMS noise in the filter's output using

$$V^2_{noise,RMS} = 2\int_0^B |NTF(f)|^2 |V_{Qe}(f)|^2 \cdot df = 2 \cdot \frac{V^2_{LSB}}{12 f_s} \cdot \int_0^B (2\pi f \cdot RC)^2 \cdot df \quad (6.13)$$

or

$$V^2_{noise,RMS} = 2 \cdot \frac{V^2_{LSB}}{12 f_s} \cdot (2\pi RC)^2 \cdot \frac{B^3}{3} \quad (6.14)$$

again noting that using a smaller digital lowpass filter bandwidth, B, reduces the noise.

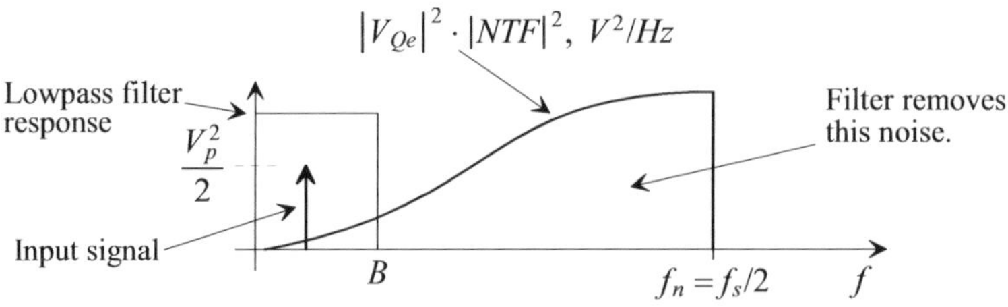

Figure 6.8 Filtering out modulation noise to calculate SNR.

We spent a great deal of time in Ch. 4 using Sinc-shaped lowpass filters for decimation. Let's use these filters here for filtering and decimating the output of the NS modulator. Let's set the bandwidth of the filter, see Fig. 4.17, to

$$B = \frac{f_s}{2K} \tag{6.15}$$

noting that we assumed, when deriving Eq. (6.14), a brickwall-shaped lowpass filter response. We can reduce the noise further, after decimation, by passing the digital data through additional lowpass filtering (perhaps using the biquad filters discussed in Ch. 4 with peaking to account for the droop introduced with the Sinc filter). Making the substitution gives

$$V_{noise,RMS}^2 = \frac{V_{LSB}^2}{12} \cdot (2\pi RC)^2 \cdot \frac{f_s^2}{12 \cdot K^3} \tag{6.16}$$

The value of RC must be much longer (at least five times) than the period of the sampling clock, T_s, to keep v_{int} from varying too much. Following the procedures used to derive Eqs. (5.10) to (5.13) we get

$$SNR_{ideal} = 20 \cdot \log \frac{V_p/\sqrt{2}}{V_{noise,RMS}} = 6.02N + 1.76 - 20 \log \frac{2\pi RC \cdot f_s}{\sqrt{12}} + 20 \log K^{3/2} \tag{6.17}$$

If we set $RC = 4.4/f_s = 4.4T_s$ then we can write

$$SNR_{ideal} = 20 \cdot \log \frac{V_p/\sqrt{2}}{V_{noise,RMS}} = 6.02N + 1.76 - 18.06 + 30 \log K \tag{6.18}$$

For $K = 4$ this equation is the same as Eq. (5.13). However, for every doubling of K beyond 4 we get an increase in resolution, N_{inc}, of 1.5 bits or an increase in SNR_{ideal} of 9 dB

$$N_{inc} = \frac{30 \log K - 18.06}{6.02} \tag{6.19}$$

$$SNR_{ideal} = 6.02(N + N_{inc}) + 1.76 \tag{6.20}$$

Figure 6.9 compares the first-order NS modulator to simple oversampling, Fig. 5.31 and Eq. (5.58). It's important to note that Bennett's criteria need not be valid using the oversampling feedback modulator discussed in this section (e.g. the input doesn't have to be busy).

6.1.2 Decimating and Filtering the Modulator's Output

It's important to note that Eq. (6.17) was derived assuming the output of the modulator was passed through a perfect lowpass filter with a bandwidth of B, $f_s/2K$. Passing the output through a Sinc averaging filter, see Fig. 4.14, will result in a poorer SNR because the higher frequency noise components will not be entirely filtered out. In this section we want to answer two questions: (1) what order, L (see Eq. [4.19]), of Sinc lowpass filter should be used in the digital filter on the output of the NS modulator, and (2) assuming we use only this filter (no additional filtering), how will the ideal SNR of the first-order NS modulator be affected?

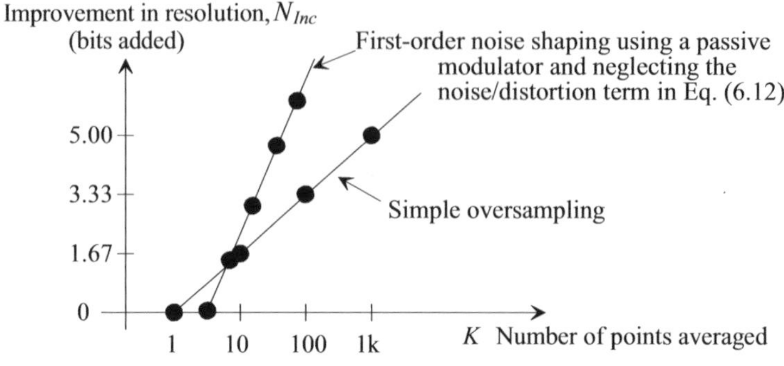

Figure 6.9 Comparing first-order noise-shaping to simple oversampling (Fig. 5.29).

We begin to answer the first question by noting that the increase in the number of bits, N_{inc}, was specified by Eq. (6.19). If our NS modulator uses a 1-bit ADC, then the final, after the digital filter, resolution of the resulting data converter is $N_{inc} + 1$ bits. (An NS modulator using a 5-bit ADC [often called a multibit NS modulator] would ideally have an output resolution of $N_{inc} + 5$ bits.) Further, we saw in Ex. 4.3 and Fig. 4.14 that the word size increased by $\log_2 K$ bits in each Sinc filter stage. For a cascade of L filters we can require

$$L \cdot \log_2 K \geq \frac{30 \log K - 18.06}{6.02} \tag{6.21}$$

For $K \leq 256$ we only need one lowpass Sinc filter or $L = 1$.

Example 6.2

Sketch the implementation of a $K = 16$ decimating filter for the NS modulator discussed in Ex. 6.1. Simulate the operation of the resulting ADC (modulator and filter). Estimate the *SNR*, number of final bits, and Nyquist frequency.

The decimating filter's transfer function is

$$\frac{1 - z^{-16}}{1 - z^{-1}} \tag{6.22}$$

and seen in Fig. 6.10 (see Sec. 4.2.5). The filter's output clock rate is 100 MHz/16 or 6.25 MHz so the Nyquist frequency is half of this or 3.125 MHz. The increase in the number of bits is 3, using Eq. (6.19), so the final output word size, based on

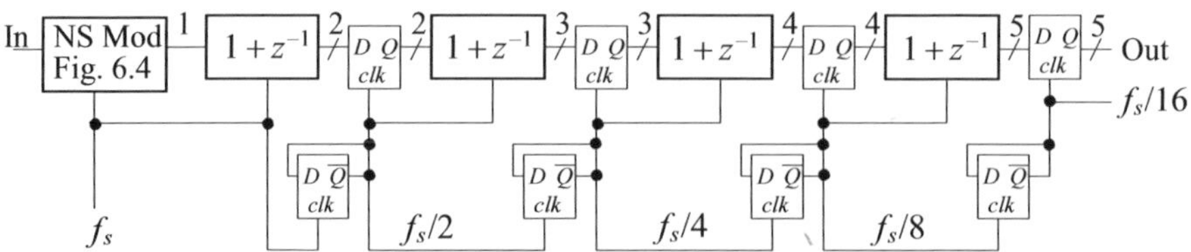

Figure 6.10 Lowpass and decimating filter used in Ex. 6.2.

the *SNR* (= 25.78 dB), is 4-bits. Our filter causes the word size to increase by 1-bit through each of the 4 stages so that our final output word size is 5-bits. We can throw the MSB of this word out; however, notice that the largest word we get out of the filter is when the filter's input remains a 1 at all times. In this case the filter's output (the output of the ADC made using the NS modulator and filter) is 16 or 1 0000. When the filter's input is a 0 at all times our filter's output is 0 0000. The output can swing from 1 0000 to 0 0000 around the common-mode code of 0 1000 (8). By removing the MSB we can make the output swing from 1111 (15) to 0000 (0) around 1000 (8). However if the NS modulator's input moves close to V_{REF+} (here *VDD*) then the output of the filter can **overflow**.

Figure 6.11 shows the simulation results. The input frequency, as in Ex. 6.1, is 500 kHz. Looking at the frequency response of our filter, Eq. (4.20) and Fig. 4.17, we should only see a minor amount of attenuation through the filter. However, looking at Fig. 6.11 we see more than a minor amount of output signal reduction. We've, up to this point, ignored the extra distortion/noise term seen in Fig. 6.12. In the next section we'll discuss this term in more detail and how matching and offsets effect the performance of the modulator. ■

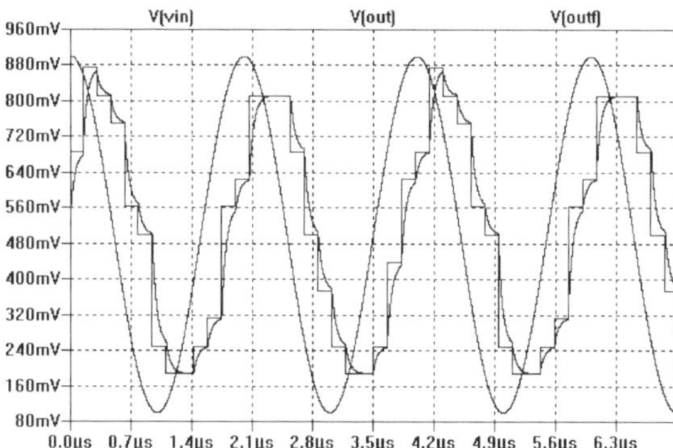

Figure 6.11 Simulating the ADC (modulator and filter) discussed in Ex. 6.2.

SNR Calculation using a Sinc Filter

Let's now consider how filtering with a Sinc filter instead of the ideal lowpass filter effects the *SNR* of the modulator/filter implementation of a data converter. Remember the SNR_{ideal} was calculated in Eq. (6.18) assuming the modulation noise was strictly bandlimited to *B*. Figure 6.12 shows the PSD of the $NTF^2(f)$ and $|V_{Qe}(f)|^2$ (the modulation noise) in a first-order passive NS modulator. Also shown in this figure is the shape of the averaging filter's magnitude response squared that is normalized to unity by dividing by *K*, see Eq. (4.20). Here we are showing the shape of a filter with $K = 16$ and a range of $f_s/2$ (see Fig. 5.12).

We can calculate the RMS quantization noise resulting from a cascade of a first-order modulator and an averaging filter, $L = 1$, using

$$V_{Qe,RMS}^2 = 2\int_0^{f_s/2} |NTF(f)|^2 \cdot |V_{Qe}(f)|^2 \cdot |H(f)|^2 \cdot df \qquad (6.23)$$

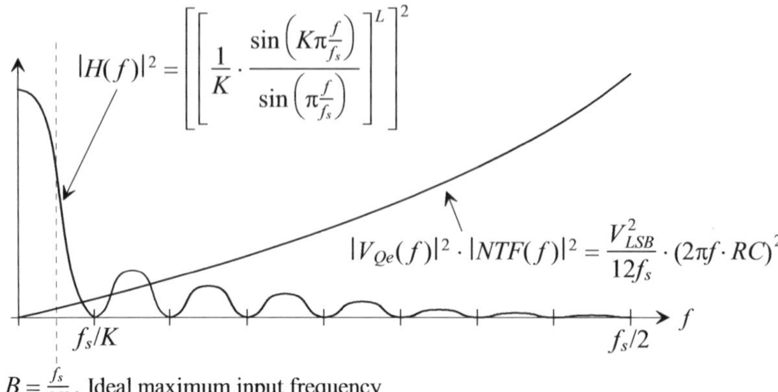

Figure 6.12 Showing modulation noise and filter response.

or

$$V_{Qe,RMS}^2 = 2\cdot\frac{V_{LSB}^2}{12f_s}\cdot\int_0^{f_s/2}(2\pi f\cdot RC)^2\frac{1}{K^2}\cdot\frac{\sin^2\left(K\pi\frac{f}{f_s}\right)}{\sin^2\left(\pi\frac{f}{f_s}\right)}\cdot df \tag{6.24}$$

For the frequency range of interest $f << f_s$ so $\sin x \approx x$ and we get

$$V_{Qe,RMS}^2 = 2\cdot\frac{V_{LSB}^2}{12f_s}\cdot(2f_s\cdot RC)^2\cdot\frac{1}{K^2}\cdot\int_0^{f_s/2}\sin^2\left(K\pi\frac{f}{f_s}\right)\cdot df \tag{6.25}$$

Let's let $\theta = \pi\cdot\frac{f}{f_s}$ so

$$V_{Qe,RMS}^2 = 2\cdot\frac{V_{LSB}^2}{12f_s}\cdot(2f_s\cdot RC)^2\cdot\frac{f_s}{K^3\pi}\cdot\overbrace{\int_0^{\frac{\pi}{2}}\sin^2 K\theta\cdot d\theta}^{=\frac{\pi}{4}} \tag{6.26}$$

and

$$V_{Qe,RMS}^2 = \frac{V_{LSB}^2}{12}\cdot(2\pi RC)^2\cdot\frac{f_s^2}{2\pi^2K^2} \tag{6.27}$$

Comparing this result to what we got in Eq. (6.16) we see that using the Sinc lowpass filter is nearly ideal for removing modulation noise. The more important concern, when using the Sinc lowpass filter, is the droop that the desired signal undergoes.

6.1.3 Offset, Matching, and Linearity

Examine the passive NS-modulator with offsets and mismatched resistors seen in Fig. 6.13. In this section we want to discuss how non-ideal behavior affects the performance of the modulator. To begin, notice that a comparator offset has the effect of causing the average value of v_{int} to be $V_{CM} \pm V_{OS}$ and thus the offset passes directly to the output of the modulator. This means that if we apply the common-mode voltage to the input the average of the output voltage is $V_{CM} \pm V_{OS}$.

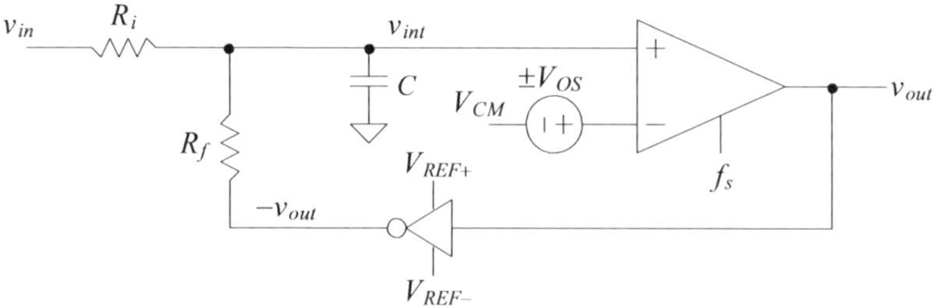

Figure 6.13 Passive modulator with mismatch and offsets.

Resistor Mismatch

In order to determine the effects of resistor mismatch we can re-write Eqs. (6.9) to (6.12), neglecting the effects of the offset voltage, as

$$\left(\frac{v_{in}-v_{int}}{R_i}+\frac{-v_{out}-v_{int}}{R_f}\right)\cdot\frac{1}{j\omega C}+V_{Qe}(f)=v_{out} \tag{6.28}$$

$$R_f\cdot(v_{in}-v_{int})-R_i\cdot(v_{out}+v_{int})+j\omega R_iR_f\cdot C\cdot V_{Qe}=j\omega R_iR_f\cdot C\cdot v_{out} \tag{6.29}$$

$$R_f\cdot v_{in}-(R_i+R_f)\cdot v_{int}+j\omega R_iR_f\cdot C\cdot V_{Qe}=(j\omega R_iR_f\cdot C+R_i)\cdot v_{out} \tag{6.30}$$

$$v_{out}=\frac{\frac{R_f}{R_i}}{1+j\omega CR_f}\cdot v_{in}+\frac{j\omega CR_f}{1+j\omega CR_f}\cdot V_{Qe}+\frac{-v_{int}\cdot\frac{R_i+R_f}{R_i}}{1+j\omega CR_f} \tag{6.31}$$

The effect of mismatched resistors is a gain error (but no non-linearity, and thus, distortion). Note that the absolute value of the capacitor isn't critical or important for precision operation.

The Feedback DAC

The feedback DAC is the most critical component in the modulator for precision operation. Consider the waveforms seen in Fig. 6.14. In this figure we show the ideal shapes of the DAC's output (remember our DAC here is a simple inverter so we get

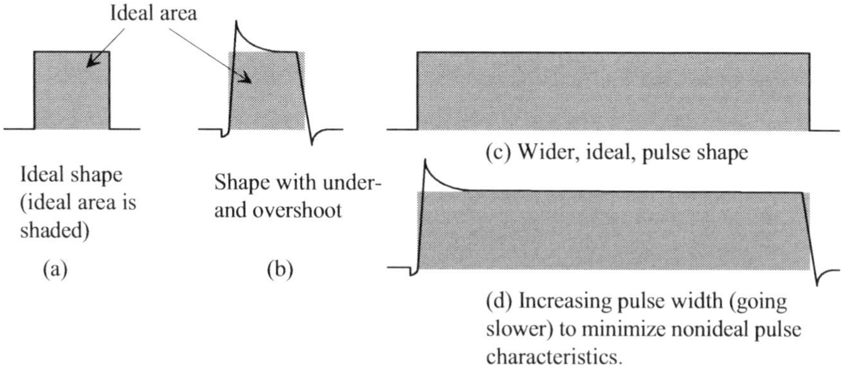

Figure 6.14 Comparator output pulse shapes, input to the integrator.

perfect linearity, Fig. 6.15). As seen in the figure, the shape of the pulse affects the amount of charge, or current, we send back to the capacitor. In part (a) we see the ideal pulse shape and the ideal area under the pulse (the shaded area). In part (b) we see how the finite rise time and fall time can affect the actual area under the curve and thus the output of the integrator. In order to minimize these unwanted effects we can use wider pulses as shown in parts (c) and (d), which means we run the modulator at a slower clocking frequency. Increasing the width of the pulses minimizes the percentage of the area affected by the transition times. Note that the feedback signal directly subtracts from the input signal so that any noise or unwanted variation in the fed back signal, such as an amplitude variation, can be considered as adding noise to the input (and thus degrading the modulator's SNR). This is important. In the coming chapters we will use switched-capacitor circuits that simply have to fully-discharge in order to make the shape of the feedback pulse less important.

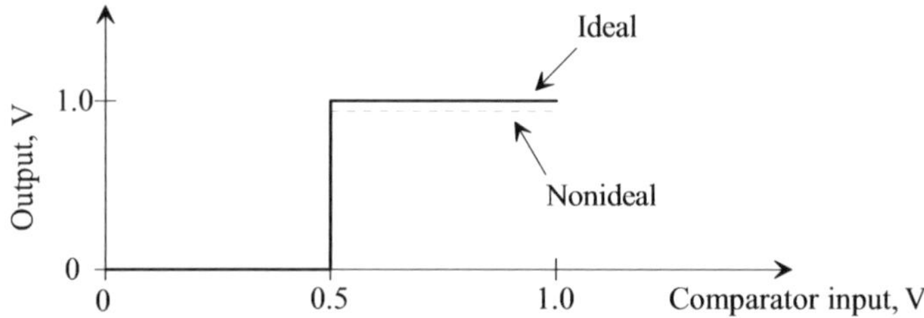

Figure 6.15 Ideal and nonideal transfer curves for a 1-bit DAC.

DAC Offset

Notice, in Fig. 6.15, that if our DAC's reference voltages are offset from their ideal values of V_{REF+} and V_{REF-} that, like mismatched resistors, all we get is a gain error (no non-linearity). Possibly the most important concern when using a single-bit inverter topology DAC is the on-resistance of the transistors used to drive the feedback resistors. The drive strength of the MOSFETs should be so high that little voltage is dropped across them (so the output of the DAC swings between V_{REF+} and V_{REF-}). This (limited DAC drive) is, again, another reason we'll soon start to focus on switched-capacitor implementations of NS modulators.

Linearity of the First-Order Modulator

We've ignored the effects of the extra noise/distortion term seen in Eq. (6.12) because it's a nonlinear error highly dependent on the input signal's characteristics (and thus hard to quantify). Let's use simulations to look at the linearity of the first-order NS modulator and digital filter seen in Fig. 6.10 (with a DAC connected to the output of the digital filter to show the digital data as an analog waveform). Figure 6.16 shows the input/output of Fig. 6.10 with a slow ramp applied to the modulator's input. Note that when the input ramp voltage is close to the power supply rails the digital output becomes nonlinear (too high when the input is close to ground and to low when the input is close to *VDD*). Also seen in this figure, as mentioned in Ex. 6.2, is the digital filter overflowing. In order to avoid this situation we can use a selector, like the one seen in Fig. 4.40, with the *Sel* input connected to the MSB of the filter (remember we do a multiply by 2 by removing the filter's MSB to scale the filter's output to full scale as discussed in Ex. 6.2). When the

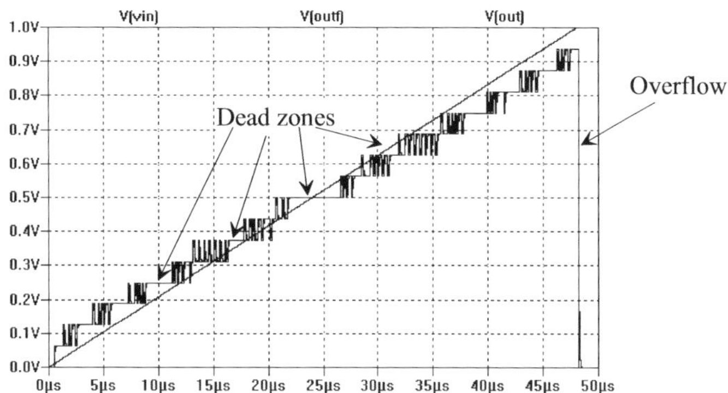

Figure 6.16 Simulating the linearity of the first-order NS modulator.

MSB of the filter's output is a 0 we simply pass the multiplied-by-2 output of the filter through the selector to the final output. When the MSB of the filter's output is a 1, we use the selector to clamp the final output to 1111 and thus avoid overflow.

Dead Zones

Besides the nonlinearity seen in Fig. 6.16, notice that for some ranges of input voltages the output doesn't move or, rather, *the output is dead.* A simple example of a dead output is seen when the input signal is close to the common-mode voltage, 500 mV. These *dead zones* are the result of a repeating modulator output code (e.g., 010101, 110110, 0000100001, etc.). The question is why does the modulator output code repeat for a relatively wide range of input voltages (e.g. 460 mV $< v_{in} <$ 540 mV)? In order to answer this question examine the modulator seen in Fig. 6.13. With an input voltage of 500 mV the output of the modulator is 101010... so that the average of the output is 500 mV. Now suppose the input voltage is increased to 510 mV. Because the voltage v_{int} varies with time, it's possible for the average current supplied from the input, $[v_{in} - v_{int}(t)]/R_i$ to equal the fed back current $[-v_{out} - v_{int}(t)]/R_f$ when the modulator's output is 1010101... By using an active integrator, Sec. 6.2.4, we can hold v_{int} constant and eliminate dead zones.

6.2 Improving SNR and Linearity

In this section we discuss techniques for improving signal-to-noise ratio and linearity. We'll start out by discussing the second-order passive (uses two capacitors) noise-shaping modulator. This topology is considerably more useful than the first-order topology discussed in the last section because it has better linearity, randomizes the output code (to avoid dead zones), and lower modulation noise in the signal bandwidth of interest. Then we'll discuss using switched-capacitors instead of resistors in the feedback paths. The big benefit of switched-capacitor circuits over continuous-time circuits is that, Fig. 6.14, we remove the effects of non-ideal pulse shapes. The drawback is the need for a non-overlapping clock generator (which consumes power and layout area). Next we'll discuss putting NS modulators in parallel. This is useful for increasing SNR while not lowering the final output clocking frequency (at all or as much). Finally, we'll end the section by showing how an active circuit, e.g. an op-amp, can be used to improve linearity and noise by removing the extra term seen, for example, in Eq. (6.12).

6.2.1 Second-Order Passive Noise-Shaping

After examining the first-order modulator seen in Fig. 6.4 we might wonder if we can get better performance by double filtering the input signal and removing the comparator's connection to V_{CM}. Figure 6.17 shows the resulting circuit. In this circuit we have two feedback paths. We've connected the largest amount of feedback to the inverting input of the comparator to ensure negative, overall, feedback. Note that, because the output is fed back to both integrating nodes v_1 and v_2, the circuit will act to drive these nodes to the same value and thus attempt to keep them equal to each other.

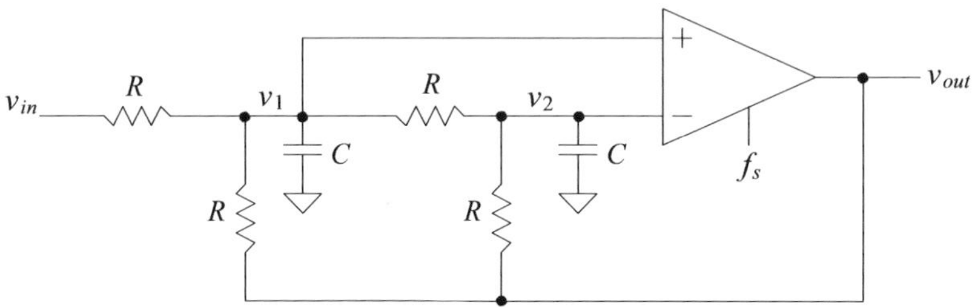

Figure 6.17 Second-order passive modulator.

To attempt to describe the operation of the circuit let's write

$$\left(\frac{v_{in}-v_1}{R}+\frac{v_{out}-v_1}{R}-\frac{v_1-v_2}{R}\right)\cdot\frac{1}{j\omega C}=v_1 \tag{6.32}$$

and

$$\left(\frac{v_1-v_2}{R}+\frac{v_{out}-v_2}{R}\right)\cdot\frac{1}{j\omega C}=v_2 \tag{6.33}$$

If the quantization noise added to v_1 and v_2 is V_{Qe1} and V_{Qe2} then $v_1+V_{Qe1}=v_{out1}$ and $v_2+V_{Qe2}=v_{out2}$. We can then write

$$v_{in}-3v_1+v_2+j\omega RC\cdot V_{Qe1}=v_{out1}(j\omega RC-1) \tag{6.34}$$

$$v_1-2v_2+j\omega RC\cdot V_{Qe2}=v_{out2}(j\omega RC-1) \tag{6.35}$$

Adding these two equations gives

$$v_{in}-2v_1-v_2+j\omega RC\cdot(V_{Qe1}+V_{Qe2})=(v_{out1}+v_{out2})\cdot(j\omega RC-1) \tag{6.36}$$

As a quick check to ensure that this equation is correct, notice that if we pass the output signal through a lowpass filter with a bandwidth approaching 0, $\omega\to 0$ and we use a large RC then for a DC input, $v_{in}\approx v_1\approx v_2$ and $v_{in}-2v_1-v_2=-2v_{out}$ (so $v_{out}\approx v_{in}$). Next, adding the random signal powers together (see, for example, Eq. [5.55] and the associated discussion) gives

$$(\omega RC)^2\cdot\left(V_{Qe1}^2+V_{Qe2}^2\right)=(\omega RC)^2\cdot\frac{V_{Qe}^2}{2} \tag{6.37}$$

and

$$\left(v_{out1}^2 + v_{out2}^2\right) \cdot \left((\omega RC)^2 + (-1)^2\right) = \left(1 + (\omega RC)^2\right) \cdot \left(v_{out,noise}^2 + v_{out,signal}^2\right) \quad (6.38)$$

Finally, we can write

$$v_{out,noise}^2 = \frac{(\omega RC)^2}{1+(\omega RC)^2} \cdot \frac{V_{Qe}^2}{2} \quad (6.39)$$

$$v_{out,signal}^2 = \frac{1}{1+(\omega RC)^2} \cdot v_{in}^2 \quad (6.40)$$

noting the quantization noise power is cut in half. Equation (6.18) can be rewritten for the second-order passive noise-shaping topology as

$$SNR_{ideal} = 20 \cdot \log \frac{V_p/\sqrt{2}}{V_{noise,RMS}} = 6.02N + 1.76 - 15.13 + 30 \log K \quad (6.41)$$

The modest increase in *SNR*, however, is not the big benefit of this topology. Rather the better linearity and reduction in dead zones are the major benefits of the topology. The extra noise/distortion term can be written as

$$\frac{2v_1 + v_2}{1+(\omega RC)^2} \quad (6.42)$$

Note that this topology attempts to drive v_1 and v_2 to the same value, that is v_{in}. The drawback of this is that the comparator must operate with an input range extending from *VDD* to ground. Figure 6.18 shows the simulation results where we repeated the simulation seen in Fig. 6.16 but used the second-order modulator. Clearly both the linearity and extent of dead-zones are much better using the second-order modulator.

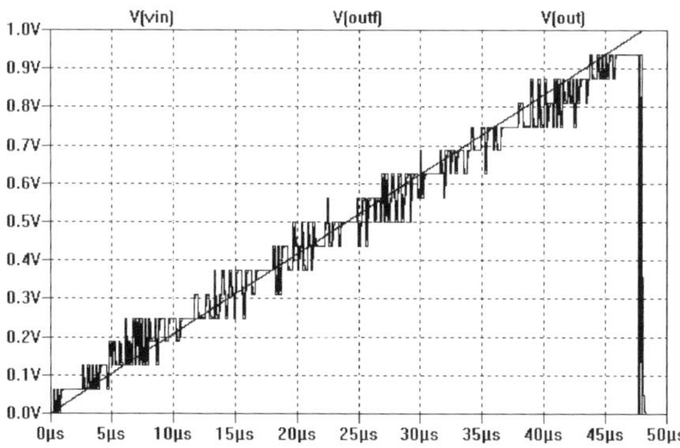

Figure 6.18 Repeating the simulation seen in Fig. 6.16 but with a second-order modulation.

Example 6.3

Repeat the simulation seen in Fig. 6.11 and Ex. 6.2 using the second-order modulator seen in Fig. 6.17.

The simulation results are seen in Fig. 6.19. The output amplitude more closely resembles the input amplitude. ■

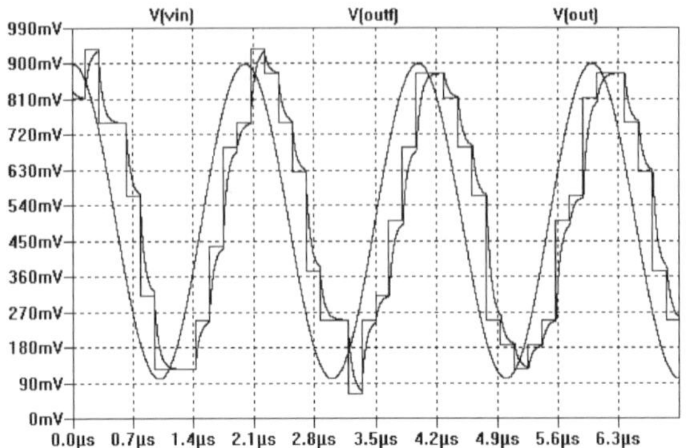

Figure 6.19 Repeating the simulation seen in Fig. 6.11 but with a second-order modulation.

6.2.2 Passive Noise-Shaping Using Switched-Capacitors

Figure 6.20 shows the first-order modulator seen in Fig. 6.4 implemented using switched-capacitor (SC) resistors, Fig. 2.35 (the resistors in Fig. 6.4 are replaced with SC resistors). Figure 6.21 shows the simulation results similar to Fig. 6.7 but using this new implementation. Again the benefit of this implementation over the continuous-time implementation seen in Fig. 6.4a is that the shape of pulse coming out of the feedback DAC, here an inverter, isn't important as long as the capacitors fully charge and discharge.

To characterize the operation of this topology let's write, again see Fig. 2.35,

$$R_{SC} = \frac{1}{f_s C} \tag{6.43}$$

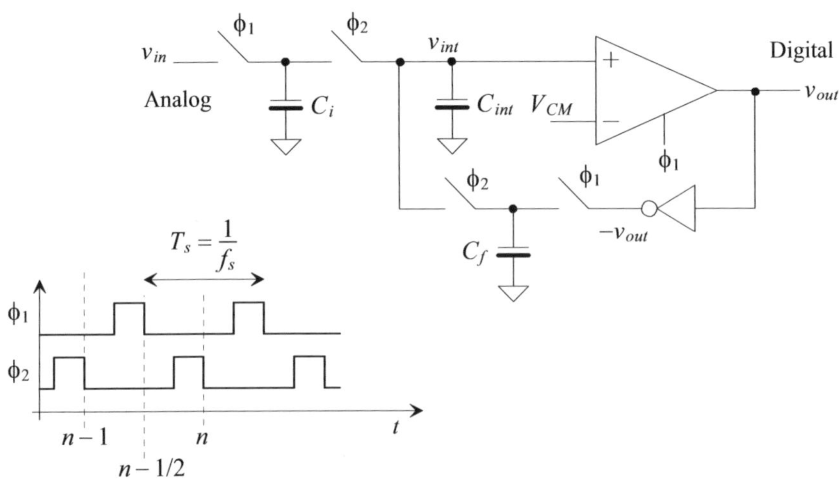

Figure 6.20 First-order passive modulator using switched-capacitors.

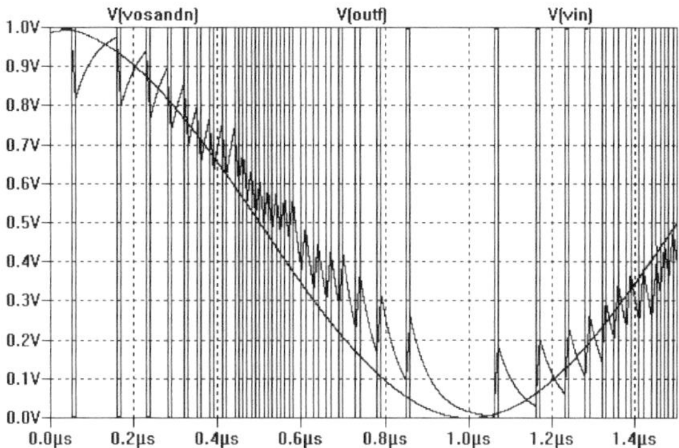

Figure 6.21 Regenerating Fig. 6.7 using a passive switched-capacitor modulator.

The operation of this topology can be described by simply substituting Eq. (6.43) into Eq. (6.31). The signal gain can then be written as

$$\frac{R_f}{R_i} = \frac{C_i}{C_f} \tag{6.44}$$

If C_i isn't exactly equal to C_f then the input signal undergoes a gain error. In order to eliminate this gain error, and to simplify the implementation of the modulator, consider sharing the feedback and input capacitors, Fig. 6.22. Simulations showing the operation of this circuit are found at CMOSedu.com. Reviewing the derivations in Sec. 2.2.3, we see that the input signal, in Fig. 6.22, is subtracted from the fed back signal and summed together using C_{int}. The inputs of the comparator are swapped, from Fig. 6.20, to ensure the polarity of the output signal matches the input signal's polarity. Notice that we connected the input and fed back signal's to the bottom plate of C_i. Remember that this is the plate with the largest parasitic capacitance. When the ϕ_1 switches are closed, the bottom plate of C_i, and its parasitic, are charged to v_{in}. Noise coupled into this bottom plate, from the substrate, sees the low-impedance input so that ideally none of the noise passes through to C_{int}. When the ϕ_2 switches close the bottom plate of C_i charges to v_{out}. Since the bottom-plate parasitic charges back and forth between v_{in} and v_{out}, none of the charge from this parasitic is transferred to C_{int}.

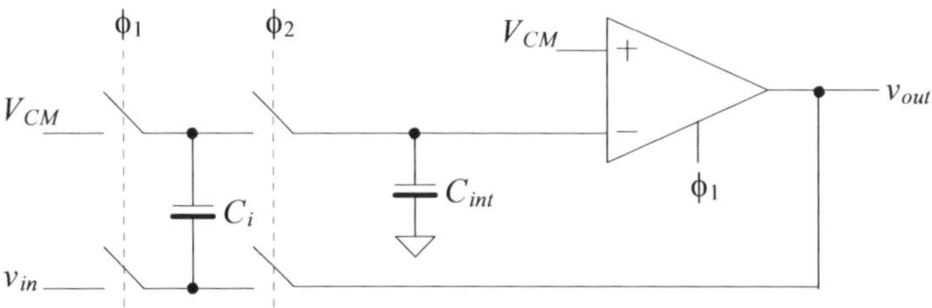

Figure 6.22 Switched-capacitor implementation of the passive modulator without gain error.

6.2.3 Increasing SNR using K-Paths

By increasing the sampling rate, f_s, we can spread the quantization noise out over a wider frequency range, Figs. 5.12 and 5.26. Thus, for a fixed digital averaging filter bandwidth (see Fig. 6.1), the amount of noise in a digital output signal can be reduced. As discussed earlier in Secs. 2.1.6 and 5.3 we can increase the effective sampling frequency by using K-paths (putting K modulators in parallel). Since we've used the same variable, K, for both the number of points averaged in the digital filter on the output of a modulator, Sec. 4.2, and the number of paths used, Sec. 2.1.6, let's, in this section, use variables with differentiating subscripts to describe the operation of a modulator implemented with K-paths and oversampling

$$K_{avg} = \text{number of samples averaged in the digital filter} \tag{6.45}$$

and

$$K_{path} = \text{number of paths used in the modulator} \tag{6.46}$$

The digital filter connected to the output of the modulator can then be described using

$$H(z) = \left[\frac{1 - z^{-K_{avg}}}{1 - z^{-1}}\right]^L \tag{6.47}$$

By using K-paths our sampling frequency changes from f_s to $K_{path} \cdot f_s$ so that the PSD of the quantization noise, Fig. 5.26, gets spread out over a wider frequency range

$$V_{Qe}^2(f) = \frac{V_{LSB}^2}{12 \cdot K_{path} \cdot f_s} \tag{6.48}$$

For the noise-shaping topologies we would replace K (for example, see Eq. [6.18]) with $K_{avg} \cdot K_{path}$ while the bandwidth of our desired signal spectrum ranges from DC to B *when decimation is used* where

$$B = \frac{f_s}{2K_{avg}} \tag{6.49}$$

The next question that we need to answer is "Is it practical to implement parallel paths of noise-shaping modulators or Nyquist-rate data converters?" Clearly there will be an increase in layout area using this approach. Is the increase in SNR worth the increase in chip size, power draw, or complexity? Well, **let's not spend any time** on Nyquist-rate converters (e.g., flash or pipeline) in a K-path configuration (called a *time-interleaved* converter) since the input signal would have to meet Bennett's criteria (pages 165-166, consider problems with a DC input). Reviewing Fig. 2.37 we see that using a SC modulator like the one seen in Fig. 6.22 may be a possibility. However, notice that we need to multiply up the clock frequency from f_s to Kf_s then pass the resulting signal through the non-overlapping clock generator seen in Fig. 2.38. Further, the SC modulators would have to settle (fully charge/discharge their capacitors) within T_s/K seconds (an even bigger concern if the modulator uses an active element like an op-amp). If this is possible why not simply use a single path clocked at the higher rate, Kf_s (or better yet two paths clocked at $Kf_s/2$ like the topology seen in Fig. 2.36)? We get the same performance as the K-paths but with a much simpler implementation. Since we have to direct, and process, the input signal to the various paths, again Fig. 2.37, at a high-rate the use of K-paths in SC circuits may be limited (more on this in a moment and in Ch. 9).

What about using the modulator seen in Fig. 6.4 in a K-path implementation? Figure 6.23 shows a 2-path, or *time-interleaved*, implementation. There are several benefits to this topology including: 1) the input can be continuously connected to the K-paths of modulators, 2) no active device (e.g. an op-amp) with settling time issues, and 3) the comparators are triggered on the rising edge of a clock signal (so the feedback operation can occur over the entire clock period, T_s for each path). The drawback of this topology is the mismatched gains through each path. Notice that we've combined the digital outputs of the modulator into a two-bit word (so the possible outputs are 00, 01, and 10). This addition, we should recognize, is a filter with a transfer function

$$H(z) = 1 + z^{-1} = \frac{1 - z^{-2}}{1 - z^{-1}} = \frac{1 - z^{-K_{path}}}{1 - z^{-1}} \text{ where } f_{s,new} = K_{path} \cdot f_s,\ z^{-1} = e^{-j2\pi \cdot \frac{f}{f_{s,new}}} \tag{6.50}$$

or an average of two filter. If we were to use 8-paths we would get a 4-bit output ranging from 0000 to 1000 (which may result in the same overflow problems we saw in Figs. 6.16 and 6.18 when the input of the topology is near V_{REF+}, or VDD, here). A simple solution to the overflow problem, other than using a selector to keep overflow from occurring as discussed earlier, is to use only 7-paths (instead of 8) so the final output code ranges from 000 to 111.

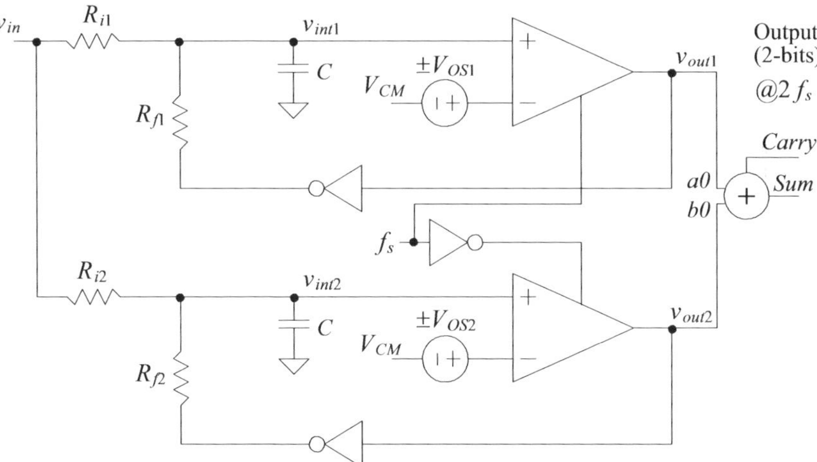

Figure 6.23 Two-path, time-interleaved, passive modulator with mismatch and offsets.

Note that each path, in Fig. 6.23, will attempt to hold the input of each comparator at a different voltage, dependent on the offset voltage of the comparator,

$$v_{int1} \rightarrow V_{CM} \pm V_{OS1} \text{ and } v_{int2} \rightarrow V_{CM} \pm V_{OS2} \tag{6.51}$$

so combining the input resistors and capacitors into a single path and the comparators and feedback paths into K-paths, to simplify the circuit and to reduce the effects of path mismatch, requires some thought. Consider the 4-path topology seen in Fig. 6.24. Here we've added an amplifier, with a gain of G, in series with the input path. Each comparator's offset is referred back to the integrating capacitor by dividing by this gain. If $G = 100$, for example, the differing offsets will have little effect on the circuit's operation.

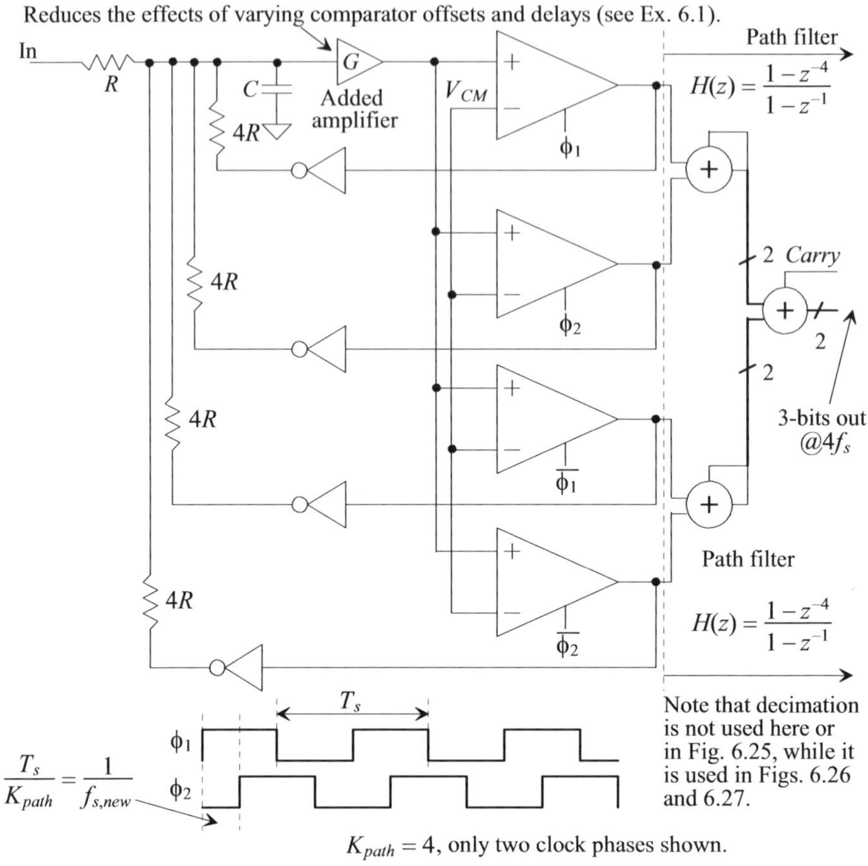

Figure 6.24 Four-path passive modulator. No longer time-interleaved since the integrator is common to all feedback paths (we'll call this the K-Delta-1-Sigma topology).

Another design concern, when using this topology, is the delay around each feedback path. If there is excessive phase shift through the forward or feedback paths, the modulator will be unstable. Using an open-loop (no feedback) amplifier, like a diff-amp, will ensure minimum delay (again noting the gain, and thus frequency response, of this amplifier aren't important as long as the gain is high enough to eliminate the offset effects of the comparators and the delay through the amplifier is comparable to T_s/K_{path}). Using an op-amp, in a feedback configuration, can ultimately limit the maximum speed of this topology. Figure 6.25 shows the output of the modulator in Fig. 6.24 with the same input signals and values used to generate Figs. 6.7 and 6.21. In Fig. 6.25 we've displayed the digital output word in analog form. Notice that the RC filtered output of this circuit is much smoother (contains less noise) than the outputs seen in Fig. 6.7 and 6.21. The extra smoothing from the inherent filtering when adding the outputs of the modulators together, (see filter transfer function in Fig. 6.24), helps to reduce the noise in the signal. Because of this inherent filtering it's easy to perform decimation directly on the output of the modulator. Figure 6.26 shows adding a register to re-time the output of the modulator (decimate) back down to f_s. Also seen in this figure is the digital filter seen in Fig. 6.1 that is used to remove the modulation noise. Regenerating the simulation results in Fig. 6.25

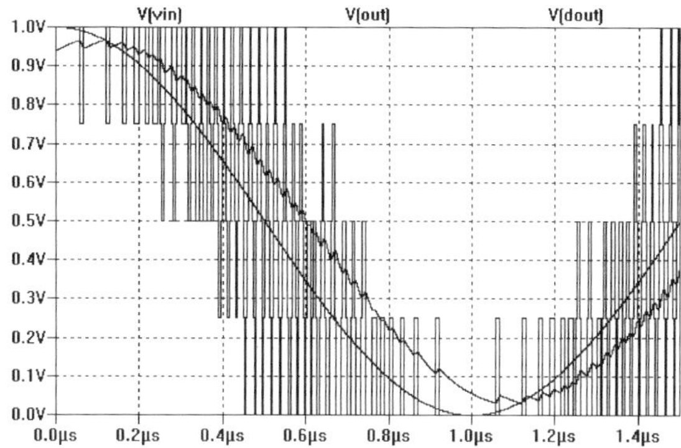

Figure 6.25 Simulating the operation of the 4-path modulator in Fig. 6.24.

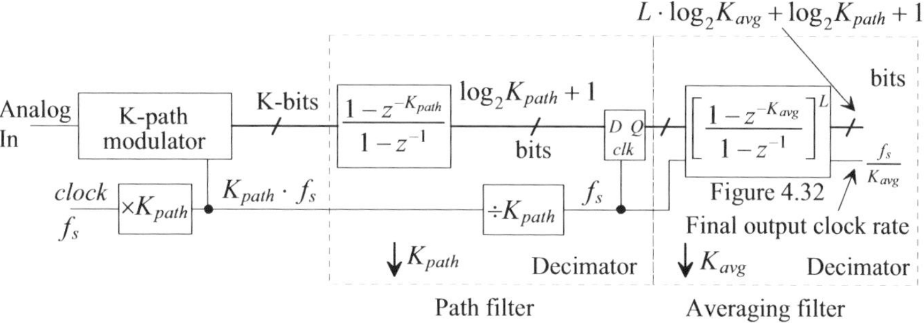

Figure 6.26 Performing decimation on the output of the modulator.

using decimation of 4 we get the results seen in Fig. 6.27. The rate of the digital word output by the path filter is f_s (just as it is when using a Nyquist-rate converter). Further, the path filter in Fig. 6.26 has the same frequency response as the S/H used in a Nyquist-rate ADC (a droop of –3.9 dB at $f_s/2$, Fig. 2.17). Thus, with some thought, we can use this topology to implement high-speed Nyquist-rate ADCs.

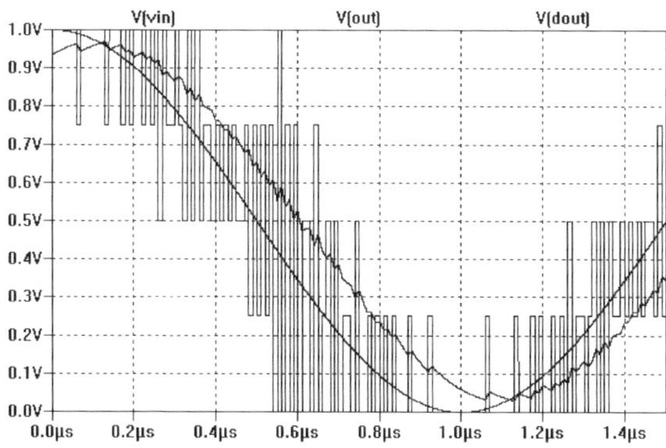

Figure 6.27 Regenerating the signals seen in Fig. 6.25 with a decimation of 4.

Revisiting Switched-Capacitor Implementations

At the beginning of the section, we dismissed the option of using switched-capacitor noise-shaping topologies in a K-path configuration, Fig. 2.37, because the capacitors would have to charge or discharge within T_s/K seconds (comparator delay $< T_s/K$ is a challenge). We said that if this were possible why not use two paths, clocked on opposite phases of a clock (Fig. 2.36), but at the higher clock frequency of $K \cdot f_s$? In the next chapter we'll show that a first-order noise-shaping delta-sigma modulator implemented using switched-capacitors has an input/output relationship given, see Eq. (7.2), by

$$v_{out}(z) = \overbrace{z^{-1}}^{STF(f)} \cdot v_{in}(z) + \overbrace{(1 - z^{-1})}^{NTF(f)} \cdot V_{Qe}(z) \tag{6.52}$$

The signal-transfer-function (*STF*) is simply one clock cycle delay, z^{-1} while the quantization noise is differentiated, Fig. 1.20. In other words the noise-transfer-function (*NTF*) is that of a differentiator, $(1 - z^{-1})$. Note the similarity to the differentiation, $j\omega RC$, the quantization noise sees at low frequencies, $1 >> j\omega RC$, in the continuous-time passive NS modulator, Eq. (6.12). If we put K of these switched-capacitor topologies described by the above equation in parallel, Fig. 2.37, we can write the corresponding output of the topology, see Eq. (2.56), as

$$v_{out}(z) = \overbrace{z^{-K}}^{STF(f)} \cdot v_{in}(z) + \overbrace{(1 - z^{-K})}^{NTF(f)} \cdot V_{Qe}(z) \tag{6.53}$$

Note that the data is changing, on the output of this topology, at a rate of $K \cdot f_s = f_{s,new}$, so that $z = e^{j2\pi \cdot f/f_{s,new}}$. While the quantization noise PSD, as indicated in Eq. (6.48), will be spread out over a wider frequency range the fact that our *NTF* has the shape of a comb filter, Fig. 1.25, instead of differentiation *limits the use of SC modulators in K-path ADCs using clock signals like those seen in Fig.* 2.37 *for high-speed conversion.*

One may wonder why the *STF* is z^{-K} or such a long delay? Looking at the timing of the clock signals in Fig. 2.37 we see that each path is only activated every T_s seconds. If $z = e^{j2\pi \cdot f/f_{s,new}}$ or $z = e^{j2\pi \cdot f/(Kf_s)}$ then to represent a single path delay of T_s seconds we have to use z^{-K}. In this scenario, Fig. 2.37 where $H(z)$ represents the transfer function of a modulator, each of the modulators is independent, that is, not sharing the integrating capacitor (integrator) as we did in the topology seen in Fig. 6.24. With some thought *we can implement a practical K-path ADC using SC circuits sharing a common integrator,* Fig. 6.24, *with clock signals, also seen in Fig.* 6.24, *having a width of* T_s *and leading edge spacing of* T_s/K *seconds so that the effective sampling frequency is* $K \cdot f_s$ (Ch. 9).

Effects of the Added Amplifier on Linearity

We added the amplifier with a gain, G, in Fig. 6.24 in order to reduce the effects of comparator offset and make combining K-paths of quantizers sharing a common integrating capacitor a practical analog-to-digital converter topology. We haven't analyzed how this added amplifier can affect the linearity and performance of the topology. Let's do that here. Rewriting Eqs. (6.8) to (6.12) with this gain included we get

$$V_{Qe}(f) + G \cdot v_{int}(f) = v_{out}(f) \tag{6.54}$$

$$\left(\frac{v_{in} - v_{int}}{R} + \frac{-v_{out} - v_{int}}{R} \right) \cdot \frac{G}{j\omega C} + V_{Qe}(f) = v_{out} \tag{6.55}$$

$$v_{in} - v_{out} - 2v_{int} + V_{Qe} \cdot j\omega R\frac{C}{G} = j\omega R\frac{C}{G} \cdot v_{out} \tag{6.56}$$

$$v_{in} - 2v_{int} + V_{Qe} \cdot j\omega R\frac{C}{G} = v_{out} \cdot (1 + j\omega R\frac{C}{G}) \tag{6.57}$$

and finally

$$v_{out} = \overbrace{\frac{1}{1 + j\omega R\frac{C}{G}}}^{STF(f)} \cdot v_{in} + \overbrace{\frac{j\omega R\frac{C}{G}}{1 + j\omega R\frac{C}{G}}}^{NTF(f)} \cdot V_{Qe} + \underbrace{\frac{-2 \cdot v_{int}}{1 + j\omega R\frac{C}{G}}}_{\text{Extra noise/distortion}} \tag{6.58}$$

It would appear that by using a very large gain, G, we can eliminate the lowpass filtering effect of the input signal (and thus pass very wideband signals to the modulator's output, useful for Nyquist-rate analog-to-digital conversion) and remove the modulation noise from the output of the modulator. The extra noise/distortion term, however, remains in the modulator's output. Again, the only way to get rid of the noise/distortion term is to hold v_{int} fixed so that it doesn't vary. This can be accomplished using relatively large f_s or an active element discussed next.

Before getting too excited about removing the modulation noise we'll find out, in the next chapter, that we can think of the comparator as also having a gain, G_c, in series with the added amplifier that varies keeping the forward gain of the modulator (a feedback system) equal to unity. In other words, if we increase the amplifier gain G we get an effective decrease in the comparator's gain G_c.

6.2.4 Improving Linearity Using an Active Circuit

Figure 6.28a shows the integrating topology we've used in our passive modulator. The input to the comparator, v_{int}, in this circuit can be written as

$$v_{int} = i_{in} \cdot \frac{1}{j\omega C} \tag{6.59}$$

In (b) we've replaced the passive integrator (the capacitor) with an active integrator. We can write

$$v_c = -G \cdot v_{int} \text{ and } i_{in} = \frac{v_{int} - v_c}{1/j\omega C} \tag{6.60}$$

If the amplifier's gain, G, is big we can write

$$v_c \approx -i_{in} \cdot \frac{1}{j\omega C} \tag{6.61}$$

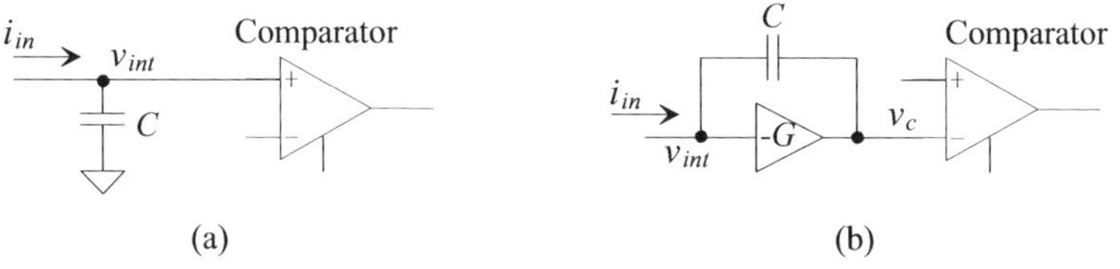

Figure 6.28 Replacing the passive integrator (a) with an active integrator (b).

where we've switched the inputs of the comparator to account for the inversion. Remember our goal is to keep v_{int} constant so that we reduce the noise/distortion term in, for example, Eq. (6.58). We can then write

$$v_{int} \approx \frac{i_{in}}{j\omega C \cdot G} \tag{6.62}$$

As $G \rightarrow \infty$ we get $v_{int} \rightarrow 0$. Figure 6.29 shows the implementation of a first-order continuous-time noise-shaping modulator using an active integrator. The same equations derived earlier, Eqs. (6.12), (6.19), and (6.20), apply to this configuration.

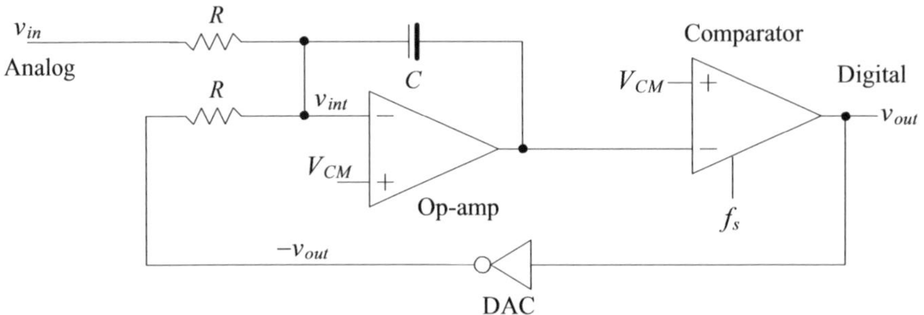

Figure 6.29 An active-integrator NS modulator.

Example 6.4
Repeat Ex. 6.1 using the modulator seen in Fig. 6.29.

The simulation results are seen in Fig. 6.30. In the simulation we can show that v_{int} doesn't vary (the point of adding the op-amp to the modulator). In order to show the better performance in linearity/distortion let's pass the 1-bit digital output through an average of 16 filter, see Ex. 6.2, and regenerate the results seen in Fig. 6.16 (see Fig. 6.31). Clearly, the active topology in Fig. 6.29 shows better linearity than the passive topology seen in Fig. 6.4. ■

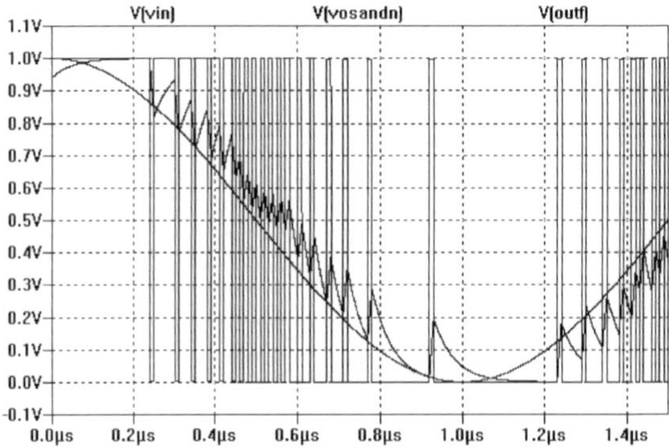

Figure 6.30 Repeating Ex. 6.1 for the first-order active NS modulator.

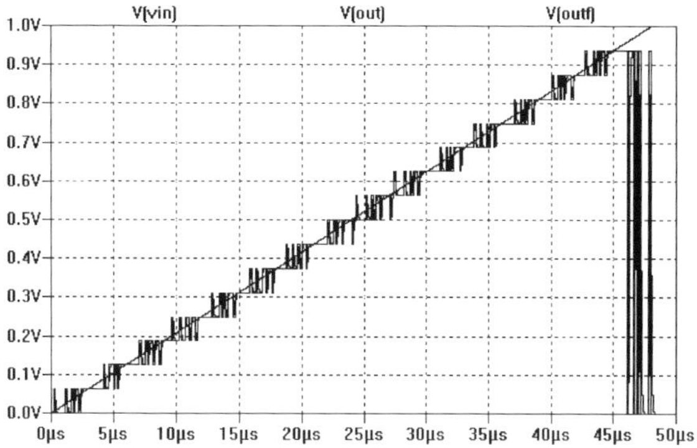

Figure 6.31 Linearity of the active NS modulator, compare to Fig. 6.16.

Second-Order Noise-Shaping

Figure 6.32 shows a second-order active noise-shaping modulator based on the passive topology seen in Fig. 6.17. Knowing that all voltages are referenced to V_{CM} (so that we don't include V_{CM} in the following derivations to keep the equations simpler) we can write

$$v_1 = -\left(\frac{v_{in} - v_{out}}{R}\right) \cdot \frac{1}{j\omega C} \tag{6.63}$$

$$v_2 = -\left(\frac{v_1 + v_{out}}{R}\right) \cdot \frac{1}{j\omega C} \tag{6.64}$$

Treating the comparator as adding noise to v_2, Fig. 6.4,

$$v_{out} = v_2 + V_{Qe}(f) \tag{6.65}$$

we can write

$$\left((v_{in} - v_{out}) \cdot \frac{1}{j\omega RC} - v_{out}\right) \cdot \frac{1}{j\omega RC} + V_{Qe} = v_{out} \tag{6.66}$$

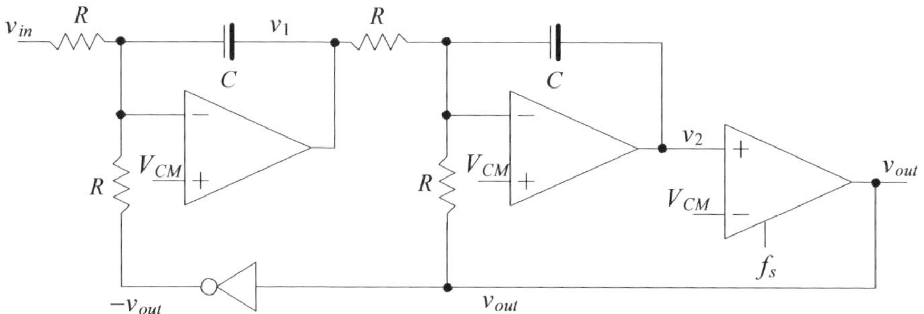

Figure 6.32 Second-order noise-shaping modulator, version I.

and finally

$$v_{out} = \overbrace{\frac{1}{(j\omega RC)^2 + j\omega RC + 1}}^{STF} \cdot v_{in} + \overbrace{\frac{(j\omega RC)^2}{(j\omega RC)^2 + j\omega RC + 1}}^{NTF} \cdot V_{Qe} \tag{6.67}$$

The *STF* has a second-order response, Sec. 3.2.2. Poor selection of the resistor and capacitor will cause instability. Note that for low frequencies the *NTF* increases as the square of frequency so we expect better *SNR*, for a given bandwidth *B* (see Eq. [6.15]), using this topology. Let's try to make this topology more robust by using only a single feedback path, Fig. 6.33. The transfer relationship for this modified topology is then

$$v_{out} = \overbrace{\frac{1}{(j\omega RC)^2 + 1}}^{STF} \cdot v_{in} + \overbrace{\frac{(j\omega RC)^2}{(j\omega RC)^2 + 1}}^{NTF} \cdot V_{Qe} \tag{6.68}$$

Reviewing Fig. 3.35, the poles fall on the y-axis and so the topology is unstable.

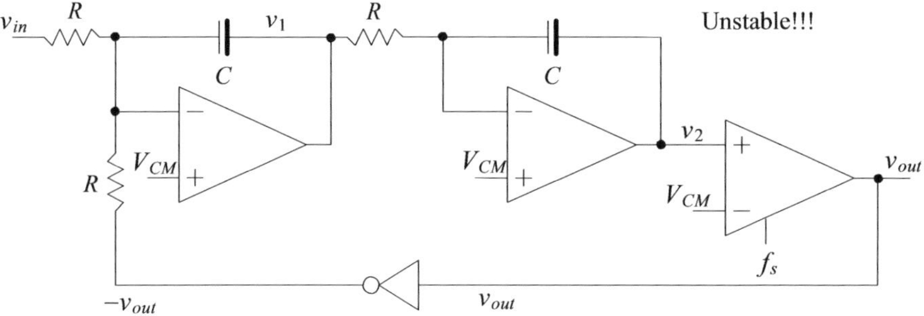

Figure 6.33 Second-order noise-shaping modulator with a single feedback path (bad).

Consider the single-feedback path modulator seen in Fig. 6.34. We've added a resistor in series with the feedback capacitor in the first integrator. We can now write

$$v_1 = -\left(\frac{v_{in} - v_{out}}{R}\right) \cdot \left(\frac{1}{j\omega C} + R\right) \tag{6.69}$$

$$v_2 = -\left(\frac{v_1}{R}\right) \cdot \frac{1}{j\omega C} = (v_{in} - v_{out}) \cdot \frac{1}{j\omega RC} \cdot \left(\frac{1}{j\omega RC} + 1\right) \tag{6.70}$$

$$v_{out} = \overbrace{\frac{j\omega RC + 1}{(j\omega RC)^2 + j\omega RC + 1}}^{STF} \cdot v_{in} + \overbrace{\frac{(j\omega RC)^2}{(j\omega RC)^2 + j\omega RC + 1}}^{NTF} \cdot V_{Qe} \tag{6.71}$$

The benefits of this topology, other than the better *SNR* and only a single feedback loop, are that we can select *RC* so that the *STF* is closer to one over a wider bandwidth and there is smaller delay between the input and output of the modulator (which is of critical importance in a *K*-path topology). Setting, for example,

$$RC = 2T_s = 2/f_s \tag{6.72}$$

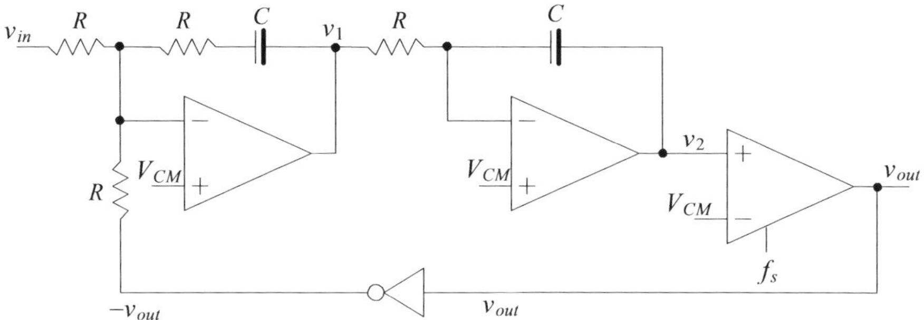

Figure 6.34 Second-order noise-shaping modulator, version II.

this equation reduces to, for frequencies of interest much less than the Nyquist frequency of $f_s/2$,

$$v_{out} \approx v_{in} + (j\omega RC)^2 \cdot V_{Qe} \tag{6.73}$$

Signal-to-Noise Ratio

Following the procedure for calculating *SNR* seen in Sec. 6.1.1

$$V^2_{noise,RMS} = 2\int_0^B |NTF(f)|^2 |V_{Qe}(f)|^2 \cdot df = 2 \cdot \frac{V^2_{LSB}}{12 f_s} \cdot \int_0^B (2\pi f \cdot RC)^4 \cdot df \tag{6.74}$$

$$V^2_{noise,RMS} = 2 \cdot \frac{V^2_{LSB}}{12 f_s} \cdot (2\pi RC)^4 \cdot \frac{B^5}{5} \tag{6.75}$$

Again with $B = f_s/2K$ we get

$$V^2_{noise,RMS} = \frac{V^2_{LSB}}{12} \cdot (2\pi RC)^4 \cdot \frac{f_s^4}{80 \cdot K^5} \tag{6.76}$$

$$SNR_{ideal} = 20 \cdot \log \frac{V_P/\sqrt{2}}{V_{noise,RMS}} = 6.02N + 1.76 - 20\log\frac{(2\pi RC \cdot f_s)^2}{\sqrt{80}} + 20\log K^{5/2} \tag{6.77}$$

Setting $RC = 2.69/f_s$ (noting that we might have to increase the *RC* to avoid saturating the integrators, discussed in more detail in the next chapter)

$$SNR_{ideal} = 6.02N + 1.76 - 30.10 + 50\log K \tag{6.78}$$

$$N_{inc} = \frac{50\log K - 30.10}{6.02} \tag{6.79}$$

$$SNR_{ideal} = 6.02(N + N_{inc}) + 1.76 \tag{6.80}$$

For every doubling in K above $K = 4$ we get 2.5 bits increase in resolution or 15 dB increase in SNR_{ideal}. To estimate the order, L, of the decimating filter let's write, see Eq. (6.21),

$$L \cdot \log_2 K \geq \frac{50\log K - 30.10}{6.02} \tag{6.81}$$

For $16 \leq K \leq 1024$ we can use a second-order filter, $L = 2$.

Example 6.5

Simulate the operation of the second-order NS modulator in Fig. 6.34 clocked at 100 MHz, with $RC = 20$ ns, and decimated with a filter having a transfer function

$$\left[\frac{1-z^{-16}}{1-z^{-1}}\right]^2$$

Estimate the bandwidth, B, of the output signal, the increase in the number of bits, N_{inc}, and SNR_{ideal}.

The simulation results are seen in Fig. 6.35. The final output clock frequency is 6.25 MHz and the bandwidth of the desired signal, B, is 3.125 MHz. Using Eq. (6.79) we estimate the increase in the number of bits as 5. Using Eq. (6.80) the ideal SNR is 37.88 dB. Note that the number of bits coming out of the filter is 9 bits, the input 1-bit word size increases 8-bits. We can throw the lower 5-bits out or throw the MSB out to divide by 2 and throw the lower 4-bits out (just not continue to pass them along in the system since they are simply noise). ■

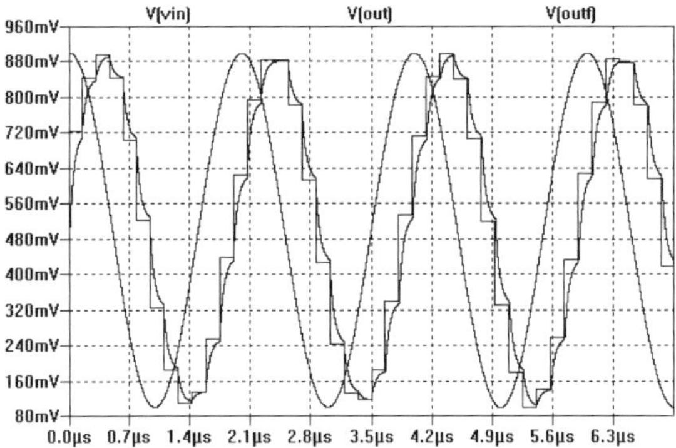

Figure 6.35 Simulation results for Ex. 6.5.

Discussion

The second-order active NS modulator is the workhorse for analog-to-digital converters using noise-shaping. The increase in resolution of 2.5 bits for every doubling in K is significant. This topology can also be used in a K-path configuration, Fig. 6.24, replacing the capacitor and amplifier (see simulation examples at CMOSedu.com). The challenge is keeping the forward delay through the modulator small (so the modulator remains stable). This dictates using simple op-amp topologies that are very fast with moderate gains, e.g., self-biased diff-amps. Finally, for high-speed we focused on using K-path topologies. The drawback of this approach is the need for several clock phases whose rising edges are spaced by T_s/K. A delay-locked loop (DLL) can be used for generating these clocks and is straightforward to design. The benefits of the K-path approach are that the matching of the resistors undergoes averaging via the K feedback paths, the pulse rising/falling edges are made less important since relatively wide pulses are used in each feedback path, and there is an inherent filtering when we combine the outputs of the paths together. Clearly further research in the design of these topologies is warranted; however, we leave this work to the refereed literature and continue to focus on the fundamentals.

ADDITIONAL READING

[1] J. C. Candy and G. C. Temes (eds.), *Oversampling Delta-Sigma Data Converters*, IEEE Press, 1992. ISBN 0-87942-285-8

[2] S. K. Tewksbury and R. W. Hallock, *Oversampled, Linear Predictive and Noise-Shaping Coders of Order N>1*, IEEE Trans. Circuits and Sys., Vol. CAS-25, pp. 436-447, July 1978.

[3] H. Inose, Y. Yasuda, *A Unity Bit Coding Method by Negative Feedback*, Proc. IEEE, Vol. 51, pp. 1524-1535, November 1963

QUESTIONS

6.1 Suggest a topology for a passive-integrator NS modulator where the input and fed back signals are currents. Derive a transfer function for your design. Does your topology have the extra noise/distortion term seen in Eq. (6.12)? Why or why not? Simulate the operation of your design.

6.2 Simulate the operation of the NS modulator seen in Fig. 6.4a but using a 4-bit quantizer (ADC). Use a 100 MHz clock frequency and an input sinewave at 500 kHz (as used to generate Fig. 6.7).

6.3 Using Eqs. (6.14) and (6.17) compare the noise performance of passive NS modulators using a 1-bit quantizer to those using a 4-bit quantizer.

6.4 Repeat Ex. 6.2 if C is changed to 1 pF and a 1 GHz clock frequency is used. Estimate the frequency where the output of the digital filter is –3 dB (0.707) from the input signal. Verify your answer with simulations.

6.5 Suppose the comparator used in the NS modulator and filter used in Ex. 6.2 has a 50 mV input-referred offset voltage. How will this offset voltage affect the conversion from analog to digital? Verify your answer with SPICE simulations.

6.6 Figure 6.36 shows the implementation of an op-amp using mixed-signal design techniques. Assuming the comparator is powered with a 1 V supply, simulate the circuit in the inverting op-amp configuration with the non-inverting input held at 0.5 V, a 10k resistor connected from the inverting input to the input source, and a feedback resistor of 100k from the op-amp's output back to the inverting input (for a closed loop gain of –10). Set the input source to have a DC offset of 500 mV, and a peak-to-peak amplitude of 20 mV at 500 kHz. Explain how the circuit operates. Note that using an active integrator, instead of the passive integrator results in more ideal behavior (less variation on the op-amp's inputs).

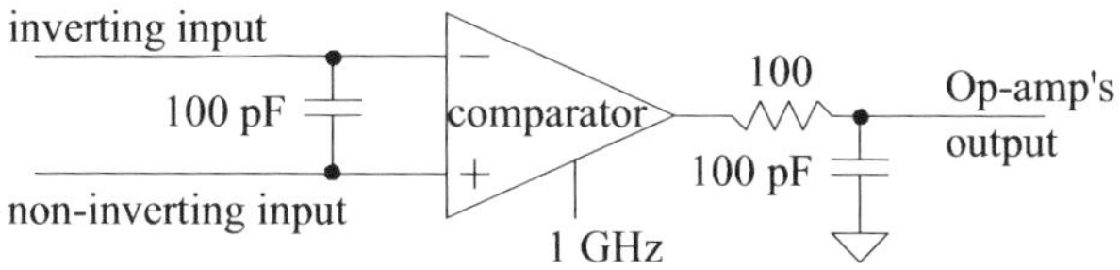

Figure 6.36 An op-amp implemented using mixed-signal techniques.

6.7 In your own words explain why dead zones in the second-order passive modulator seen in Fig. 6.17 are less of a problem than the first-order modulator seen in Fig. 6.4a.

6.8 Verify, using simulations, that the modulator seen in Fig. 6.20 suffers from capacitor mismatch while the one in Fig. 6.22 does not.

6.9 What is a time-interleaved data converter? Why is a time-interleaved converter different from the converter seen in Fig. 6.24?

6.10 Show the details of how to derive the transfer function of the path filter seen in Fig. 6.24.

6.11 Repeat question 6.7 if an active integrator, Fig. 6.28, is used in place of the passive integrator.

6.12 Repeat Ex. 6.5 if K is changed from 16 to 8.

Chapter

7

Noise-Shaping Data Converters

In this chapter we continue our discussion of first-order noise-shaping data converters. We start out discussing first-order topologies and then move on to higher-order topologies including second-order noise-shaping and cascaded modulators.

7.1 First-Order Noise Shaping

The block diagram of a NS feedback modulator is shown in Fig. 7.1. At the end of Ch. 5 we showed, but in the s domain, that the output of the modulator, $v_{out}(z)$, can be related to the input, $v_{in}(z)$, and the ADC's quantization noise, $V_{Qe}(z)$, by

$$v_{out}(z) = \overbrace{\frac{A(z)}{1+A(z)}}^{STF(z)} \cdot v_{in}(z) + \overbrace{\frac{1}{1+A(z)}}^{NTF(z)} \cdot V_{Qe}(z) \tag{7.1}$$

where, as in the last chapter, $STF(z)$ is the signal's transfer function and $NTF(z)$ is the noise's transfer function.

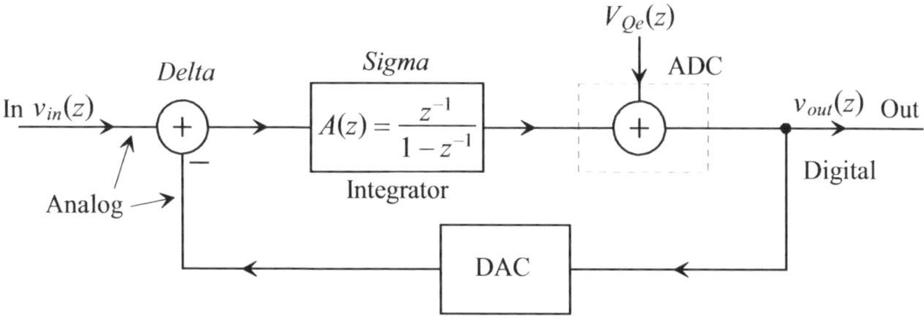

Figure 7.1 Block diagram of a noise-shaping (NS) modulator.

Consider what happens if $A(z)$ is an integrator (implemented using a DAI, Sec. 2.2.3) as shown in the figure. Equation 7.1 becomes

$$v_{out}(z) = z^{-1} v_{in}(z) + (1 - z^{-1}) \cdot V_{Qe}(z) \tag{7.2}$$

This equation is important! It shows that the input signal simply passes through the modulator with a delay while the quantization noise is differentiated (see Fig. 1.20 for the magnitude response of a digital differentiator with a transfer function $1-z^{-1}$). We can think of the noise differentiation as pushing the quantization noise to higher frequencies. We'll come back to how NS affects the quantization noise spectral density, $V_{Qe}(f)$, in a moment. But first let's attempt to understand what's happening here.

In Fig. 7.1 the summer takes the difference (*Delta*) between the input signal and the fed back signal. The integrator accumulates or sums (*Sigma*) this difference and feeds the result back, via the ADC and DAC, to the summer. This forces the output of the modulator to track the average of the input. Sometimes the fed back signal will have a value greater than the input signal, while at other times the fed back signal will be less than the input signal. The *average* signal fed back, however, should ideally be the same as the input signal. Note that this type of NS modulator is often called a *Delta-Sigma* or *Sigma-Delta* modulator. Also, at this point, we should see the need for the averaging filters discussed earlier.

A circuit implementation of a first-order NS modulator is shown in Fig. 7.2. For the moment we use a single-bit ADC and DAC (both implemented using the clocked comparator) for gain linearity reasons (discussed in more detail later). The analog voltage coming out of the integrator is compared to the common-mode voltage (this is our 1-bit ADC) using the comparator. For the 1-bit DAC a logic-0 has an analog voltage of 0 *V*, while a logic-1 has an analog voltage of *VDD* (= 1 *V* here) so that the comparator's output can be used directly (fed back to the DAI).

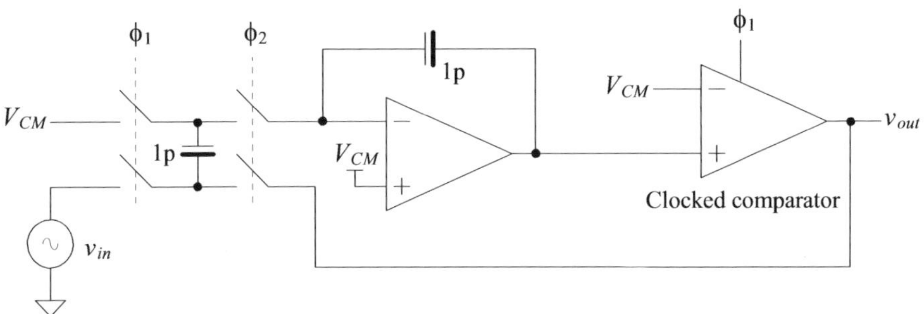

Figure 7.2 Circuit implementation of a first-order NS modulator.

The comparator is clocked on the rising edge of ϕ_1 resulting in a T_s delay (z^{-1}) in series with the fed back signal and a delay of $T_s/2$ ($z^{-1/2}$) in series with the input signal. To understand this statement, remember that the nonoverlapping clock dead time (the time both ϕ_1 and ϕ_2 are low) is short and, practically, the falling edge of ϕ_2 occurs at the same instance as the rising edge of ϕ_1 (and so we could also use $\overline{\phi_2}$ to clock the comparator). This results in a transfer function, see Table 2.2, to the input of the comparator (which can also be thought of as the ADC output since the fed back signal and modulator output are the same signal) of

$$\text{Desired ADC input/output} = \frac{z^{-1}}{1-z^{-1}}(v_{in} - v_{out}) \tag{7.3}$$

After careful review we should see that the circuit implementation of Fig. 7.2 corresponds to the NS modulator represented by the block diagram shown in Fig. 7.1.

We can use the SPICE models developed earlier to demonstrate the operation of the NS modulator of Fig. 7.2. Assuming our sampling frequency is 100 MHz, the input is a 500 kHz sinewave centered around V_{CM} (= 0.5 V) with a peak amplitude of 0.5 V. The input and output of the modulator are shown in Fig. 7.3.

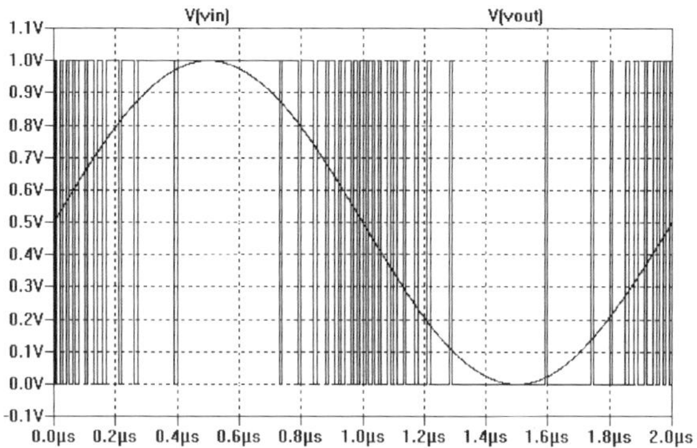

Figure 7.3 Simulating the operation of the modulator in Fig. 7.2.

A Digital First-Order NS Demodulator

So far our noise-shaping discussion has centered around analog-to-digital conversion. Figure 7.4 shows a first-order NS demodulator-based topology for digital-to-analog conversion. While the discussion concerning selection of the digital interpolating and reconstruction filters is given in Ch. 4, we are interested here in the topology of the first-order NS demodulator for use in a DAC.

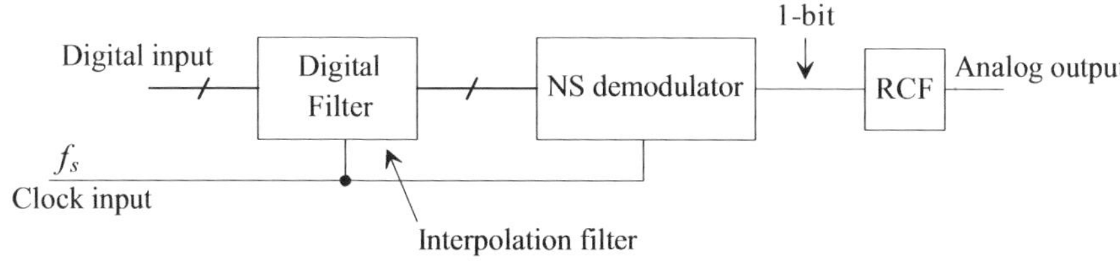

Figure 7.4 DAC using a NS modulator and digital filter.

Figure 7.5a shows a block diagram of a first-order NS demodulator for use in a DAC. Figure 7.5b shows the practical implementation. The only differences between this circuit and the circuit of Fig. 7.1 is that the DAI is replaced with an all-digital integrator (see Fig. 1.28), and the quantizer (comparator) is replaced with a circuit that performs quantization by selecting (using the MSB of the accumulator output word) digital V_{REF+} (= 011111 in two's complement) or V_{REF-} (= 100000 in two's complement) which, for our current discussion, are *VDD* and ground. Note that this topology isn't the preferred way to implement a DAC using noise-shaping. Using the error-feedback topology, discussed later, is the preferred method.

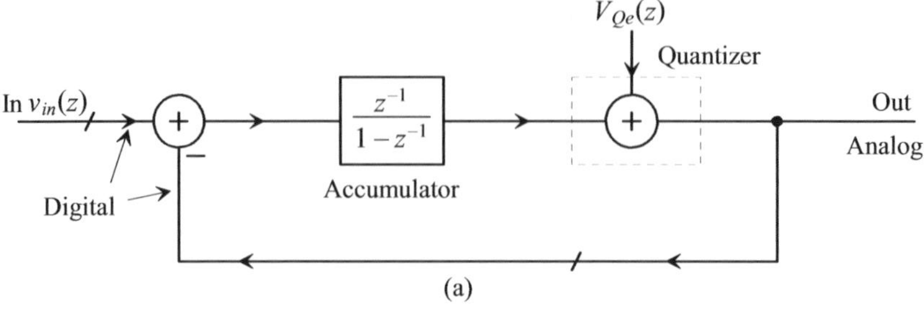

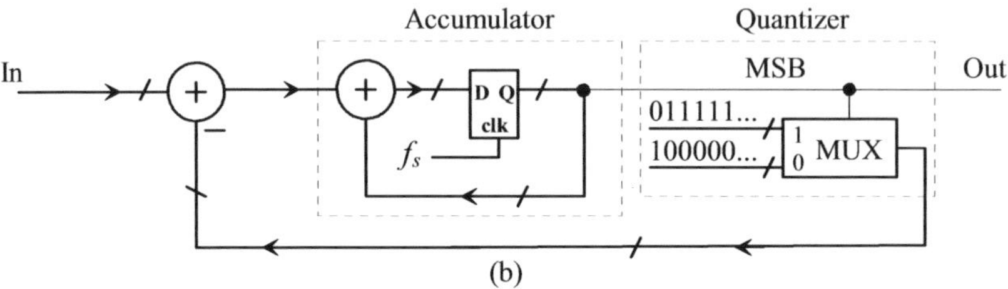

Figure 7.5 Block diagram of (a) a NS demodulator and, (b) a more detailed implementation for use in a DAC.

7.1.1 Modulation Noise in First-Order NS Modulators

Here we present a more detailed discussion of the quantization noise spectrum for a first-order NS modulator. To begin, let's write Eq. (7.1) in the time domain,

$$v_{out}[nT_s] = v_{in}[(n-1)T_s] + v_{Qe}[nT_s] - v_{Qe}[(n-1)T_s] \tag{7.4}$$

which shows the output is a function of the first difference (order) of the quantization noise $v_{Qe}[nT_s] - v_{Qe}[(n-1)T_s]$. Intuitively note that the smaller we make T_s (the faster we sample since $T_s = 1/f_s$), the closer our digital output $v_{out}[nT_s]$ approaches the analog input $v_{in}[nT_s]$.

Next, using Eq. (7.1), let's write the product of the noise transfer function and $v_{Qe}[z]$ (the modulation noise) of the first-order NS modulator in the frequency domain as

$$NTF(z)V_{Qe}(z) = (1-z^{-1})V_{Qe}(z) \to NTF(f)V_{Qe}(f) = \left(1 - e^{-j2\pi\frac{f}{f_s}}\right) \cdot \frac{V_{LSB}}{\sqrt{12f_s}} \tag{7.5}$$

where we have used, see Fig. 5.12,

$$V_{Qe}(f) = \frac{V_{LSB}}{\sqrt{12f_s}} \left(\text{units, V}/\sqrt{\text{Hz}}\right) \text{ for } 0 \le f \le f_s/2 \tag{7.6}$$

and, once again (see Eq. [4.1] or, for the 1-bit ADC [comparator], Eq. [6.1]),

$$V_{LSB} = \frac{V_{REF+} - V_{REF-}}{2^N} \tag{7.7}$$

where N is the number of bits used in the low-resolution ADC/DAC in the modulator. Using a single-bit ADC/DAC in a NS modulator, $N = 1$, results in this book in $V_{LSB} = 1$ V. This again shows that we are not reducing the quantization noise, but are pushing it to higher frequencies so that it can be filtered out. Using Eq. (1.46), we can write the PSD of the NTF as

$$|NTF(f)|^2 \cdot |V_{Qe}(f)|^2 = \frac{V_{LSB}^2}{12f_s} \cdot 4\sin^2\pi\frac{f}{f_s} \quad \left(\text{units, V}^2/\text{Hz}\right) \tag{7.8}$$

Figure 7.6 shows the PSD of the first-order NS modulation noise for $V_{LSB} = 1$ V and $f_s = 100$ MHz. Note that we are discussing modulation noise instead of quantization noise. *The modulation noise is the quantization noise after being differentiated by the NS modulator.* The modulation noise is the unwanted signal added to the input signal. After reviewing Fig. 7.6 we see that the magnitude of the modulation noise is significant. However, after passing this signal through a lowpass filter, we can remove the higher frequency noise resulting in a lower value of data converter RMS quantization noise, $V_{Qe,RMS}$. By restricting the bandwidth of the modulation noise we can drive the RMS quantization noise in our signal to zero. Of course, by lowering the bandwidth of the digital filter on the output of the modulator we also limit the possible bandwidth, B, of the input signal. Notice that we have violated Bennett's criteria, Sec. 5.1.1, by utilizing a quantizer with an LSB that is comparable to the input signal. Now, however, we are using feedback that adds or subtracts a signal from the input and ultimately affects the quantizer input.

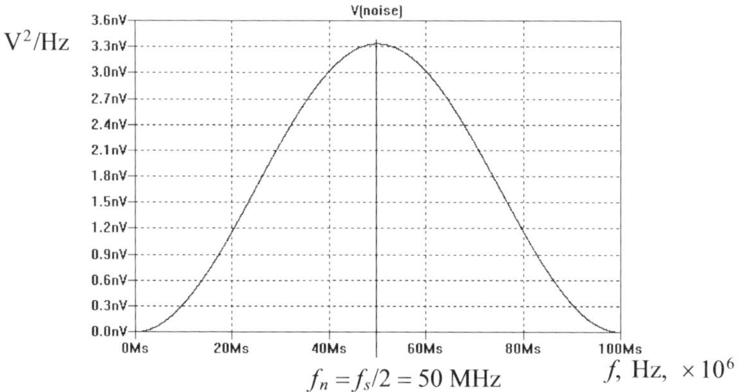

Figure 7.6 Modulation noise for a first-order NS modulator.

7.1.2 RMS Quantization Noise in a First-Order Modulator

The RMS quantization noise present in a bandwidth, B, can be calculated, see Eq. (6.13), using

$$V_{Qe,RMS}^2 = 2\int_0^B |NTF(f)|^2 |V_{Qe}(f)|^2 \cdot df = 2 \cdot \frac{V_{LSB}^2}{12f_s} \cdot 4 \cdot \int_0^B \sin^2\pi\frac{f}{f_s} \cdot df \tag{7.9}$$

Remembering that the maximum bandwidth of our input signal is related to the sampling frequency, f_s, and the oversampling ratio, K, by

$$B = \frac{f_s}{2K} \tag{7.10}$$

and, for small values of x,

$$\sin x \approx x \tag{7.11}$$

then

$$V_{Qe,RMS} \approx \frac{V_{LSB}}{\sqrt{12}} \cdot \frac{\pi}{\sqrt{3}} \cdot \frac{1}{K^{3/2}} \tag{7.12}$$

This equation should be compared to Eq. (5.56). Further, we can describe the ideal data converter SNR using first-order NS, see Eqs. (5.10) - (5.13) as

$$SNR_{ideal} = 20 \cdot \log \frac{V_p/\sqrt{2}}{V_{Qe,RMS}} = 6.02N + 1.76 - 20\log\frac{\pi}{\sqrt{3}} + 20\log K^{3/2} \text{ (in dB)} \tag{7.13}$$

or

$$SNR_{ideal} = 6.02N + 1.76 - 5.17 + 30\log K \quad \text{(in dB)} \tag{7.14}$$

This equation should be compared to Eq. (5.58) where we saw every doubling in the oversampling ratio, K, results in a 0.5-bit increase in resolution (called simple oversampling). Here we see that *every doubling in the oversampling ratio results in* 1.5 *bits increase in the resolution* (or a 9 dB increase in SNR_{ideal}). A first-order NS modulator's performance is compared to simple oversampling in Fig. 7.7 (see also Sec. 6.1.1 and Fig. 6.9).

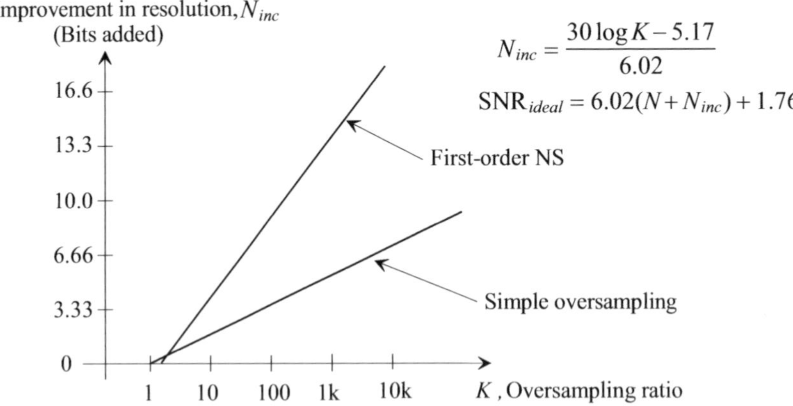

Figure 7.7 Comparing simple oversampling to first-order noise-shaping.

Example 7.1

Determine the ideal signal-to-noise ratio and the maximum signal bandwidth allowed, B, for the first-order NS modulator of Fig. 7.1 if 16 of its output samples are averaged ($K = 16$).

Because the sampling frequency, f_s, is 100 MHz, we can use Eq. (7.10) to determine the maximum input signal bandwidth, B, is 3.125 MHz. Using Eq.

(7.14) we can solve for the SNR_{ideal} as (knowing that the NS modulator of Fig. 7.1 uses a 1-bit quantizer) 38.73 dB. This corresponds to an equivalent data converter (ADC) resolution, using Eq. (5.13), of 6.14 bits (number of bits added is 5.14). Note that the ADC is made with the components, modulator, and digital filter, shown in Fig. 6.1. ■

7.1.3 Decimating and Filtering the Output of a NS Modulator

It's important to note that Eq. (7.14) was derived assuming the output of the modulator was passed through a perfect lowpass filter with a bandwidth of B. Passing the output through a Sinc averaging filter, see Figs. 4.10 and 4.13, will result in a poorer SNR because the higher frequency noise components will not be entirely filtered out. In this section we want to answer two questions: (1) what order, L (see Eq. [4.19]), of Sinc averaging filter should be used in the digital filter on the output of the NS modulator, and (2) assuming we use only this filter (no additional filtering), how will the ideal SNR of the first-order NS modulator be affected.

We begin to answer to the first question by writing the increase in the number of bits, N_{inc}, as

$$N_{inc} = \frac{30\log K - 5.17}{6.02} \tag{7.15}$$

If our NS modulator uses a 1-bit ADC, then the final, after the digital filter, resolution of the resulting data converter is N_{inc} + 1 bits. (An NS modulator using a 5-bit ADC [often called a multibit NS modulator] would ideally have an output resolution of N_{inc} + 5 bits.) Further, we saw in Sec. 4.2.5 that the word size increased by $\log_2 K$ bits in each stage so that we can require

$$L \cdot \log_2 K \geq \frac{30\log K - 5.17}{6.02} \tag{7.16}$$

Solving this equation results in L being greater than or equal to 2. In general, we can write

$$L = 1 + M \tag{7.17}$$

where M is the order of the modulator. For a first-order modulator we use two stages in the averaging filter, or,

$$H(z) = \left[\frac{1}{K}\frac{1 - z^{-K}}{1 - z^{-1}}\right]^2 \tag{7.18}$$

In the next section we discuss second-order NS modulators ($M = 2$). For these modulators we use a Sinc averaging filter with $L = 3$.

Example 7.2

Comment on the implementation of the digital decimation filter for the modulator described in Ex. 7.1. Assume the final output clocking frequency is 100 MHz/16 or 6.25 MHz.

The transfer function of the digital filter is

$$H(z) = \left[\frac{1 - z^{-16}}{1 - z^{-1}}\right]^2$$

The block diagram of the filter is seen in Fig. 4.32 using 4 stages and $L = 2$. The increase in resolution through each $(1+z^{-1})^2$ stage is 2 bits (see Fig. 6.10 for the filter when $L = 1$). The resolution calculated in Ex. 7.1 was 6.14 bits, which we round up to 7-bits. Because the output of the digital filter is 9-bits (8-bits through the four stages plus the 1-bit coming out of the modulator), we drop the lower two bits (divide by 4) to get our final 7-bit resolution. ■

Next let's examine how filtering with a Sinc filter affects the SNR of the data converter. Remember the SNR_{ideal} was calculated in Eq. (7.14) assuming the modulation noise was strictly bandlimited to B. Figure 7.8 shows the PSD of the $NTF^2(f) \cdot |V_{Qe}(f)|^2$ (the modulation noise) of the first order NS modulator. Also shown in this figure is the shape of the averaging filter's magnitude response squared, Eq. (4.20). Here we are showing the shape of a filter with $L = 2$ (set by Eq. [7.17] for a first-order modulator) and $K = 16$. We limit our range to $f_s/2$.

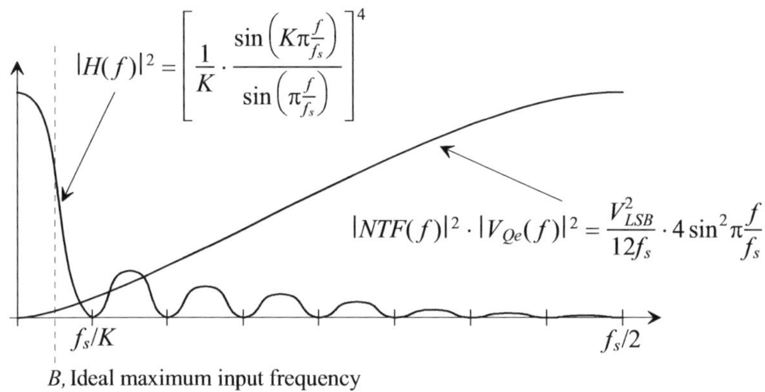

Figure 7.8 Showing modulation noise and filter response.

We can calculate the RMS quantization noise on the output of the Sinc filter in a cascade of a first-order modulator and an averaging filter using (see Sec. 6.1.2)

$$V_{Qe,RMS}^2 = 2 \int_0^{f_s/2} |NTF(f)|^2 \cdot |V_{Qe}(f)|^2 \cdot |H(f)|^2 \cdot df \tag{7.19}$$

$$V_{Qe,RMS}^2 = \frac{V_{LSB}^2}{12 f_s} \cdot \frac{8}{K^4} \cdot \int_0^{f_s/2} \frac{\sin^4\left(K\pi \frac{f}{f_s}\right)}{\sin^2\left(\pi \frac{f}{f_s}\right)} \cdot df \tag{7.20}$$

If we let $\theta = \pi \frac{f}{f_s}$, then this equation can be written as

$$V_{Qe,RMS}^2 = \frac{V_{LSB}^2}{12 f_s} \cdot \frac{8}{K^4} \cdot \frac{f_s}{\pi} \cdot \overbrace{\int_0^{\frac{\pi}{2}} \frac{\sin^4(K\theta)}{\sin^2\theta} d\theta}^{= K \cdot \frac{\pi}{4}} \tag{7.21}$$

and finally,

$$V_{Qe,RMS} = \frac{V_{LSB}}{\sqrt{12}} \cdot \frac{\sqrt{2}}{K^{3/2}} \tag{7.22}$$

This equation should be compared to Eq. (7.12), which was derived assuming the digital filter was ideal with a bandwidth of B. The SNR resulting from using a first-order ($M = 1$) NS modulator and a second-order ($L = 2$) Sinc averaging filter is

$$SNR_{Sinc} = 6.02N + 1.76 - 3.01 + 30 \log K \text{ (in dB)} \tag{7.23}$$

Comparing this to SNR_{ideal} given in Eq. (7.14), we see that using a Sinc filter for averaging results in only a 2.16 dB difference (increase) in SNR over the ideal filter. If we remember that using a Sinc filter results in a droop in the desired signal, Fig. 2.31, the SNR will be lower than what is predicted by Eq. (7.23). (Note that an analysis of higher order modulators using Sinc averaging filters would show that as long as Eq. [7.17] is valid the deviation from SNR_{ideal} is negligible.)

7.1.4 Pattern Noise from DC Inputs (Limit Cycle Oscillations)

In the ideal Nyquist-rate ADC, a DC input signal results in a single output code. In other words the output code doesn't vary. In a NS-based ADC the output code varies as seen in Fig. 7.3 (the average of the output is equal to the input signal). When this code is applied to a digital filter the output of the filter shows a ripple or variation. Unfortunately, this ripple on the output of the filter can cause noise in the spectrum of interest. The frequency of the ripple and the amplitude of the ripple depend on the DC input value. Figure 7.9 shows an example of this ripple. The input signal for the modulator in Fig. 7.2 is 0.5 V (the common-mode voltage) while the modulator is clocked at 100 MHz. The modulator's output is a square wave of alternating ones and zeroes. The first harmonic of this signal is at half the clocking frequency or, in this example, 50 MHz. Since the ripple frequency lies outside our base spectrum (which, from Ex. 7.2, is from DC to 3.125 MHz) it will not, in a significant way, affect the SNR. (The digital filter will likely have a zero in its transfer function that falls at half the clocking frequency eliminating the ripple altogether and resulting in a constant filter output value.) If we increase the input signal to 510 mV we get the digital filter output ripple seen in Fig. 7.10. A component of the ripple is at a

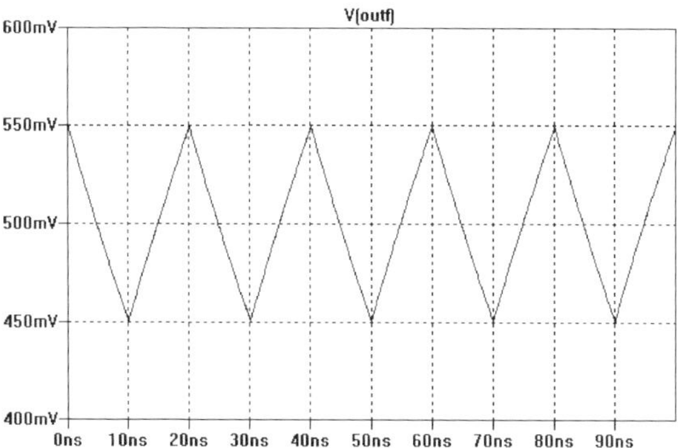

Figure 7.9 Showing how filtering the output of a modulator results in ripple.

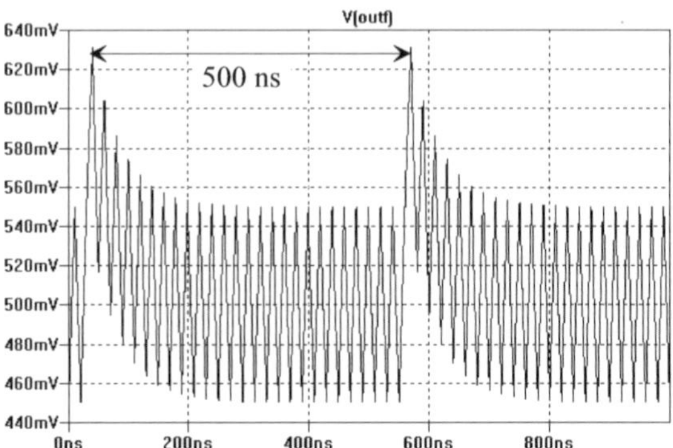

Figure 7.10 Showing a lower frequency ripple in the modulator's output.

frequency that is roughly 1/500 ns or 2 MHz, which is well within our base spectrum. The resulting tone will lower the SFDR and the SNR of the data converter. The question now becomes, "How do we minimize the possibility of unwanted tones appearing in the data converter's output spectrum?" Looking at the digital output of the modulator we see that it would be better to spread or flatten the repeating data out so that we don't get repeating tones (as discussed in the dead zone discussion in Sec. 6.1.3). Although we may still have a tone, or repeating sequence, at a frequency in the base spectrum, the amplitude of the tone will be well below the $V_{Qe,RMS}$ of the data converter (and so it won't affect the SFDR of the data converter). In order to accomplish this spread or randomization we can add a noise dither source (see Sec. 5.3.3) to our basic NS modulator, as seen in Fig. 7.11. By applying the dither to the input of the comparator (quantizer) the dither will be noise-shaped like the quantization noise (the spectral content of the dither, Eq. [5.62], is less important).

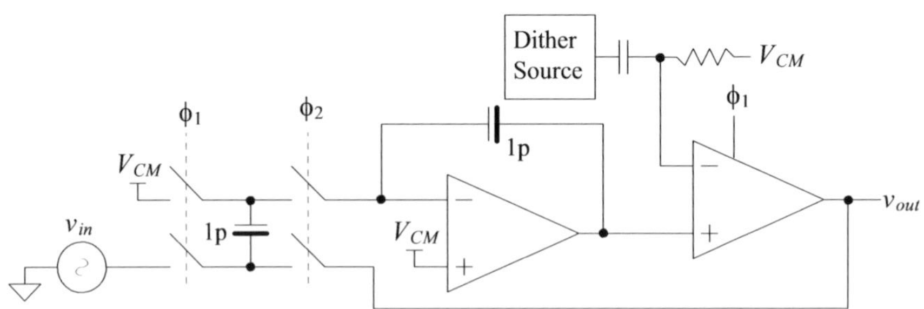

Figure 7.11 Adding a dither source to a first-order NS modulator.

Finally, note that unwanted tones are usually not a problem if the input signal is busy and random (not DC as discussed in this section). Later in the chapter, we discuss second-order modulators that utilize two integrators. The second integration helps to spread the repeating sequences out over a longer period of time so that, hopefully, negligible unwanted tone energy is present in the base spectrum.

7.1.5 Integrator and Forward Modulator Gain

So far we haven't discussed the shape or amplitude of the integrator's output. Because we are using near-ideal components in our simulations, we haven't seen any limitations due to the finite op-amp output swing. Figure 7.12 shows the integrator's output for the input and output signals shown in Fig. 7.3 (using the modulator of Fig. 7.2). Clearly the output swing of the op-amp is beyond the power supply rails. If the transistor-level model of the op-amp were to replace the ideal op-amp, the integrator's output would *saturate* at voltages less than *VDD* (= 1.0 V here) or greater than ground. While in some situations op-amp saturation is not necessarily bad (the gain of the integrator goes to zero), it is desirable to understand how decreasing forward loop gain affects the performance of the modulator. Note also, in Fig. 7.12, that the output of the integrator makes the largest change when it passes through the comparator reference voltage, $V_{CM} = 0.5$ V, since the fed back signal, the comparator output (a full-scale signal), is input to the integrator.

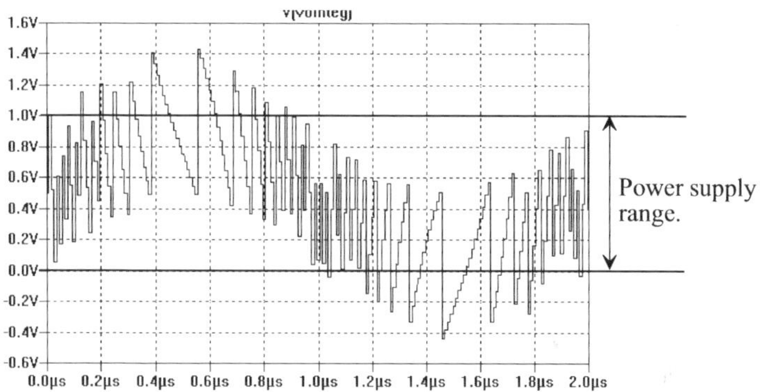

Figure 7.12 Output swing limitations in the op-amp (integrator).

Consider the linearized model of our first-order NS modulator shown in Fig. 7.13. The gain of the integrator, see Eq. (2.103) or Fig. 2.55, is given by

$$G_I = \frac{C_I}{C_F} \tag{7.24}$$

We have also drawn the comparator with a gain. Up until this point we have assumed the gain of the comparator is unity. We'll comment on this more in a moment. Let's define the modulator's forward gain as

$$G_F = G_I \cdot G_c \tag{7.25}$$

We can rewrite Eq. (7.2) using this gain as

$$v_{out}(z) = \frac{z^{-1} \cdot G_F}{1 + z^{-1}(G_F - 1)} \cdot v_{in}(z) + \frac{1 - z^{-1}}{1 + z^{-1}(G_F - 1)} \cdot V_{Qe}(z) \tag{7.26}$$

If G_F approaches zero (the integrator saturates while the comparator gain stays finite), then the output of the modulator is the sum of the integrated input and the quantization noise. (This is bad.) Since the quantization noise is not spectrally shaped it will be difficult to filter the modulator's output to recover the input signal. If the forward gain is

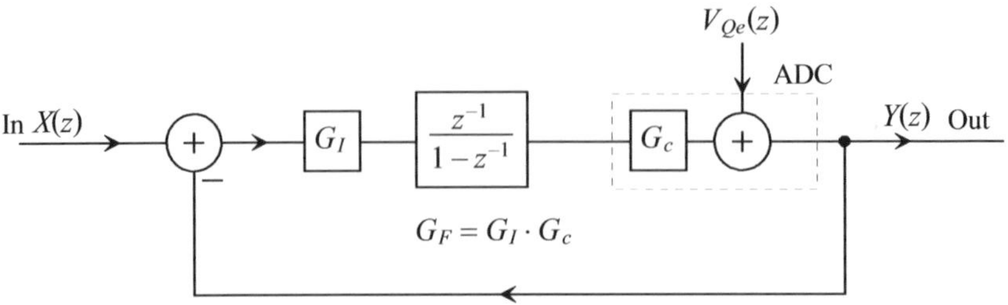

Figure 7.13 Block diagram of a NS modulator showing forward gains.

greater than two, then, as seen in Fig. 4.38 and the associated discussion, the poles of the transfer function reside outside the unit circle and the modulator will be unstable. We can restrict the values of the forward gain to

$$0 \le G_F \le 2 \tag{7.27}$$

Ideally, however, the gain is one.

Example 7.3

Show, using SPICE simulations and the modulator of Fig. 7.2, that an integrator gain of 0.4 will result in an op-amp output range well within the power supply range.

Figure 7.14a shows a schematic of the modulator with $G_I = 0.4$. Figure 7.14b shows the output of the integrator (the output of the op-amp) in the modulator of part (a) with the input sinewave shown in Fig. 7.3. The output swing is limited to roughly 80% of the supply range. *For general design it is desirable to set our integrator gain to 0.4.* This ensures our integrator doesn't saturate unless the input to the modulator goes outside the supply voltage range.

It's interesting to note that in both modulators, Fig. 7.2 and Fig. 7.14a, the forward gain is unity. This is a result of the effective gain of the comparator changing forcing the forward gain, controlled by the fed back signal, to unity. What this means is that our modulator functions as expected with a signal gain of one (Eq. [7.2] is valid) whether G_I is 1 or 0.4. Next we discuss how this change in comparator gain occurs. ■

Figure 7.15 shows the transfer curves for the comparator. The x-axis, the comparator input, is the output of the integrator in our modulator. Shrinking the integrator's output swing while holding the output swing of the comparator at the supply rails (1 V) results in an increase in effective comparator gain. This gain variation, with the integrator output swing, helps to set the forward gain of the modulator to precisely 1. We can write this using equations as

$$\overbrace{\frac{\text{Integrator output}}{\text{Modulator input}}}^{\text{Integrator gain, } G_I} \cdot \overbrace{\frac{\text{Comparator(modulator) output}}{\text{Integrator output}}}^{\text{Comparator gain, } G_c} = \overbrace{\frac{\text{Modulator output}}{\text{Modulator input}}}^{G_F} \tag{7.28}$$

If the modulator is functioning properly, then the average value of the modulator output will be equal to the modulator input and thus $G_F = 1$. It's interesting to note that this result

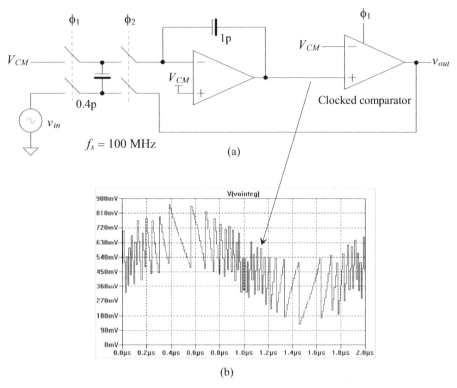

Figure 7.14 (a) First-order NS modulator with an integrator gain of 0.4, and (b) the output of the op-amp.

(*precise integrator gain isn't important*) will apply to any integrator that is directly followed by an ADC.

Before leaving this section, let's point out a couple of problems with a noise-shaping modulator that uses a multibit ADC, Fig. 7.16. Since the output of the integrator is the input signal to the ADC, the limited integrator output swing will directly effect the range of ADC output codes. Limiting the range of ADC output codes will then limit the allowable range of modulator inputs unless scaling is used (shifting the output codes or

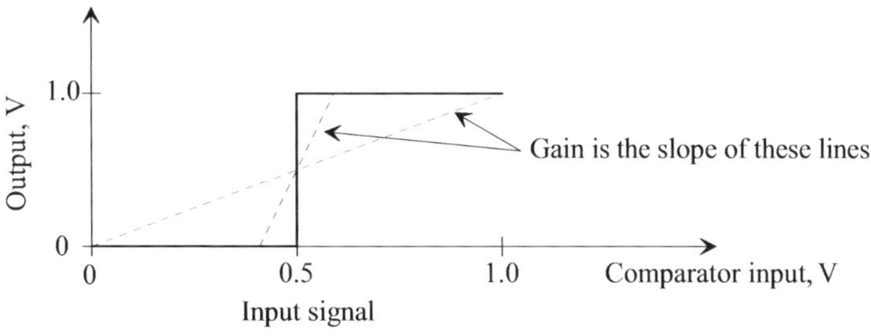

Figure 7.15 Comparator gain as a function of input voltage.

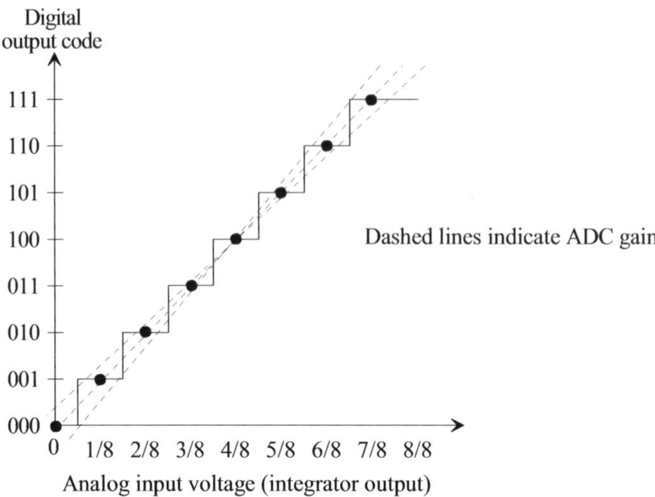

Figure 7.16 A 3-bit ADC.

sizing of capacitors in the DAI). Next, notice in Fig. 7.16 that the variation in the gain of the ADC, with input signal, is more limited than the gains attainable with the simple comparator seen in Fig. 7.15. Limiting the range of ADC gains can result in modulator forward gains that are not exactly unity. This is especially true at high input frequencies where the gain of the integrator is low. However, if the integrator gain is high, the effective gain of the ADC is not important. The point here is that using a multibit ADC will increase the open-loop gain requirements of the op-amp used in the integrator.

7.1.6 Comparator Gain, Offset, Noise, and Hysteresis

It's of interest to determine how the performance of the comparator influences the operation of the modulator. Both the comparator's offset and input-referred noise, Fig. 7.17a, can be referred back to the modulator's input, Fig. 7.17b. By doing so we can determine how they effectively change the input signal seen by the modulator. As seen in Fig. 7.17, the high gain of the integrator, $A(f)$, reduces the effect of the comparator's noise and offset on the input signal. For example, if the gain of the integrator at DC is 1,000 and the offset voltage of the comparator is 50 mV, then the input-referred offset is only 50 μV.

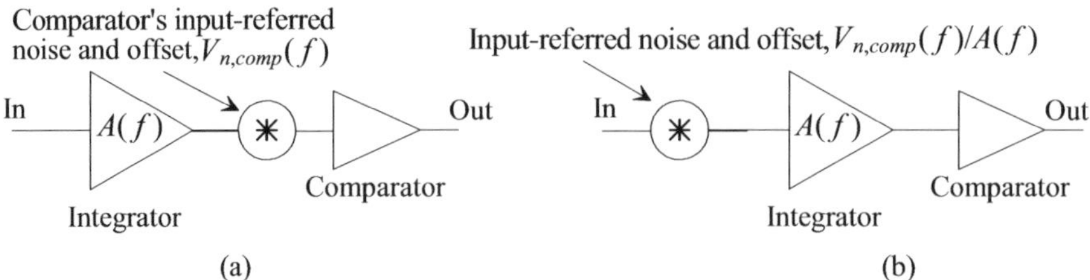

Figure 7.17 (a) Referring the comparator offset and noise to (b) the input of the modulator.

In order to determine the minimum gain and maximum allowable hysteresis requirements of the comparator, let's review Fig. 7.14. We see that when the output of the comparator changes states, the output of the integrator changes by at least

$$\text{Change in integrator output} = G_I \cdot (VDD - V_{CM}) = \frac{C_I}{C_F} \cdot \frac{V_{REF+} - V_{REF-}}{2} \tag{7.29}$$

For the modulator of Fig. 7.14 this equation can be evaluated as 0.2 V. As long as the hysteresis is much less than this value and the gain of the comparator (1/0.2 or 5) is large enough so that the comparator can make a full output transition with this input difference, then the modulator will function properly. Very simple, low-performance comparator designs can be used while not affecting the modulator's performance.

7.1.7 Op-Amp Gain (Integrator Leakage)

Now that we've discussed the gain of the comparator, let's determine how high the open-loop gain of the op-amp must be for proper integrator action. With low op-amp gain, some of the charge stored on the integrator's input capacitor, C_I, is not transferred to the feedback capacitor, C_F. This loss of charge is sometimes referred to as *integrator leakage*. The charge on the input capacitance effectively leaks off when it is transferred to the feedback capacitance.

We can write the open-loop, frequency-dependent gain of the op-amp as $A_{OL}(f)$. The output voltage of the op-amp is then $v_{out} = A_{OL}(f)(v_+ - v_-)$, where v_+ is our common-mode voltage V_{CM} (the noninverting terminal of the op-amp), see Fig. 2.54, and v_- is the op-amp's inverting input terminal. Following the procedure to derive Eq. (2.102), we can rewrite Eq. (2.100) with finite op-amp gain as

$$Q_2 = C_I\left(V_{CM} - \frac{v_{out}[nT_s]}{A_{OL}(f)} - v_2[nT_s]\right) \tag{7.30}$$

or, rewrite Eq. (2.102) to include the effects of finite op-amp gain to get

$$v_{out}(z) = \frac{C_I}{C_F} \cdot \frac{v_1(z) \cdot z^{-1/2} - v_2(z)}{\left(1 + \frac{C_I}{C_F}\frac{1}{A_{OL}(f)}\right) - z^{-1}} \tag{7.31}$$

Using this result in Eq. (7.1) and, as discussed in the last section, assuming the forward gain of the modulator, G_F, is one gives

$$v_{out}(z) = \frac{z^{-1}}{1 + \frac{C_I}{C_F}\frac{1}{A_{OL}(f)}} \cdot v_{in}(z) + \frac{1 + \frac{C_I}{C_F}\frac{1}{A_{OL}(f)} - z^{-1}}{1 + \frac{C_I}{C_F}\frac{1}{A_{OL}(f)}} \cdot V_{Qe}(z) \tag{7.32}$$

The gain error term

$$\varepsilon_{gain} = \frac{C_I}{C_F} \cdot \frac{1}{A_{OL}(f)} \tag{7.33}$$

is ideally zero so that Eq. (7.32) reduces to Eq. (7.2). Note that reducing the integrator's gain, C_I/C_F, reduces the gain error while increasing the gain required of the comparator. Note also that the denominator term is common in both the signal and the noise. This term results in a data converter gain error (it behaves as if it were an op-amp offset voltage that is a function of the integrator's output amplitude [which results in the gain error] and frequency), but it will not affect the modulator's SNR. In order to determine the increase in the modulator's output noise (the change in the shape of the modulation noise) we need to look at the noise transfer function including the effects of the gain error

$$NTF_{\varepsilon}(z) = (1 + \varepsilon_{gain}) - z^{-1} \quad (7.34)$$

or, in the frequency domain,

$$|NTF(f)|^2 = 2(1 + \varepsilon_{gain})\left(1 - \cos 2\pi \frac{f}{f_s}\right) + \varepsilon_{gain}^2 \quad (7.35)$$

Following the same procedure used to arrive at Eq. (7.12) and assuming constant op-amp gain, $A_{OL}(f)$, from DC to B, results in

$$V_{Qe,RMS}^2 = 2 \cdot \frac{V_{LSB}^2}{12f_s} \cdot \left[4(1 + \varepsilon_{gain})\frac{\pi^2}{f_s^2} \cdot \frac{1}{3} \cdot \left(\frac{f_s}{2K}\right)^3 + \varepsilon_{gain}^2 \cdot \frac{f_s}{2K}\right] \quad (7.36)$$

noting that if $\varepsilon_{gain} = 0$, this equation reduces to Eq. (7.12).

If we assume the contribution to the noise from the error term squared, ε_{gain}^2, is small, which is valid for op-amp gain

$$A_{OL}(f) > K \text{ (the oversampling ratio)} \quad (7.37)$$

over the frequency range of DC to B, then we can rewrite Eq. (7.13), to include the effects of finite op-amp gain, as

$$SNR_{gerr} = 6.02N + 1.76 - 20 \log \frac{\pi}{\sqrt{3}} + 20 \log K^{3/2} - 10 \log(1 + \varepsilon_{gain}) \quad (7.38)$$

The largest degradation in the *SNR*, resulting from integrator leakage, can be estimated as 0.5 dB if $K \geq 8$ ($\varepsilon_{gain} \approx 1/8$ neglecting G_I). The minimum gain·bandwidth product of the op-amp is estimated as

$$\text{Op-amp unity gain frequency}, f_{un} = K \cdot B = f_s/2 \quad (7.39)$$

assuming the op-amp is rolling off at 20 dB/decade at B (a dominant pole compensated op-amp). Otherwise, the minimum gain of the op-amp can be estimated simply as the oversampling ratio, K.

In order to illustrate typical op-amp requirements, let's consider the modulator of Fig. 7.14 with K = 16 and B = 3.125 MHz (see Ex. 7.1). The f_{un} of the op-amp is estimated, using Eq. (7.39), as 50 MHz. If the open-loop response of the op-amp starts to roll off at 10 kHz, then the DC gain of the op-amp must be at least 5,000. However, we could also use an op-amp with a DC gain of 100 (remembering low integrator gain increases the undesirable effects [noise and offset] of the comparator on the performance of the modulator) that rolls off at 500 kHz.

7.1.8 Op-Amp Settling Time

Equation (7.39) can be used, for the moment, to provide an estimate for the settling time requirements of the op-amp in a first-order modulator. *Assuming* the settling time is linear, and not slew-rate limited, we can write the change in the op-amp's output assuming a dominant pole compensated op-amp as

$$v_{out} = V_{outfinal}(1 - e^{-t/\tau}) \text{ where } \tau = \frac{1}{2\pi f_u \cdot \beta} \quad (7.40)$$

where, for the DAI (see Fig. 7.18), the feedback factor is

$$\beta = \frac{C_F}{C_F + C_I} \tag{7.41}$$

The feedback factor is 0.714 in the modulator of Fig. 7.14. The output of the DAI, v_{out}, must settle in a time, t ($< T_s/2$), to some percentage of an ideal value, $V_{outfinal}$. Solving for this percentage using Eqs. (7.39), (7.40), and (7.41) and assuming $T_s/2 = t$ results in

$$\frac{v_{out}}{V_{outfinal}} \times 100\% = 1 - \exp\left[-\frac{\pi}{2} \cdot \frac{C_F}{C_F + C_I}\right] \times 100\% \tag{7.42}$$

The output will only reach 67% of its ideal final value in the modulator of Fig. 7.14 when the op-amp used has a unity gain frequency of $f_s/2$.

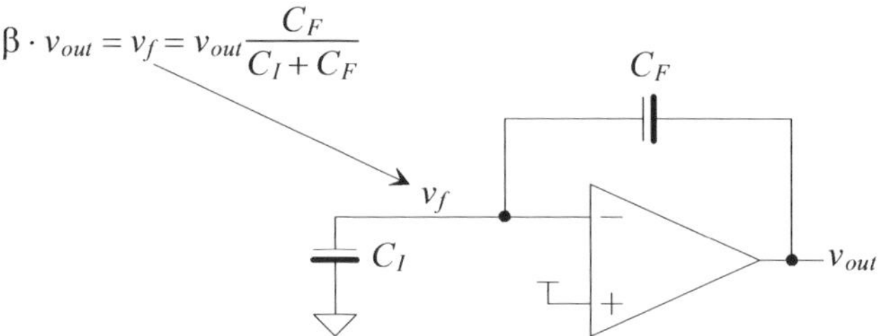

Figure 7.18 The feedback factor in the DAI.

In deriving Eq. (7.42), we used an op-amp unity gain frequency specified by Eq. (7.39) to determine the settling response of the integrator. If the settling is linear then incomplete settling will result in a constant DAI gain error (0.67 above). Every time the output changes it will change by some constant percentage of its ideal value. Rewriting the transfer function of our DAI to include this constant gain error results in

$$V_{out}(z) = \frac{C_I}{C_F} \cdot \overbrace{\left(1 - e^{-\pi\beta\cdot(f_{un}/f_s)}\right)}^{\text{Settling gain error, } G_s} \cdot \frac{V_1(z) \cdot z^{-1/2} - V_2(z)}{1 - z^{-1}} \tag{7.43}$$

Full, or complete, settling requires that the op-amp's unity gain frequency, f_{un} , be much larger than the sampling frequency, f_s (in other words we can't use Eq. [7.39] to specify the required bandwidth of the op-amp if settling time is important). The constant gain error, resulting from incomplete settling, can be tolerated in the first-order modulator because the integrator is directly followed by a comparator, as discussed earlier. In some of the modulator topologies, though, the integrator is not followed by a comparator so settling time becomes more important. In order to determine to what percentage the integrator output must settle in these topologies, a gain term, say G_s, is added to the linearized block diagram of the modulator (integrator). The transfer function of the modulator is then evaluated to determine the allowable values of G_s for the application.

It's important to realize that we are assuming the op-amp doesn't experience slew-rate limitations. If slewing is present, then the added gain term, in Eq. (7.43), will not be a constant and will introduce *distortion* into the modulator's output spectrum (whether a comparator follows the integrator or not).

7.1.9 Op-Amp Offset

The operation of the DAI is subject to the op-amp's offset. It can be shown that this offset will effectively add (or subtract) from the common-mode voltage, V_{CM}, and thus effectively shift the input signals upwards or downwards. The resulting modulator output will then show an offset equal to the op-amp's offset. In order to circumvent this problem, offset storage can be used in the integrator.

7.1.10 Op-Amp Input-Referred Noise

Here we discuss how the DAI's unwanted noise contributions affect the *SNR* of the modulator, assuming we know the DAI's input-referred noise PSD $V^2_{n,DAI}(f)$. Figure 7.19 shows the modulator's input-referred noise source, $V_{n,ckt}(f)$, in series with the input signal. This noise source, with units of $\text{V}/\sqrt{\text{Hz}}$, includes both the integrator's and the comparator's contributions. However, as discussed earlier, the noise contributions from the comparator are usually negligible because of the high-gain of the integrator.

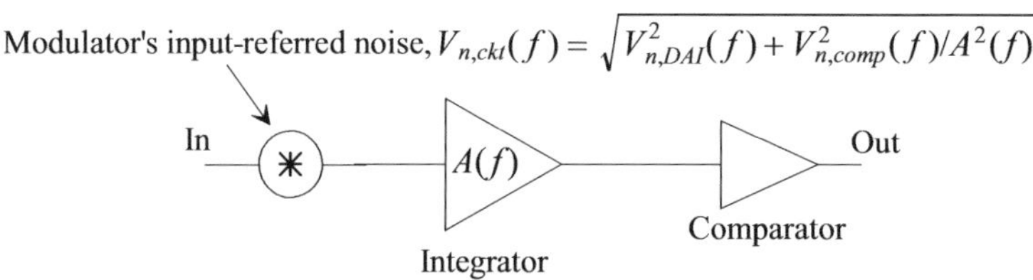

Figure 7.19 The modulator's input-referred noise contributions from both the comparator and the integrator.

Because the modulator's input-referred noise adds directly to the input signal, we can use the derivations developed earlier in the chapter. As specified in Eq. (7.2), the modulator's input, and thus its input-referred noise, pass through the modulator with a delay of z^{-1}. If we assume the modulator's input-referred noise is white and bandlimited to $f_s/2$ such that

$$V_{n,ckt}(f) = \frac{V_n}{\sqrt{f_s}} \text{ for } f < f_s/2 \tag{7.44}$$

then passing the output of the modulator through an ideal lowpass filter with a bandwidth of $B\ (=f_s/[2K])$ results in

$$V_{ckt,RMS} = \sqrt{2 \cdot \int_0^B \frac{V_n^2}{f_s} \cdot df} = \frac{V_n}{\sqrt{K}} \tag{7.45}$$

Noting that not passing the output of the modulator through a lowpass filter results in an RMS output noise of V_n, we see that the averaging filter (the lowpass filter) reduces the noise by the root of *K*. We could also think of the filtering as reducing the PSD of the modulator's input-referred noise by *K*. Remembering the jitter discussion from Ch. 5, we see a direct parallel in the derivations of how averaging affects the RMS value of a random signal (noise or jitter).

Finally, as used in Ex. 5.13, we can estimate the finite *SNR* of a data converter from quantization noise, jitter, and circuit noise using

$$V_{n,RMS} = \sqrt{V_{Qe,RMS}^2 + V_{jitter,RMS}^2 + V_{ckt,RMS}^2} \qquad (7.46)$$

and

$$SNR = 20 \cdot \log \frac{V_p / \sqrt{2}}{V_{n,RMS}} \qquad (7.47)$$

where V_p is, again, the peak amplitude of an input sinewave and

$$V_{jitter,RMS} = \sqrt{P_{AVG,jitter}} \text{ (see Eq. [5.51])} \qquad (7.48)$$

7.1.11 Practical Implementation of the First-Order NS Modulator

Switched-capacitor circuits suffer from the problems of capacitive feedthrough and charge injection. To reduce these effects, fully-differential circuit topologies are used. It could be stated that if reasonable size capacitors and dynamic range are required, fully-differential topologies are a necessity simply because they subtract out, to a first-order, the voltage changes on the switched-capacitors resulting from these problems. In addition fully-differential topologies are used because they improve power supply and substrate-coupled noise rejection and improve distortion (even-order harmonics cancel).

Figure 7.20 shows the fully-differential implementation of the DAI of Fig. 2.54. The inputs are now differential, that is, now $v_1 = v_{1+} - v_{1-}$ and $v_2 = v_{2+} - v_{2-}$, as is the output of the integrator, $v_{out} = v_{out+} - v_{out-}$. The fully-differential DAI has the same transfer function as the single-ended DAI assuming the input signals are differential.

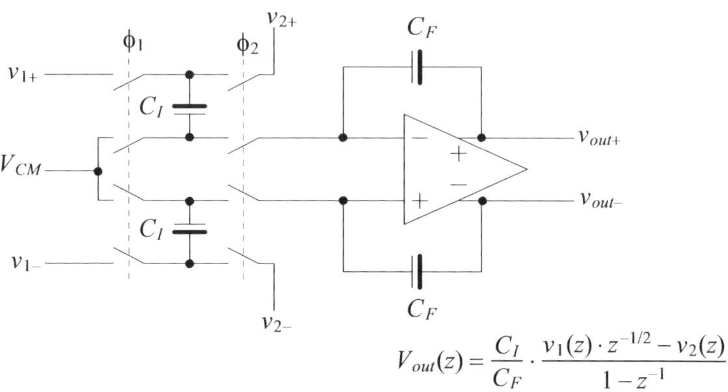

Figure 7.20 Fully-differential discrete-analog integrator (DAI) implementation.

It's important to understand the signal levels in the fully-differential DAI. Let's assume $V_{CM} = 0.5$ V and the input voltages can range in amplitude from 0 to 1 V. Assuming the input is balanced correctly if $v_{1+} = 0.85$ V, then v_{1-} must equal 0.15 V. The maximum input voltage is $v_{1max} = 1 - 0 = 1$ V. The minimum input signal, on the other hand, is $v_{1min} = 0 - 1 = -1$ V. The range of inputs, or outputs, is then 2 V or twice the range of the single-ended DAI.

Figure 7.21 shows the implementation of a first-order NS modulation utilizing a fully-differential DAI. Let's simulate the operation of this modulator with the input signals and capacitor sizes used in Fig. 7.14 (f_s = 100 MHz, C_I = 0.4 pF, and C_F = 1 pF, f_{in} = 500 kHz, and a 0.5 V peak input sinewave [the input sinewave, $v_{in+} - v_{in-}$, has a peak amplitude of 1 V]).

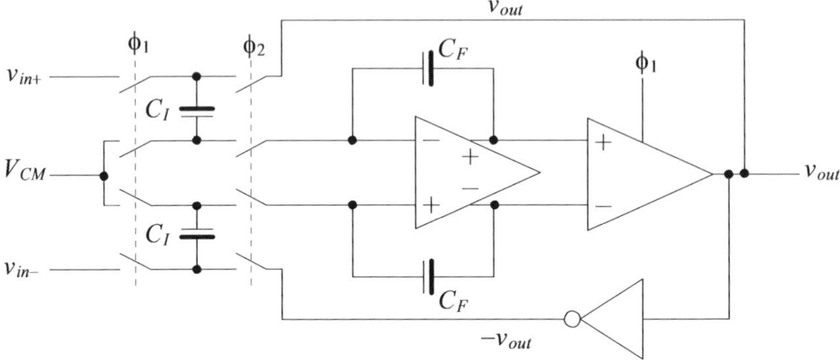

Figure 7.21 Fully-differential implementation of a first-order NS modulator.

Figure 7.22 shows the simulation results for the outputs of the modulator of Fig. 7.21 after being passed through two RC filters with time constants of 100 ns. Passing a single modulator output to the decimating filter would result in an output that is half the input signal amplitude, which can be compensated for at the output of the filter by a shift-left operation (multiply by two). Note that the input common-mode voltage of the op-amp remains at 0.5 V. This is important as the design of the op-amp becomes more challenging if the common-mode voltage is not constant. The finite op-amp common-mode rejection ratio (CMRR) can introduce distortion into the output of the modulator.

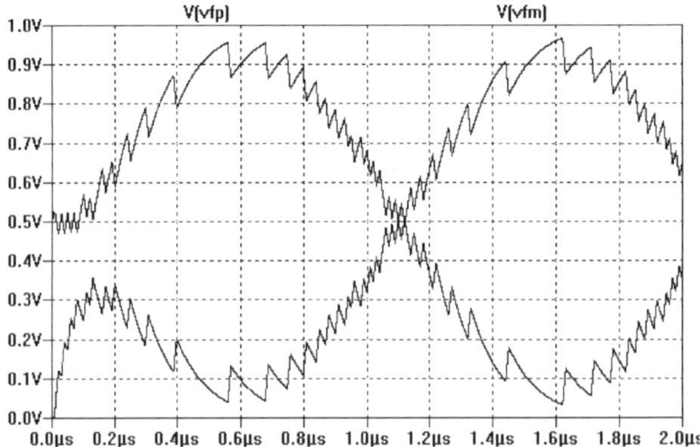

Figure 7.22 Outputs of the fully-differential first-order modulator after RC filtering.

7.2 Second-Order Noise-Shaping

If we review Eq. (7.2), we might wonder if further filtering of the quantization noise, $V_{Qe}(z)$, can result in an improvement in the data converter's *SNR* over an input signal bandwidth *B*. The second-order modulator's output shows a double differentiation of the quantization noise

$$v_{out}(z) = z^{-1} v_{in}(z) + (1 - z^{-1})^2 V_{Qe}(z) \tag{7.49}$$

The modulation noise may then be written, see Eq. (7.8), as

$$|NTF(f)|^2 \cdot |V_{Qe}(f)|^2 = \frac{V_{LSB}^2}{12 f_s} \cdot 16 \sin^4 \pi \frac{f}{f_s} \tag{7.50}$$

Figure 7.23 shows a comparison between the modulation noise of first- and second-order NS modulators. Notice that the modulation noise is "flatter" in the bandwidth of interest.

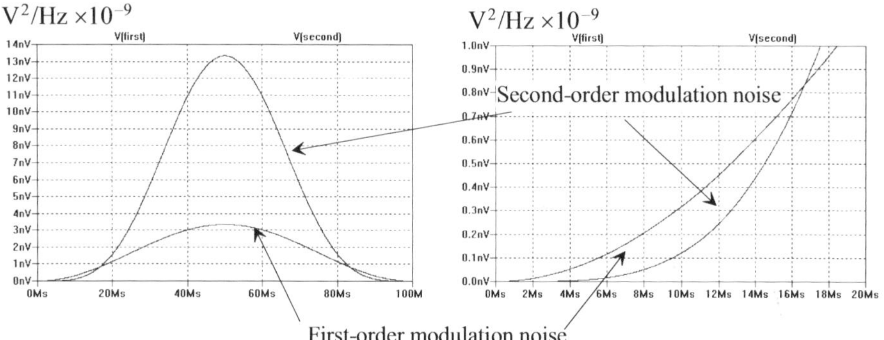

Figure 7.23 Comparing first- and second-order NS modulator's modulation noise.

Following the procedure used earlier to calculate the RMS quantization noise in a bandwidth *B* results in

$$V_{Qe,RMS} \approx \frac{V_{LSB}}{\sqrt{12}} \cdot \frac{\pi^2}{\sqrt{5}} \cdot \frac{1}{K^{5/2}} \tag{7.51}$$

with an increase in the SNR of

$$SNR_{ideal} = 6.02N + 1.76 - 12.9 + 50 \log K \tag{7.52}$$

Every doubling in the oversampling ratio results in an increase in SNR of 15 *dB or* 2.5 *bits increase in resolution!* Figure 7.24 shows a comparison between simple oversampling, first-order NS, and second-order NS-based data converters. Note that, as discussed earlier, the oversampling ratio is generally greater than or equal to eight.

7.2.1 Second-Order Modulator Topology

Consider the block diagram of a NS modulator shown in Fig. 7.25 (see Fig. 5.39). The transfer function of this modulator may be written as

$$v_{out}(z) = \frac{A(z)}{1 + A(z)B(z)} \cdot v_{in}(z) + \frac{1}{1 + A(z)B(z)} \cdot V_{Qe}(z) \tag{7.53}$$

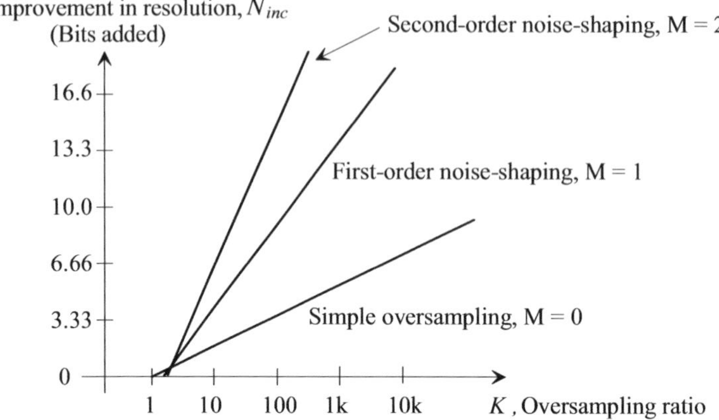

Figure 7.24 Comparing improvement in modulator resolution.

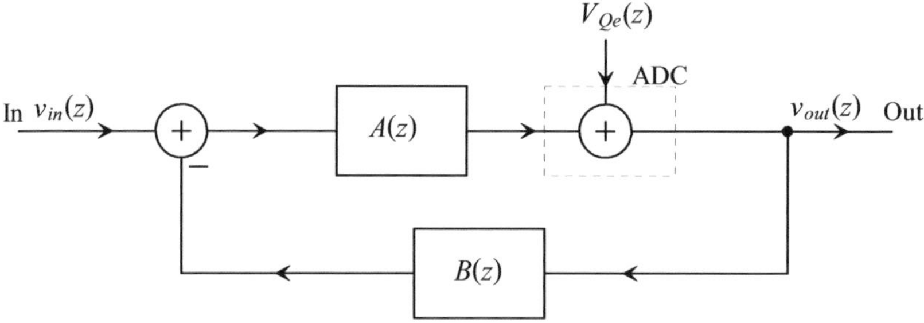

Figure 7.25 Block diagram of a feedback modulator.

Comparing this equation to Eq. (7.49), we can solve for the forward and fed-back circuit blocks, $A(z)$ and $B(z)$, by equating coefficients

$$STF(z) = \frac{A(z)}{1 + A(z)B(z)} = z^{-1} \tag{7.54}$$

and

$$NTF(z) = \frac{1}{1 + A(z)B(z)} = (1 - z^{-1})^2 \tag{7.55}$$

The results are

$$A(z) = \frac{z^{-1}}{(1 - z^{-1})^2} \tag{7.56}$$

and

$$B(z) = 2 - z^{-1} \tag{7.57}$$

The second-order modulator can be implemented using the topology shown in Fig. 7.26a. The output of $B(z)$ is the sum of the modulator output and the differentiated, $(1 - z^{-1})$,

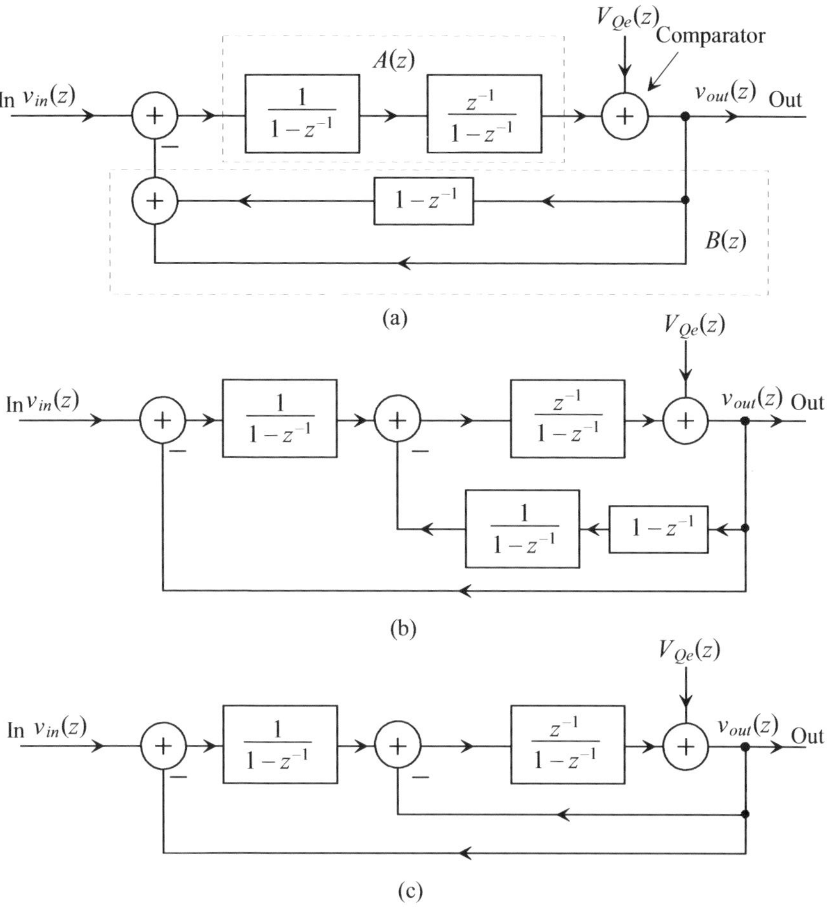

Figure 7.26 Block diagrams of second-order modulators.

modulator output. We can redraw the block diagram in (a), as shown in Fig. 7.26b, resulting in the implementation of a second-order NS modulator shown in Fig. 7.26c.

The second-order (de) modulator topology of Fig. 7.26c can be used directly to implement a NS DAC (see Figs. 7.4 and 7.5). However, this topology doesn't lend itself directly to implementation using the DAI. The major concern, as discussed in the last section, is the op-amp's output going to the power-supply rails (integrator saturation). This is more of a concern in the second-order modulator since the output of the first integrator isn't connected directly to a comparator.

Figure 7.27a shows how we can add an integrator gain to the block diagram of Fig. 7.26c without changing the system's transfer function. Figure 7.27b shows pushing the gain, $1/G_I$, through the second summer so that it is directly preceding the second integrator. Notice that in Fig. 7.27b this (the second integrator's gain) is in series with the comparator's gain (not shown; see Fig. 7.15 and the associated discussion). This means we can arbitrarily change the second integrator's gain because the comparator gain

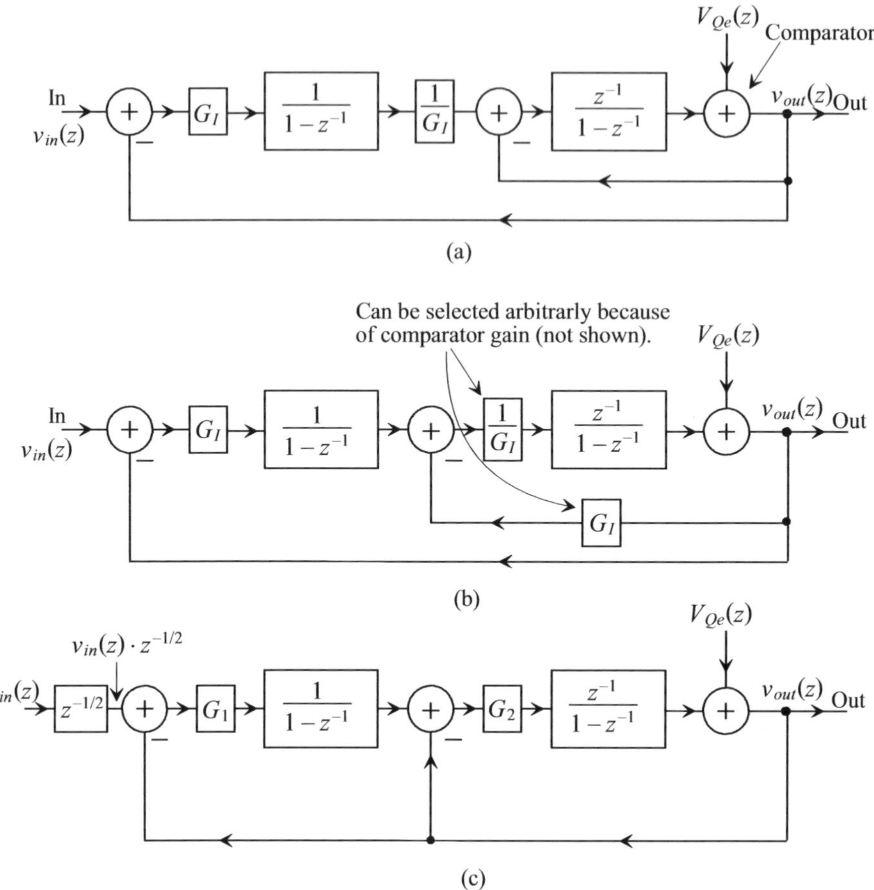

Figure 7.27 Block diagrams of second-order modulator introducing integrator gains.

changes to force the loop gain to unity. Figure 7.27c shows the resulting configuration where the second integrator has the a gain of G_2 and the first integrator has a gain of G_1. Also notice that we have added a delay in series with the input signal. This delay was added to show how using a DAI results in an added delay in series with the input signal (see Fig. 2.55 and the associated discussion). The delay doesn't affect the magnitude of modulator's transfer function but rather indicates the input signal arrives half a clock cycle later.

Figure 7.28 shows the DAI implementation of the second-order modulator of Fig. 7.27c. Note how the output of the modulator is fed back and immediately passes through the first integrator and is applied to the second integrator (no delay as seen in Fig. 7.27c). This is a result of switching the phases of the clock signals in the first integrator. We should also see how the input signal sees an added half-clock cycle delay. Note that at this point it should be straightforward to sketch the circuit implementation of the fully-differential, second-order modulator (see Fig. 7.21).

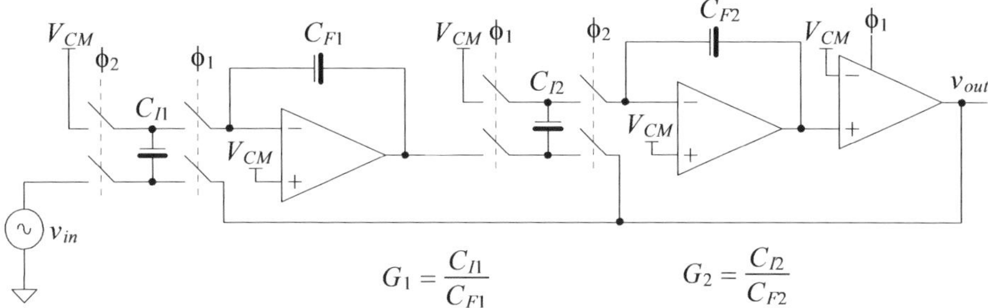

Figure 7.28 Implementation of the second-order modulator of Fig. 7.27c.

7.2.2 Integrator Gain

As we showed in Eq. (7.28) for the first-order modulator, the forward gain of a second-order modulator will also be unity when the modulator is functioning properly. We now need to discuss how to select the integrator gains to avoid harmful integrator saturation. If noise and offsets were not a concern, as shown in Fig. 7.17 and the associated discussions, then we could make our integrator gains very small (ultimately limited by imperfections in the switches such as clock feedthrough and charge injection). In a practical modulator, integrator saturation (the integrator's gain going to zero) can also lead to modulator instability, as shown in Eq. (7.27), and the associated discussion.

Figure 7.29 shows the integrator outputs for the modulator of Fig. 7.28 if both integrator gains are set to 0.4. Notice that both outputs go outside the supply voltage range. If we replace the ideal op-amps in the simulation with transistor-based op-amps, the integrator outputs will saturate at some voltage within the supply range. This saturation can be thought of as noise and ultimately limits the data converter's SNR. Integrator saturation can be avoided by limiting the input signal range, designing with small integrator gain, and using op-amps that have a wide output swing.

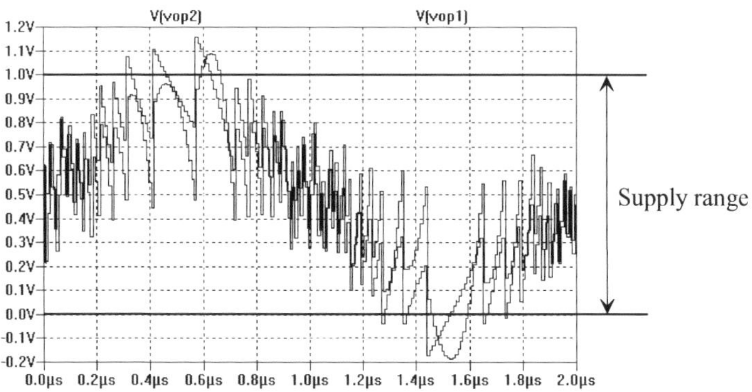

Figure 7.29 Showing integrator outputs in a second-order modulator with the integrator gains both set to 0.4.

For a more quantitative view of how the gains in a second-order NS modulator affect performance, let's consider a couple of different topologies. Figure 7.30 shows the block diagram of the second-order NS modulator topology of Fig. 7.25, with an integrator gain coefficient, G_I, and a comparator gain, G_c, added. Deriving the transfer function of this linearized model with $G_F = G_I \cdot G_c$ results in

$$v_{out}(z) = \frac{G_F \cdot z^{-1} \cdot v_{in}(z)}{1+z^{-1} \cdot 2(G_F - 1) + z^{-2} \cdot (1 - G_F)} + \frac{(1-z^{-1})^2 \cdot V_{Qe}(z)}{1+z^{-1} \cdot 2(G_F - 1) + z^{-2} \cdot (1 - G_F)} \tag{7.58}$$

The poles of this transfer function are located at

$$z_{p1,p2} = (1 - G_F) \pm \sqrt{(1-G_F)^2 - (1 - G_F)} \tag{7.59}$$

We know that for the modulator to remain stable the poles must reside within the unit circle. This means that our values of forward gain are restricted to

$$0 \leq G_F \leq 1.333 \tag{7.60}$$

Again, if the modulator is functioning properly, $G_F = 1$ (because of the comparator's gain variation as seen in Fig. 7.15 and the associated discussion).

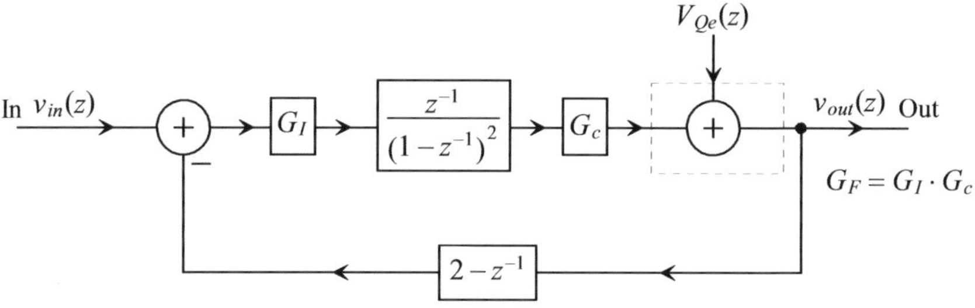

Figure 7.30 Block diagram of a second-order feedback modulator with gains.

We should make some observations at this point. Reviewing Eq. (7.27), we see that the allowable range of forward gain, in the first-order modulator, is larger than the allowable range in the second-order modulator. However, as long as the integrators don't saturate (G_I doesn't approach zero), stability for either modulator is easy to attain. An analysis of the stability of higher-order modulators show that the range of allowable forward gains decreases with the order of the modulator. For example, a third-order modulator can have a forward gain of at most 1.15. Finally, notice that the input signal range is more restricted for the second-order modulator, in order to avoid integrator saturation, as seen in Fig. 7.29. We'll discuss methods to attain wider input signal range and more robust stability criteria by adjusting the feedback gains later.

Notice that we are treating our modulator as a linear system even though it isn't linear; the comparator gain is a nonlinear variable. The linear approximation is useful in order to give an idea of the stability of the modulator under certain operating conditions. Generally, a DC input is applied to the modulator in the simulation, while lowpass filters are added to determine the average comparator gain, G_c. Figure 7.31 shows this schematically. Assuming we know G_I (the gain coefficient of the integrators), we can then look at the stability and forward gain of the modulator for varying DC input signal voltages.

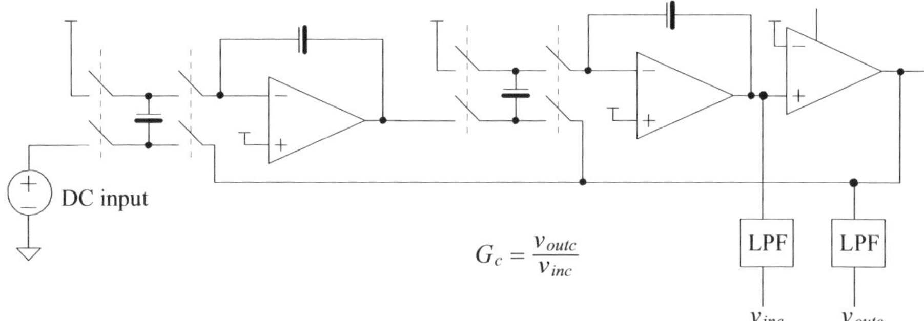

Figure 7.31 Simulating the gain of the comparator.

Next, consider the more generic block diagram of the second-order NS modulator shown in Fig. 7.32. In a moment we'll discuss how to implement the feedback gain, G_3, using the DAI. The transfer function of this topology can be written as

$$v_{out}(z) = \frac{G_1 G_2 G_c \cdot z^{-1} v_{in}(z) + (1 - z^{-1})^2 \cdot V_{Qe}(z)}{1 + z^{-1} \cdot (G_1 G_2 G_c + G_2 G_3 G_c - 2) + z^{-2} \cdot (1 - G_2 G_3 G_c)} \tag{7.61}$$

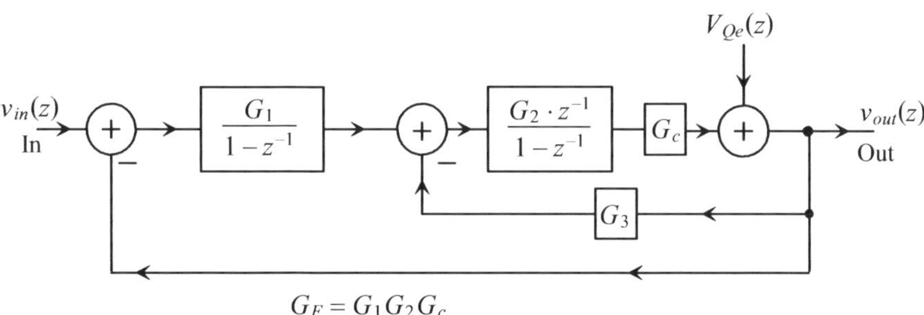

Figure 7.32 Generic block diagram of a second-order NS modulator.

Notice that if $G_1 = G_2 = G_3 = G_c = 1$ (where $G_1 G_2 G_c = G_F$), then this equation reduces to Eq. (7.49). The poles of this equation are located at

$$z_{p1,p2} = \frac{2 - G_1 G_2 G_c - G_2 G_3 G_c \pm \sqrt{(2 - G_1 G_2 G_c - G_2 G_3 G_c)^2 - 4(1 - G_2 G_3 G_c)}}{2} \tag{7.62}$$

When the modulator is functioning properly we require the (linearized) coefficient of the input, $v_{in}(z)$ in Eq. (7.61), to be unity

$$\left| \frac{G_1 G_2 G_c}{(z - z_{p1})(z - z_{p2})} \right| = 1 \tag{7.63}$$

Again, if we set $G_1 = G_2 = G_3 = 1$ (and $G_c = 1$), then the poles are located at DC, that is,

$$z_{p1,p2} = 0 \tag{7.64}$$

Equation (7.62) is useful to estimate the modulator's stability when scaling amplitudes by adjusting the integrator gain coefficients, G_1, G_2, and G_3.

Implementing Feedback Gains in the DAI

Consider the modified DAI shown in Fig. 7.33. Notice that if $C_{I2} = C_{I3}$, this topology reduces to the DAI shown in Fig. 2.54. Also note that some of the switches can be combined to simplify the circuitry. Assuming that the output is connected through the ϕ_2 switches (or that there are no switches connected to the output of the op-amp, see Eq. [2.102]) we can write the transfer function of the integrator as

$$v_{out}(z) = v_1(z) \cdot \frac{C_{I2}}{C_{F2}} \cdot \frac{z^{-1/2}}{1-z^{-1}} - v_2(z) \cdot \frac{C_{I3}}{C_{F2}} \cdot \frac{1}{1-z^{-1}} \tag{7.65}$$

The block diagram of this topology is shown in Fig. 7.34a. We want to implement a block diagram like the one shown in Fig. 7.34b. Because we have already defined

$$G_2 = \frac{C_{I2}}{C_{F2}} \tag{7.66}$$

we define our feedback gain, G_3, as

$$G_3 = \frac{C_{I3}}{C_{F2}} \cdot \frac{1}{G_2} = \frac{C_{I3}}{C_{I2}} \tag{7.67}$$

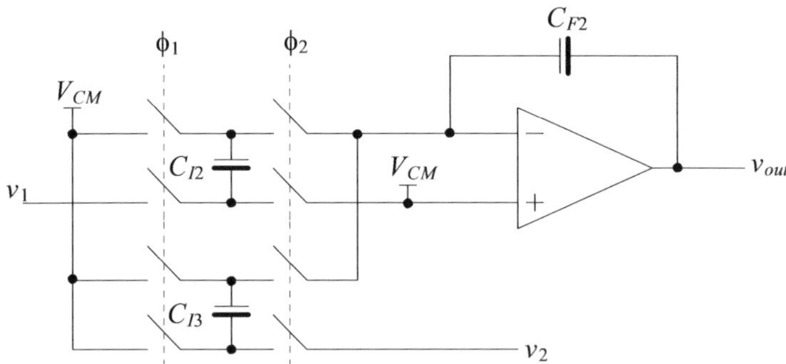

Figure 7.33 Adding an additional gain setting to our DAI.

Example 7.4

Sketch the circuit implementation of a second-order NS modulator based on the topology of Fig. 7.32, where $G_1 = G_2 = G_3 = 0.4$. Comment on the stability of the resulting configuration. Simulate the design and show the integrator output swing.

The implementation of the modulator is shown in Fig. 7.35. We could dissect Eq. (7.62) at this point to determine the transient properties of the modulator. However, before discussing the transient characteristics of the modulator, let's look at the integrator output swing.

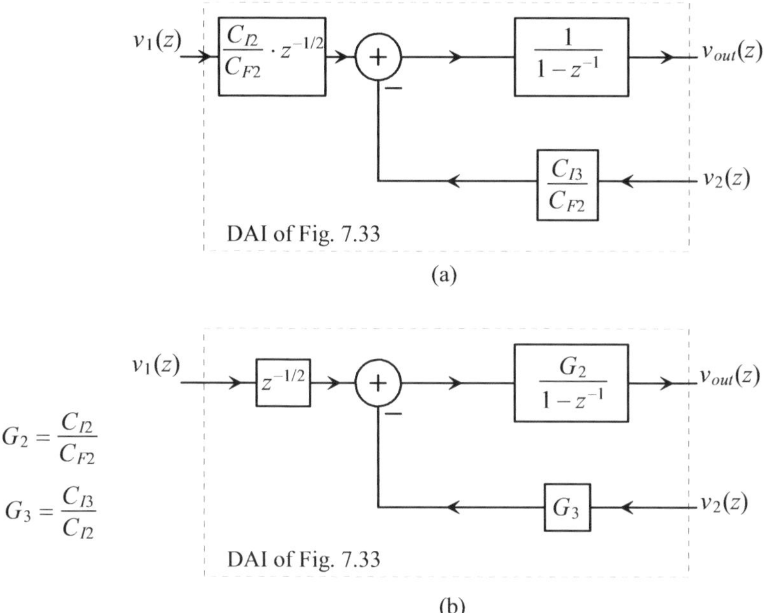

Figure 7.34 Block diagram of a DAI.

Figure 7.36 shows the output swing of the integrators. This figure should be compared with Fig. 7.29. The output of the first integrator now falls within the power supply range. The output of the second integrator is reduced but still exceeds the power supply range. This, as discussed earlier, has less impact on performance in the actual transistor-based modulator because the integrator is followed by a comparator.

Let's attempt to get an idea for the stability of the modulator by adding LPFs, as seen in Fig. 7.31, to the simulation (with a DC input) to measure G_c. Figure 7.37 shows how we will implement the LPFs. The voltage-controlled voltage source is used to keep from loading the modulator with the RC circuit when it is added into the general simulation. In our ideal modulator shown in Fig. 7.35 both the comparator output and integrator outputs are ideal voltage sources, so we don't

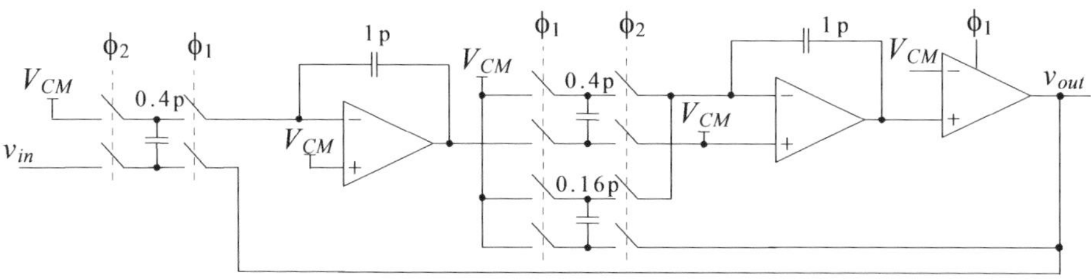

Figure 7.35 Implementation of a second-order modulator with feedback gain.

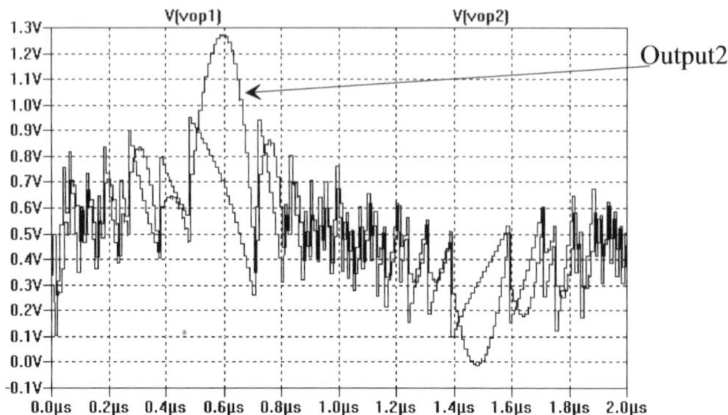

Figure 7.36 Integrator output signals in the modulator shown in Fig. 7.35.

need the isolation (and therefore we can add the RC LPF directly into the simulation).

Figure 7.38 shows the comparator input and output, after lowpass filtering, for the modulator of Fig. 7.35 when the input signal is 0.1 V (DC). The average comparator input voltage is roughly 0.4 V. The resulting comparator gain is then only 0.25. Using Eq. (7.62) to calculate the location of the poles results in $z_{p1,p2} = 0.96 \pm j \cdot 0.195$. These poles are very close to the unit circle. Small shifts in the DAI gains can result in an unstable modulator. Increasing the input signal amplitude makes the modulator more stable. Increasing G_3 also increases the modulator's stability.

The simulation that generated Fig. 7.38 can be very useful in understanding basic second-order modulator's stability criteria. Changing the simulation variables and looking at the resulting simulation outputs can be very instructional. Note that increasing the simulation time in the netlist that generated Fig. 7.38 would reveal that the comparator input actually has small amplitude oscillations. Also note that Fig. 7.36 shows the output of the second integrator going outside the power supply rails with variations in the input signal. This is related to the stability of the modulator being a function of the input voltage. An input signal close to *VDD*, for example, causes the modulator to be less stable than an input signal close to V_{CM}. ■

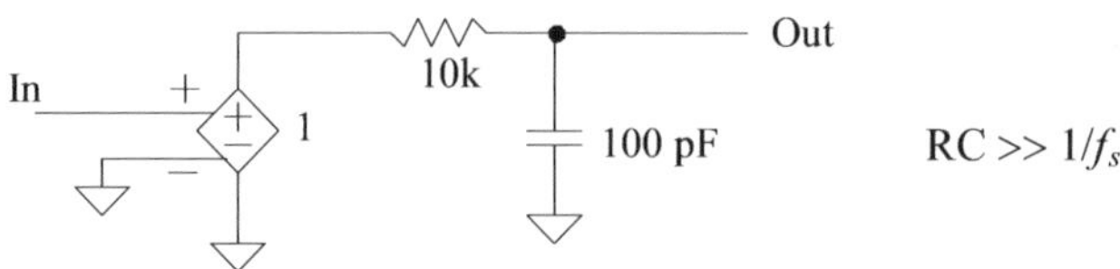

Figure 7.37 SPICE implementation of a LPF for determining comparator gain.

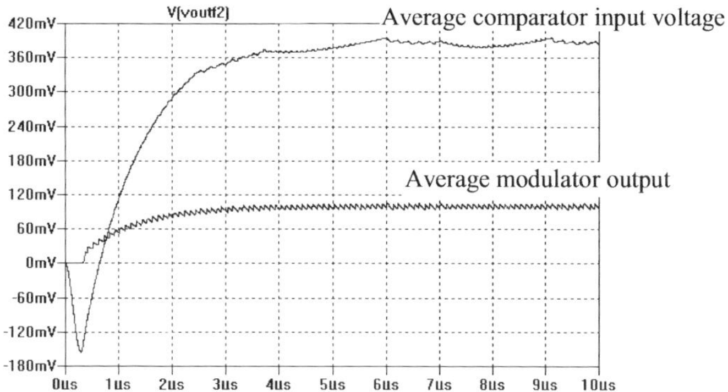

Figure 7.38 Average comparator input and output when using the modulator of Fig. 7.35 with an input signal of 0.1 V.

Using Two Delaying Integrators to Implement the Second-Order Modulator

Consider the second-order modulator topology shown in Fig. 7.39. This topology can be implemented using the circuits in either Fig. 7.28 or Fig. 7.35 by simply switching the phases of the clocks in the first integrator (by making both integrators delaying). Without too much thought we know that adding gratuitous delay to the forward or feedback paths of a feedback system moves the system towards instability, however let's show this mathematically. The transfer function of this topology is

$$v_{out}(z) = \frac{G_1 G_2 G_c \cdot z^{-2} v_{in}(z) + (1 - z^{-1})^2 \cdot V_{Qe}(z)}{1 + z^{-1} \cdot (G_2 G_3 G_c - 2) + z^{-2} \cdot (1 - G_2 G_3 G_c + G_1 G_2 G_c)} \tag{7.68}$$

The poles are located at

$$z_{p1,p2} = \frac{2 - G_2 G_3 G_c \pm \sqrt{(2 - G_2 G_3 G_c)^2 - 4(1 - G_2 G_3 G_c + G_1 G_2 G_c)}}{2} \tag{7.69}$$

This equation should be compared to Eq. (7.62). Remembering that for a stable modulator the poles must be inside the unit circle, we see that using two delaying integrators will not result in a modulator that has as robust stability criteria as the general implementation of Fig. 7.27. The extra delay won't cause instability in modulators with large oversampling ratios (since the delay is then relatively small); however, for high-speed topologies or bandpass modulators excess forward (or feedback) delay can be a problem. Note that the benefit of this topology, and why it's used in many applications, is that it gives the comparator an extra half-clock cycle to make a decision.

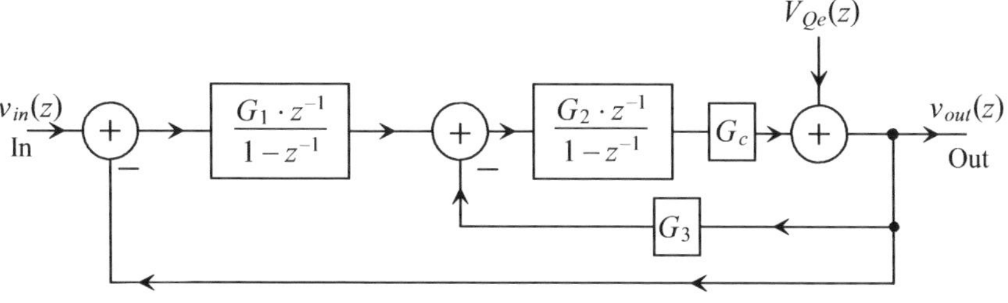

Figure 7.39 Second-order NS modulator using two delaying integrators.

7.2.3 Selecting Modulator (Integrator) Gains

Before leaving this section, let's discuss the general selection of modulator gains. In general, for good stability, the inner loop feedback gain, G_3, should be made as large as possible. For general design, set $G_3 = 1$. This simplifies the design of the modulator circuitry and provides good flexibility when selecting the values of G_1 and G_2. If $G_3 = 1$, then Eq. (7.62) may be rewritten to show the location of the poles as

$$z_{p1,p2} = \frac{2 - G_1 G_2 G_c - G_2 G_c \pm \sqrt{(2 - G_1 G_2 G_c - G_2 G_c)^2 - 4(1 - G_2 G_c)}}{2} \tag{7.70}$$

Keeping in mind that the reason we are not setting all gains to one is to avoid integrator saturation, we can look at Eq. (7.70) as a guide to determine how we can reduce G_1 and G_2. Since G_2 is directly followed by the comparator, we can set its gain to 0.4 as discussed earlier. Practically then, we can reduce the value of G_1 to a very small number and still have a stable modulator. At the same time, using a small G_1 avoids integrator saturation. The practical problem with a small G_1, as discussed earlier, is the increase in the input-referred noise. Again trade-offs must be made for given design criteria. Figure 7.40 shows the integrator outputs for the modulator of Fig. 7.32 when $G_1 = 0.2$, $G_2 = 0.4$, and $G_3 = 1$. Note that, when compared to Figs. 7.29 and 7.36, the outputs are very well behaved. We don't have the abnormal transitions above the power-supply rails indicating that the modulator stability is becoming marginal with input signal values close to the power-supply rails.

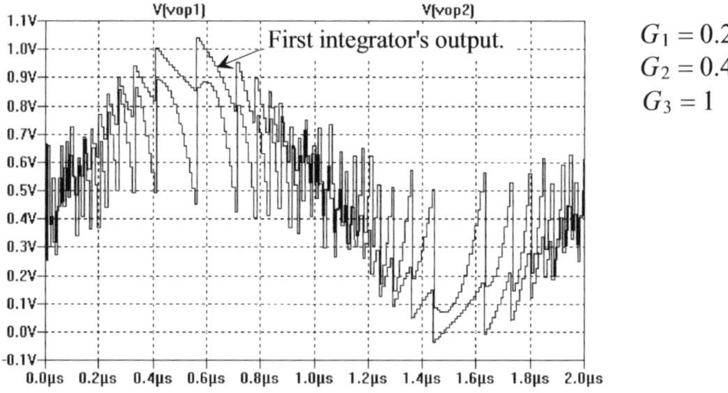

Figure 7.40 Integrator outputs for a modulator with first integrator gain of 0.2.

7.3 Noise-Shaping Topologies

The last section presented the fundamentals of NS data converters. It's important to understand this fundamental material before proceeding with the topics presented in this section.

In this section we cover (1) higher-order NS modulators, (2) NS modulators using multibit ADCs and DACs (multibit modulators), and (3) cascaded modulators (higher-order modulators built with a cascade of first- and/or second-order modulators).

7.3.1 Higher-Order Modulators

We can take the theory developed for our first- and second-order modulators in the last section and generalize it for an Mth-order modulator (a modulator having M integrators and M feedback loops). Rewriting Eqs. (7.8) and (7.50) for the general Mth-order modulator results in

$$|NTF(f)| \cdot |V_{Qe}(f)| = \frac{V_{LSB}}{\sqrt{12 f_s}} \cdot \left[2 \sin \pi \frac{f}{f_s}\right]^M \tag{7.71}$$

The RMS noise in a bandwidth, B, can be written, see Eqs. (7.12) and (7.51), as

$$V_{Qe,RMS} = \frac{V_{LSB}}{\sqrt{12}} \cdot \frac{\pi^M}{\sqrt{2M+1}} \cdot \frac{1}{K^{M+1/2}} \tag{7.72}$$

The ideal increase in the SNR can be written as

$$\text{SNR}_{ideal} = 6.02N + 1.76 - 20\log\left[\frac{\pi^M}{\sqrt{2M+1}}\right] + [20M+10] \cdot \log K \tag{7.73}$$

or

$$\text{SNR}_{ideal} = 6.02(N + N_{inc}) + 1.76 \tag{7.74}$$

The increase in resolution, N_{inc}, is given by

$$N_{inc} = \frac{1}{6.02}\left[(20M+10) \cdot \log K - 20\log\left(\frac{\pi^M}{\sqrt{2M+1}}\right)\right] \tag{7.75}$$

This equation shows that for every doubling in the oversampling ratio, K, the resolution increases by $M + 0.5$ bits.

Mth-Order Modulator Topology

Reviewing the general NS modulator topology of Fig. 7.25, we want to determine the forward transfer function, $A(z)$, and the feedback transfer function, $B(z)$, for an Mth-order NS modulator. The transfer function of a general Mth-order modulator is

$$v_{out}(z) = v_{in}(z) \cdot (z^{-1}) + V_{Qe}(z) \cdot (1 - z^{-1})^M \tag{7.76}$$

Using this equation together with Eq. (7.53) results in a forward modulator transfer function of

$$A(z) = \frac{z^{-1}}{(1 - z^{-1})^M} \tag{7.77}$$

and a feedback filter transfer function of

$$B(z) = \frac{1 - (1 - z^{-1})^M}{z^{-1}} \tag{7.78}$$

The block diagram of an Mth-order NS modulator is shown in Fig. 7.41. Note, as we'll see shortly, it's impossible (in an analog-to-digital converter) to implement this topology using only one delaying integrator.

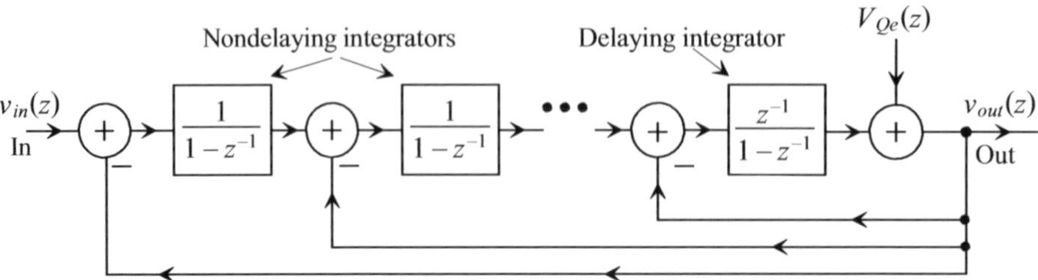

Figure 7.41 Generic block diagram of an Mth-order NS modulator.

7.3.2 Filtering the Output of an Mth-Order NS Modulator

Let's revisit the derivation of Eq. (7.17). This equation states that the number of Sinc stages, L, used in cascade, for near optimum removal of the modulation noise, is one more than the order of the modulator ($L = M + 1$). Rewriting Eq. (7.19)

$$V_{Qe,RMS}^2 = 2\int_0^{f_s/2} |NTF(f)|^2 \cdot |V_{Qe}(f)|^2 \cdot |H(f)|^2 \cdot df \tag{7.79}$$

where the decimation filter's transfer function is given by

$$|H(f)|^2 = \left[\frac{1}{K}\cdot\frac{\sin\left(K\pi\frac{f}{f_s}\right)}{\sin\left(\pi\frac{f}{f_s}\right)}\right]^{2(M+1)} \tag{7.80}$$

The mean-squared quantization noise is calculated by evaluating

$$V_{Qe,RMS}^2 = 2\cdot\overbrace{\frac{V_{LSB}^2}{12f_s}}^{|V_{Qe}(f)|^2}\cdot\int_0^{f_s/2}\overbrace{\left[2\sin\pi\frac{f}{f_s}\right]^{2M}}^{|NTF(f)|^2}\cdot\left[\frac{1}{K}\cdot\frac{\sin\left(K\pi\frac{f}{f_s}\right)}{\sin\left(\pi\frac{f}{f_s}\right)}\right]^{2(M+1)}\cdot df \tag{7.81}$$

or

$$V_{Qe,RMS}^2 = 2\cdot\frac{V_{LSB}^2}{12f_s}\cdot 2^{2M}\cdot\left[\frac{1}{K}\right]^{2(M+1)}\cdot\int_0^{f_s/2}\frac{\sin^{2(M+1)}\left(K\pi\frac{f}{f_s}\right)}{\sin^2\left(\pi\frac{f}{f_s}\right)}\cdot df \tag{7.82}$$

If we let $\theta = \pi\frac{f}{f_s}$, then we get

$$V_{Qe,RMS}^2 = \frac{V_{LSB}^2}{12f_s}\cdot\left[\frac{2}{K}\right]^{2(M+1)}\cdot\frac{f_s}{\pi}\cdot\overbrace{\int_0^{\frac{\pi}{2}}\frac{\sin^{2(M+1)}(K\theta)}{\sin^2\theta}\cdot d\theta}^{=\frac{K}{2}\pi\cdot\prod_{m=1}^{M}\frac{2m-1}{2m}} \tag{7.83}$$

Finally, the RMS quantization noise associated with an Mth-order modulator followed by an $M+1$ ($=L$) Sinc averaging filter is

$$V_{Qe,RMS} = \frac{V_{LSB}}{\sqrt{12}} \cdot \left[\frac{2}{K}\right]^{M+1/2} \cdot \prod_{m=1}^{M} \frac{2m-1}{2m} \tag{7.84}$$

The change in *SNR*, when using the Sinc averaging filter decimator instead of the ideal filter with bandwidth, *B*, is given by looking at the ratio of Eq. (7.72) to Eq. (7.84)

$$\text{Increase in } SNR = -20\log\left[2^{M+1/2} \cdot \prod_{m=1}^{M} \frac{2m-1}{2m} \cdot \frac{\sqrt{2M+1}}{\pi^M}\right] \tag{7.85}$$

For first-, second-, and third-order modulators, the improvement in the *SNR*s is 2.16, 6.35, and 10.39 dB, respectively. This shows that using a Sinc averager, theoretically, increases the *SNR* if we neglect the decrease in the desired signal amplitude because of the droop in the Sinc-filters response, Fig. 4.17. To avoid the droop, as discussed earlier, the desired signal content is often limited to frequencies well below $f_s/2K$ (= *B*). When the droop (reduction in the desired signal amplitude) is taken under consideration, the *SNR*, when using the Sinc averaging filter, is worse than the ideal filter with bandwidth *B*.

7.3.3 Implementing Higher-Order, Single-Stage Modulators

The single-stage, higher-order modulator of Fig. 7.41 can be difficult to implement directly. It is impossible to implement a higher-order modulator, when using DAIs, where all but the last integrator are nondelaying. However, as we saw with the second-order modulator using two delaying integrators in Fig. 7.39 and Eqs. (7.68) and (7.69), the stability criteria of a modulator using only delaying integrators is poorer than the criteria of the topology shown in Fig. 7.41 (where only the last integrator is delaying). While we can help the situation by staggering delaying and nondelaying integrators in a modulator, the point is that implementing a higher-order modulator without modifying our basic NS topology will result in an unstable circuit. Intuitively, we can understand this by noting that if the modulator's forward gain is too high and the delay through the forward path is too long (because of the large number of integrators), the signal fed back may add to the input signal instead of subtracting from it.

In order to help with the stability of a higher-order modulator a topology that feeds the input signal forward into additional points in the modulator (thereby reducing the forward gain and delay) and feeds the output signal back as discussed earlier (allowing scaling of amplitudes) is needed. In order to move towards this goal, consider the modified NS topology for higher- order modulators shown in Fig. 7.42. The forward and feedback transfer functions can be written as

$$A(z) = \frac{a_1 \cdot z^{-M}}{(1-z^{-1})^M} + \frac{a_2 \cdot z^{-(M-1)}}{(1-z^{-1})^{M-1}} + \frac{a_3 \cdot z^{-(M-2)}}{(1-z^{-1})^{M-2}} + \ldots + \frac{a_M \cdot z^{-1}}{1-z^{-1}} = \sum_{i=1}^{M} a_i \cdot \left(\frac{z^{-1}}{1-z^{-1}}\right)^{M-i+1} \tag{7.86}$$

$$A(z) = (z-1)^{-M} \cdot \left[a_1 + a_2(z-1)^1 + a_3(z-1)^2 + \ldots + a_M(z-1)^{M-1}\right] \tag{7.87}$$

and

$$-A(z)B(z) = \frac{b_1 \cdot z^{-M}}{(1-z^{-1})^M} + \frac{b_2 \cdot z^{-(M-1)}}{(1-z^{-1})^{M-1}} + \frac{b_3 \cdot z^{-(M-2)}}{(1-z^{-1})^{M-2}} + \ldots + \frac{b_M \cdot z^{-1}}{1-z^{-1}} = \sum_{i=1}^{M} b_i \cdot \left[\frac{1}{z-1}\right]^{M-i+1} \tag{7.88}$$

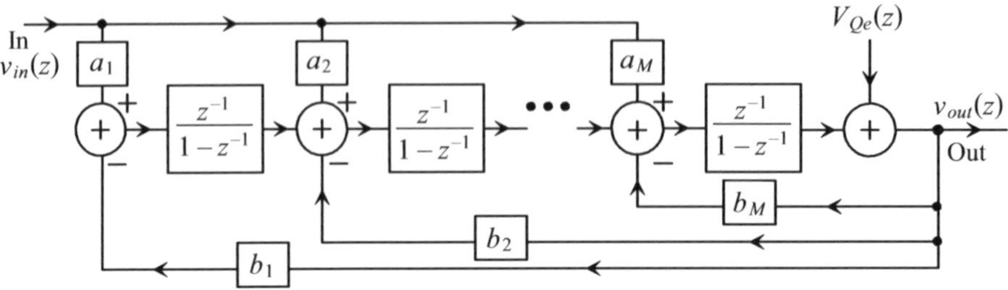

Figure 7.42 Block diagram of a modified M[th]-order NS modulator.

or

$$-A(z)B(z) = (z-1)^{-M} \cdot \left[b_1 + b_2(z-1)^1 + b_3(z-1)^2 + \ldots + b_M(z-1)^{M-1} \right] \tag{7.89}$$

Before going any further, let's explain what we are trying to do with the modified, higher-order, NS topology of Fig. 7.42. We know that the $NTF(z)$, for a general modulator, is of the form $(1-z^{-1})^M$ with a shape seen in Fig. 7.43. At high frequencies the modulation noise will get very large. At $f_s/4$, for example, the magnitude of the noise transfer function, $|NTF(f)|$, is $\left(\sqrt{2}\right)^M$ (see Fig. 1.20). For the modified NS modulator we will try to reduce the modulation noise at higher frequencies by changing the shape of the $NTF(z)$. Our modified $NTF(z)$ will be of the form

$$NTF(z) = LPF(z) \cdot (1-z^{-1})^M = LPF(z) \cdot \left(\frac{z-1}{z}\right)^M = HPF(z) \tag{7.90}$$

where $LPF(z)$ [$HPF(z)$] is a lowpass [highpass] filter implemented with the feedback coefficients b_x. The goal is to flatten out the higher frequency modulation noise (keep the noise from getting too large) thereby reducing the $NTF(f)$ at high frequencies and keeping the modulator stable. One drawback of using this technique is that the signal no longer sees just a delay in its transfer function but rather it sees the lowpass response. The modified STF will be of the form

$$STF(z) = NTF(z) \cdot A(z) = LPF(z) \cdot \sum_{i=1}^{M} a_i \cdot (z-1)^{i-1} \tag{7.91}$$

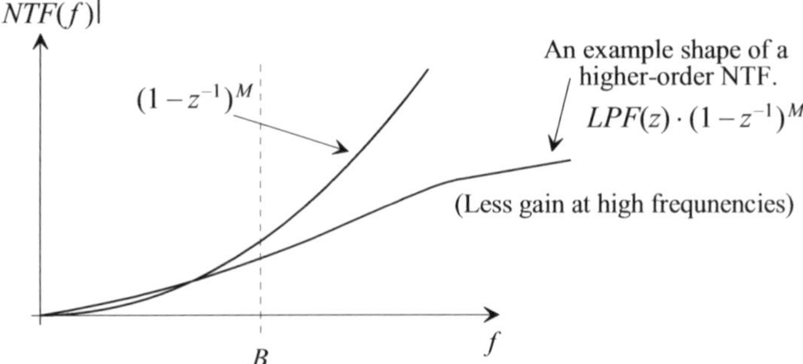

Figure 7.43 Showing the change in the NTF in a higher-order modulator.

so that the feed forward coefficients, a_x , can be used to help make the $STF(f)$ constant over the region of interest (the *STF* can be made to have an overall lowpass response). The *NTF* is given by

$$NTF(z) = \frac{1}{1 + A(z)B(z)} \tag{7.92}$$

or

$$NTF(z) = \frac{1}{1 - (z-1)^{-M} \cdot \sum_{i=1}^{M} b_i \cdot (z-1)^{i-1}} = HPF(z) \tag{7.93}$$

The coefficients, b_x, are selected for a highpass response. Note also that our coefficients are positive since the feedback paths, as seen in Fig. 7.42, are subtracting. The design of the modulator is performed by determining the feed-forward and feedback coefficients using basic digital-signal processing filter design (and, to keep the algebra simple, a computer program of some sort), then to simulate the design to see if it exceeds specifications. One challenge, among others, is to meet a given *SNR* without causing harmful integrator saturation.

7.3.4 Multi-Bit Modulators

Throughout this chapter we have assumed $N = 1$; that is, we have used a comparator for our quantizer in the forward path of our NS modulator. The main advantage of single-bit modulators, as discussed earlier, is the inherent linearity of the 1-bit feedback DAC. Feedback DAC linearity is important because the output of the DAC is directly subtracted from the input signal. Any distortion or nonlinearity (or noise) in the output of the DAC will directly affect the modulator's performance and, ultimately, limit the modulator's SNR. The benefits of using a multibit ($N > 1$) quantizer in a NS modulator are increased SNR (see Eq. 7.73), better stability (the modulator behaves closer to the linearized theory developed in this chapter), fewer spectral tones, and simpler digital-decimation filter. The drawbacks of using multibit topologies, are the increase in ADC complexity (the ADC must be a flash converter) and the need for the DAC to be accurate to the final accuracy of the modulator. The ADC errors, like gain errors in the integrators, are less important since they are in the forward, high-gain path of the modulator.

Simulating a Multibit NS Modulator Using SPICE

Figure 7.44 shows a circuit-level implementation of a first-order, multibit, NS modulator using a 4-bit ADC and DAC. Figure 7.45 shows the SPICE simulation outputs of this modulator. Most of the design effort, when developing multi-bit modulators, goes into the design of the feedback DAC. Because it is nearly impossible to design highly accurate (say 12-bits resolution or better) DACs without trimming, or some sort of error correction, methods have been developed that attempt to randomize DAC errors. If the errors appear as a random variable, they may appear as white noise in the output spectrum and not affect the SNR of the data converter.

Figure 7.46 shows one possible implementation of a DAC. This DAC utilizes resistive unit elements, resistors laid out in a square that connect, on one side, between *VDD* and ground while the other side connects to a rotating switch connected to the analog output. In other words, in one case we connect *VDD* to one corner of the resistor

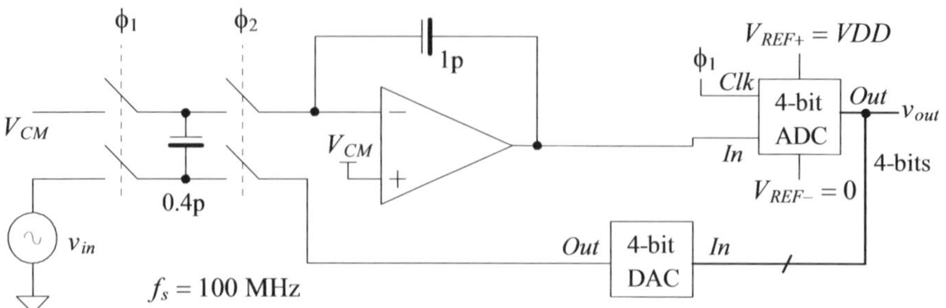

Figure 7.44 Circuit implementation of a first-order multi-bit NS modulator.

cube and ground to the opposite corner (assuming $V_{REF+} = VDD$ and $V_{REF-} = 0$). There exist two voltage dividers along each of the sides of the resistor square. The output of the DAC can change from zero, to $(1/8)VDD$, to $(2/8)VDD$, ... up to $(8/8)VDD$. Depending on the output of the decoder, one tap from each side is fed to the analog output. Because there are two sides, the outputs from each side are combined and effectively averaged.

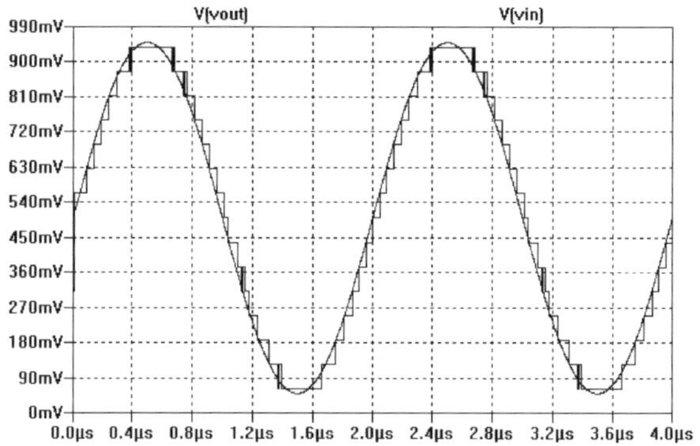

Figure 7.45 Output of the multi-bit modulator in Fig. 7.44.

The purpose of the counter is to vary the connections of *VDD* and ground around the outside of the resistive divider to randomize variations in the output voltage due to resistor mismatch. In order to understand this in more detail, consider a constant DAC output voltage of *VDD*/2. As the counter changes output values, so do the connections to *VDD* and ground in the resistor string. In order to keep a constant output voltage of *VDD*/2 the switches in the center of the DAC move accordingly based on the output of the counter and the input to the decoder. In this way variations in the resistors, hopefully, average out to a constant value.

With a little thought the reader can think of other schemes to attempt to randomize the errors that are fed back and subtracted from the input. In all cases it's presumed, for these techniques to work correctly, that the DAC errors average to zero or a very small value.

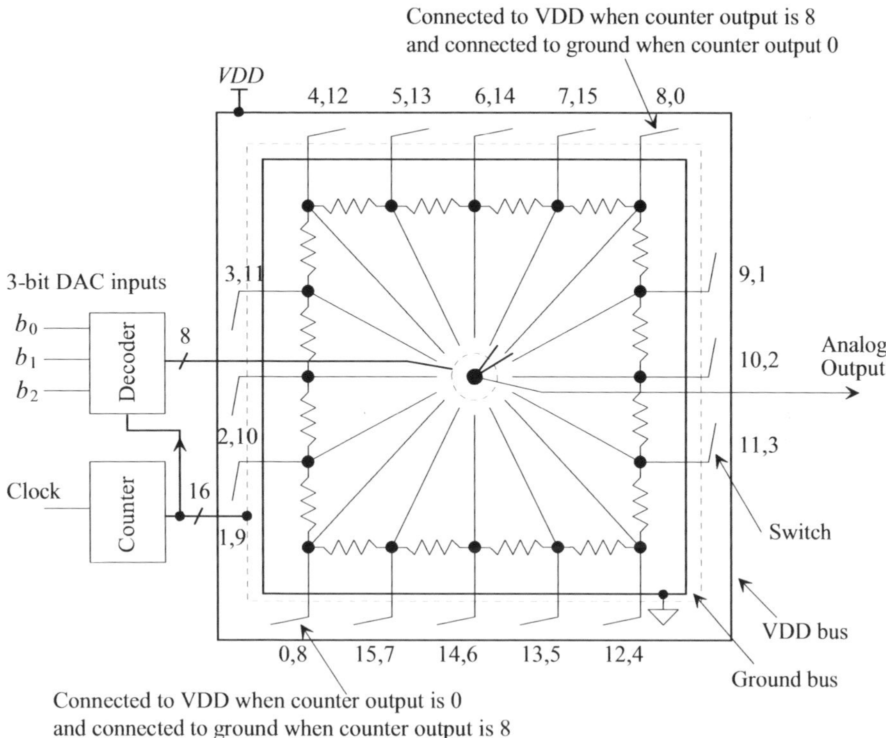

Figure 7.46 Implementation of a DAC for use in a multibit NS modulator.

7.3.5 Error Feedback

The NS topologies we've discussed so far are sometimes called *interpolative modulators* since the signal fed back is the average of the input signal interpolated between known values of the modulator output (the average of the modulator outputs is the input signal). However, NS modulators were first introduced (see C. C. Cutler, "Transmission systems employing quantization," 1960, U.S. Patent No. 2,927,962 [filed 1954]) using the error feedback topology shown in Fig. 7.47. Error feedback topologies are not used in analog input modulators because errors in the analog subtraction directly add to the input signal. We can use this topology, however, in the implementation of a digital input demodulator (sometimes also called a modulator), as the subtraction is digital.

Looking at Fig. 7.47 we note that by definition the difference between the input and the output of the quantizer is the quantization noise, $V_{Qe}(z)$. This noise is subtracted from the input after a delay (for a first-order modulator) resulting in

$$v_{out}(z) = v_{in}(z) - F(z) \cdot V_{Qe}(z) + V_{Qe}(z) = v_{in}(z) + V_{Qe}(z)[1 - F(z)] \tag{7.94}$$

Note that the signal transfer function for an error feedback-based modulator is simply one; that is, $STF(f) = 1$. For a first-order NS modulator we set $F(z) = z^{-1}$ (a register), which results in

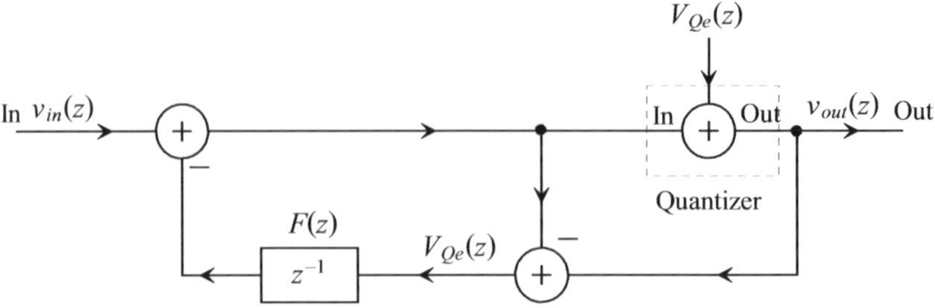

Figure 7.47 Block diagram of an error feedback modulator.

$$v_{out}(z) = v_{in}(z) + V_{Qe}(z) \cdot (1 - z^{-1}) \tag{7.95}$$

A second-order modulator with a $NTF(f) = (1 - z^{-1})^2$ would use a feedback filter, noticing from Eq. (7.94) that $NTF(f) = 1 - F(z)$, of

$$F(z) = 1 - (1 - z^{-1})^2 = z^{-1} \cdot (2 - z^{-1}) \tag{7.96}$$

Implementation of a second-order NS modulator is shown in Fig. 7.48. Note that when trying to implement higher-order modulators using error feedback we run into the same problem we encountered when using an interpolative modulator, namely, instability resulting from a *NTF* that is too large at higher-frequencies. As with interpolative modulators, we can design the *NTF* to have a highpass response.

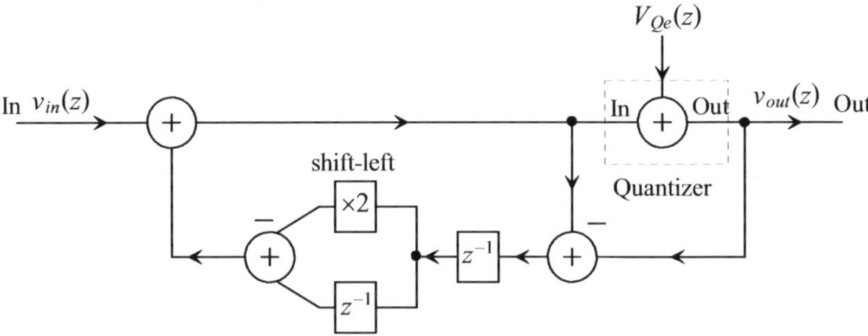

Figure 7.48 Block diagram of a second-order error feedback modulator.

We've introduced the error feedback topology with the idea that it can be used in a modulator (demodulator) that performs digital-to-analog conversion. We first introduced a modulator for use in a DAC back in Fig. 7.5. At this point we need to answer the question, "Why is the NS topology of Fig. 7.47 a better choice for DAC implementation, in general, than the topologies of Figs. 7.5 and 7.41?" The answer to this comes from the realization that the quantizer and difference block in Fig. 7.47 can be implemented by simply removing lower bits from the digital input words. This is illustrated in Fig. 7.49. The resulting error feedback modulator will be simpler to implement than the modulators based on interpolative topologies. Figure 7.50 shows Fig. 7.47 redrawn to show the simpler implementation.

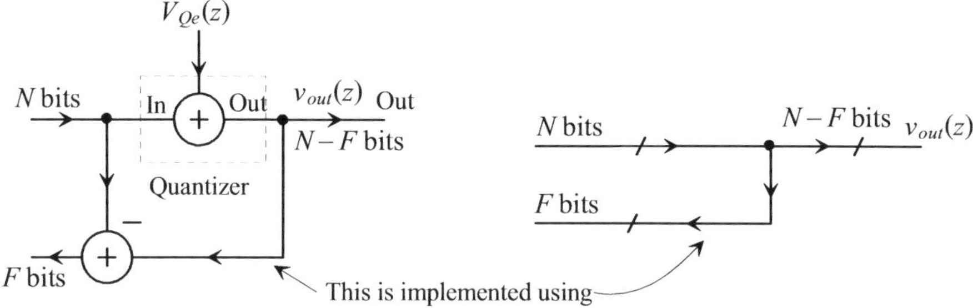

Figure 7.49 Showing how the quantizer and difference block are implemented.

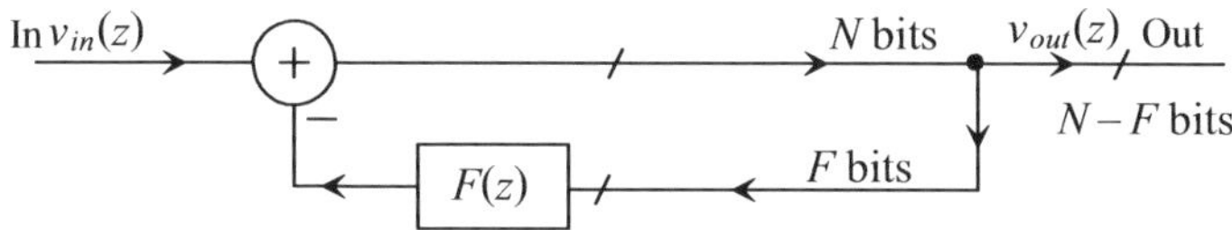

Figure 7.50 Block diagram of an error feedback modulator.

The number of bits used in the modulator, N, is selected to avoid overflow when the maximum input signal and fed-back signal are subtracted. When using two's complement numbers, the words input to the adder must be the same length. The smaller word's MSB is used to increase the smaller word's size until the word lengths match. We'll comment more on this important concern in a moment.

Figure 7.51 shows the block diagram of an NS-based DAC. As we saw in Fig. 7.4, if a 1-bit output is used, the modulator can be connected directly to the reconstruction filter (RCF). The 1-bit DAC is perfectly linear so distortion concerns are reduced. Using a multibit modulator and DAC gives a better SNR, for a given oversampling ratio and modulator order, as well as easing the requirements placed on the RCF. The drawback, as discussed earlier, is that the DAC must be accurate to the final desired output resolution since it is in series with the output signal path.

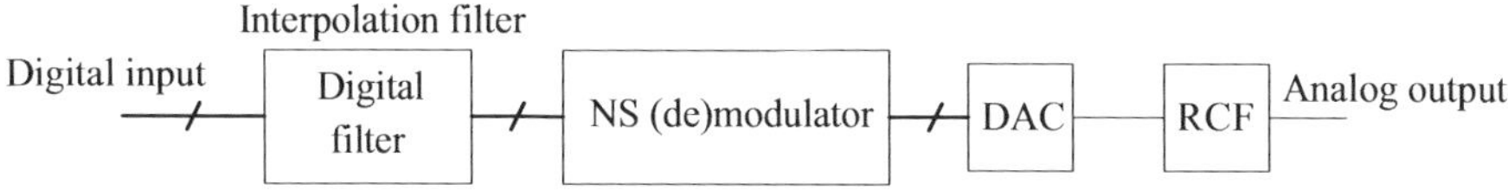

Figure 7.51 DAC using a NS modulator and digital filter.

Let's look at how to estimate the quantization noise added to the signal from the error feedback quantization process. Assuming we are using two's complement numbers we know that

$$\overbrace{0111111...}^{N\text{ bits}} = V_{REF+} - \overbrace{\frac{V_{REF+} - V_{REF-}}{2^N}}^{1\text{ LSB}} \tag{7.97}$$

and

$$1000000... = V_{REF-} \tag{7.98}$$

For the DAC to function properly we must change the numbers back to binary offset (complement the left-most or most-significant bit) unless the DAC input uses two's complement format. This is easy to see if the output of the modulator is a single bit ($N-F=1$) since an MSB of $1 = V_{REF+}$ and a $0 = V_{REF-}$ where $V_{LSB} = V_{REF+} - V_{REF-}$. By dropping F bits the voltage weighting of an LSB in the modulator output can be written as

$$V_{LSB} = \frac{V_{REF+} - V_{REF-}}{2^{N-F}} \text{ if } N-F > 1 \tag{7.99}$$

This result is used in Eq. (7.6) to estimate the quantization noise spectrum in a NS modulator.

Implementation Concerns

We know from our discussions in the last chapter that most digital additions and subtractions utilize two's complement numbers because of the simplicity (see Sec. 4.1.3 and the associated discussion) in implementing the hardware. However, consider the two's complement N-bit input in Fig. 7.49. If we drop the lower F bits, the resulting number fed back to $F(z)$ (the quantization noise) is not in two's complement format.

In order to circumvent these types of problems, the topology shown in Fig. 7.52 can be used. The input to the quantizer/subtractor is changed from two's complement format into binary offset format. (See Figs. 4.4 and 4.5 for a comparison of the formats.) Quantization is then performed; the lower bits are dropped from the output and fed back. The fed-back word (the quantization error) is then changed from a binary offset number back into a two's complement number. The size of the word fed back is adjusted to match the size of the modulator's input (knowing that the words used in two's complement arithmetic must be the same size so that the sign bit is in the same location in each word, see also Fig. 4.7).

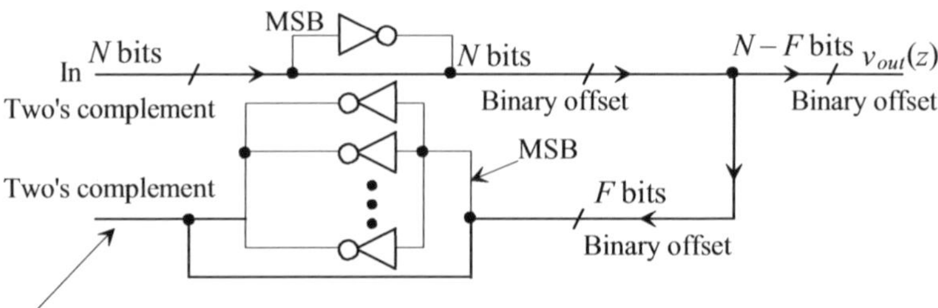

Figure 7.52 Implementation of the quantizer and difference blocks.

7.3.6 Cascaded Modulators

The NS modulators discussed up to this point, in this chapter, have been single feedback loop topologies with the general form seen in Fig. 5.38. This includes the higher-order topologies discussed in Sec. 7.3.1 and the error-feedback topologies of the last section. In this section we discuss cascaded or multistage NS modulators. The cascaded modulators discussed here are sometimes called *MultistAge noise SHaping or MASH modulators.* While our focus in this section is on modulators for ADCs, it is easy to extend the theory to modulators used in DACs.

We indicated, in the last section, that feeding back the quantization noise, $V_{Qe}(z)$, to the input isn't practical in analog implementations of NS modulators. The output of the analog subtraction, $V_{Qe}(z)$, would be added directly to the input signal. Instead of feeding $V_{Qe}(z)$ back to the input, cascaded modulators feed it forward to the input of another modulator. The second modulator's output is then a delayed version of $V_{Qe}(z)$ as well as its own unwanted modulation noise. If this output is subtracted from the output of the first modulator, we can effectively reduce the resulting overall quantization noise.

The major benefit of a cascaded topology is stability. Unconditionally stable first- and second-order loops can be cascaded to implement higher-order modulators. In addition, as we'll briefly discuss, modulators consisting of a first-stage modulator using a 1-bit ADC and DAC followed by a multibit modulator can provide reasonable resolutions with low oversampling ratio K.

Second-Order (1-1) Modulators

A second-order NS modulator can be implemented using a cascade of two first-order modulators (called a 1-1 modulator), as seen in Fig. 7.53. The output of the first modulator is given by

$$v_1(z) = z^{-1} v_{in}(z) + (1 - z^{-1}) V_{Qe1}(z) \tag{7.100}$$

while the output of the second modulator is

$$v_2(z) = -z^{-1} V_{Qe1}(z) + (1 - z^{-1}) V_{Qe2}(z) \tag{7.101}$$

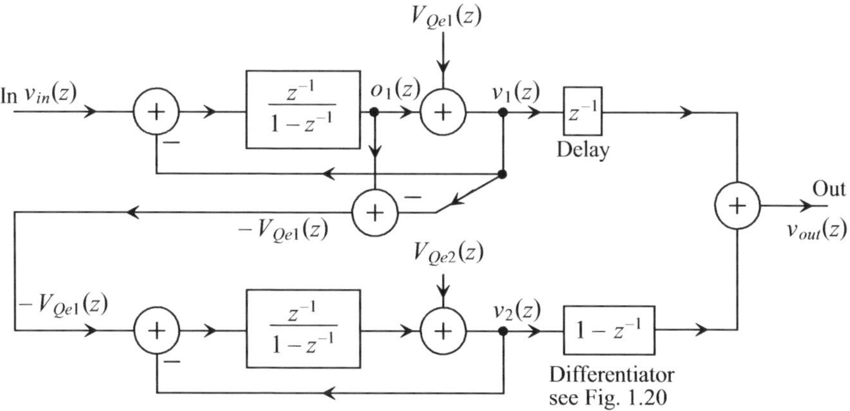

Figure 7.53 Second-order (1-1) cascaded modulator.

The overall modulator output is given by

$$v_{out}(z) = z^{-1}v_1(z) + (1 - z^{-1})v_2(z)$$
$$= z^{-2}v_{in}(z) + z^{-1}(1 - z^{-1})V_{Qe1}(z) - z^{-1}(1 - z^{-1})V_{Qe1}(z) + (1 - z^{-1})^2 V_{Qe2}(z)$$
$$= z^{-2}v_{in}(z) + (1 - z^{-1})^2 V_{Qe2}(z) \quad (7.102)$$

Note that the second modulator is used to subtract the first's quantization noise, $V_{Qe1}(z)$, from the final output. If all of the components are ideal, the resulting modulator has second-order noise shaping. In practice, however, the coefficients of $V_{Qe1}(z)$ in Eq. (7.102) will not exactly cancel. When this occurs, $V_{Qe1}(z)$ is said to *leak to the output of the modulator*. Differences in the coefficients are caused by gain errors in the first modulator's analog integrator when compared to the output of the digital differentiator.

Let's attempt to characterize the performance of the 1-1 modulator if the integrators have gain coefficients, G_I, other than one, as seen in Fig. 7.13. We can write the output of the first modulator's integrator in Fig. 7.53 as

$$o_1(z) = \frac{G_{I1} \cdot z^{-1}}{1 + (G_F - 1)z^{-1}} \cdot v_{in}(z) - \frac{G_{I1} \cdot z^{-1}}{1 + (G_F - 1)z^{-1}} \cdot V_{Qe1}(z) \quad (7.103)$$

Using Eq. (7.26) with this equation we can write

$$V_{Qe1out}(z) = v_1(z) - o_1(z) = \frac{(G_F - G_{I1}) \cdot z^{-1}}{1 + (G_F - 1)z^{-1}} \cdot v_{in}(z) + \frac{1 - (1 - G_{I1}) \cdot z^{-1}}{1 + (G_F - 1)z^{-1}} \cdot V_{Qe1}(z) \quad (7.104)$$

where, ideally, the output quantization noise of the first modulator, $V_{Qe1out}(z)$, is $V_{Qe1}(z)$. If the modulator is functioning properly, then $G_F = 1$ independent of G_I as discussed earlier. Equation (7.104) can then be written as

$$V_{Qe1out}(z) = (1 - G_{I1}) \cdot z^{-1} \cdot v_{in}(z) + [1 - (1 - G_{I1}) \cdot z^{-1}] \cdot V_{Qe1}(z) \quad (7.105)$$

Using this equation in Eq. (7.101) while assuming the second modulator uses an integrator scaling factor, G_{I2}, and G_{F2} is one results in (rewriting Eq. [7.102])

$$v_{out}(z) = z^{-2}v_{in}(z) + z^{-1}(1 - z^{-1})V_{Qe1}(z) - z^{-1}(1 - z^{-1})V_{Qe1out}(z) + (1 - z^{-1})^2 V_{Qe2}(z)$$
$$= \overbrace{z^{-2}v_{in}(z) + (1 - z^{-1})^2 V_{Qe2}(z)}^{\text{Desired output}} + \overbrace{[V_{Qe1}(z) - v_{in}(z)] \cdot z^{-2}(1 - z^{-1}) \cdot (1 - G_{I1})}^{\text{Unwanted term}} \quad (7.106)$$

While we can set the second modulator's integrator gain coefficient, G_{I2}, to 0.4 to avoid integrator saturation, as discussed earlier, we must set G_{I1} as close to unity as possible. Using a unity gain coefficient results in a reduction in the modulator's overall dynamic range (see Fig. 7.12 and the associated discussion). Note that the input signal appears in the unwanted term in Eq. (7.106). It should be obvious at this point that we can add scaling parameters at various points in the modulator to attempt to maximize the modulator's dynamic range. Also note that the number of bits in the 1-1 modulator's output will be more than one bit (two bits if comparators are used in each first-order modulator).

Third-Order (1-1-1) Modulators

By adding a third first-order modulator to our 1-1 modulator of Fig. 7.53, we get a 1-1-1 or third order modulator, Fig. 7.54. The output of the added third modulator can be written as

$$v_3(z) = -z^{-1}V_{Qe2}(z) + (1-z^{-1})V_{Qe3}(z) \tag{7.107}$$

while the ideal output of the 1-1-1 cascade is given by

$$v_{out}(z) = v_1(z)\cdot z^{-2} + v_2(z)(1-z^{-1}) + v_3(z)(1-z^{-1})^2 = z^{-3}v_{in}(z) + (1-z^{-1})^3 V_{Qe3}(z) \tag{7.108}$$

Again, as we saw in Eq. (7.106), noise from the first modulator can leak through to the output and spoil the overall cascade's *SNR*. Indeed, if the leakage from the first modulator is large enough, we get no benefit from adding the third modulator. Notice, in Eq. (7.106), that the unwanted term exhibits first-order differentiation, $(1-z^{-1})$. We might expect better overall performance, that is, less leakage if the first modulator is second order. The unwanted term would then exhibit second-order differentiation.

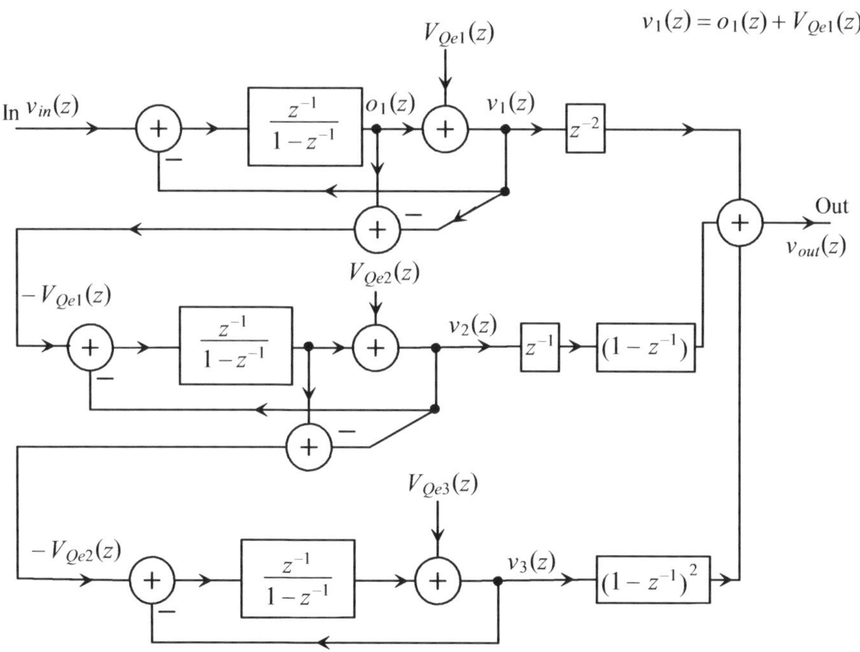

Figure 7.54 Third-order (1-1-1) cascaded modulator.

Third-Order (2-1) Modulators

A third-order modulator formed by using a second-order modulator followed by a first-order modulator is shown in Fig. 7.55. The output of the first modulator is given by

$$v_1(z) = z^{-1}v_{in}(z) + (1-z^{-1})^2 V_{Qe1}(z) \tag{7.109}$$

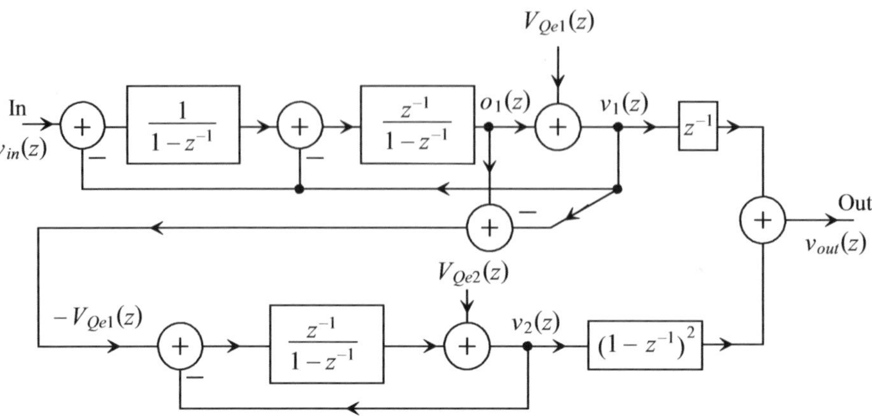

Figure 7.55 Third-order (2-1) cascaded modulator.

while the output of the second modulator is

$$v_2(z) = -z^{-1}V_{Qe1}(z) + (1-z^{-1})V_{Qe2}(z) \tag{7.110}$$

The output of the 2-1 modulator is then, ideally,

$$v_{out}(z) = v_1(z)z^{-1} + (1-z^{-1})^2 v_2(z) = z^{-2}v_{in}(z) + (1-z^{-1})^3 V_{Qe2}(z) \tag{7.111}$$

Let's attempt to characterize the leakage to the output by first determining the output of the second integrator $o_1(z)$ (the input to the comparator). We'll use the topology shown in Fig. 7.32, with $G_3 = 1$, to define our gains. The output, $o_1(z)$, is (assuming that $G_F = G_1G_2G_c = 1$)

$$o_1(z) = \frac{G_1G_2 \cdot z^{-1}v_{in}(z) - [(G_1G_2 + G_2) - G_2z^{-1}] \cdot z^{-1}V_{Qe1}(z)}{1 + z^{-1}\cdot(G_2G_c - 1) + z^{-2}(1 - G_2G_c)} \tag{7.112}$$

Again, writing the input to the second modulator as (using Eq. [7.61])

$$V_{Qe1out}(z) = v_1(z) - o_1(z)$$

$$= \frac{(1-G_1G_2)\cdot z^{-1}v_{in}(z) + \left[(1-z^{-1})^2 + (G_1G_2+G_2)\cdot z^{-1} - G_2z^{-2}\right]V_{Qe1}(z)}{1 + z^{-1}\cdot(G_2G_c - 1) + z^{-2}(1 - G_2G_c)} \tag{7.113}$$

noting that if $G_1 = G_2 = G_c = 1$ then $V_{Qe1out}(z) = V_{Qe1}(z)$. If we write the output of the cascade as

$$v_{out}(z) = z^{-2}v_{in}(z) + z^{-1}(1-z^{-1})^2V_{Qe1}(z) - z^{-1}(1-z^{-1})^2V_{Qe1out}(z) + (1-z^{-1})^3V_{Qe2}(z) \tag{7.114}$$

then

$$v_{out}(z) = \overbrace{z^{-2}v_{in}(z) + (1-z^{-1})^3V_{Qe2}(z)}^{\text{Desired output}} +$$

$$z^{-1}(1-z^{-1})^2\overbrace{\left[\frac{(1-G_1G_2)\cdot z^{-1}v_{in}(z) + [(1-G_2) - (G_2G_c - G_2)z^{-1}]\cdot z^{-1}V_{Qe1}(z)}{1 + z^{-1}\cdot(G_2G_c - 1) + z^{-2}(1-G_2G_c)}\right]}^{\text{Undesired term}} \tag{7.115}$$

When this equation is compared to Eq. (7.106), we see that the undesired term is second-order differentiated. Also, we have more control over the integrator gains. Third-order modulators using the 2-1 topology are much more robust than the 1-1-1-based topology and can provide output signals free of unwanted tones. Again, if integrator saturation (and thus dynamic range) isn't a concern, then we can set $G_1 = G_2 = 1$.

One of the interesting uses of the 2-1 modulator is the configuration where the first (second-order) modulator utilizes a 1-bit ADC and DAC, while the second (first-order) modulator utilizes a multibit ADC and DAC. The overall linearity of this topology is dominated by the second-order modulator, while the multibit modulator provides an enhancement in dynamic range for a given oversampling ratio. These very interesting data converters are discussed in greater detail in [7].

Implementing the Additional Summing Input

Before leaving our introduction to cascaded converters, let's discuss the implementation of the extra summing block used to generate the quantization noise, $V_{Qe}(z)$. Figure 7.56 shows the topology of the two summing blocks and how they can be combined.

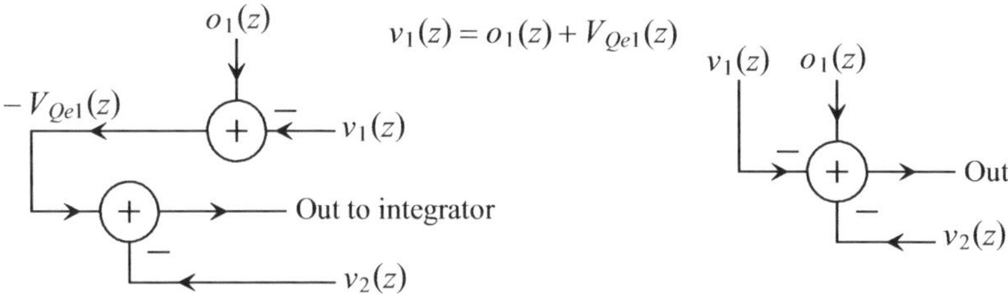

Figure 7.56 Showing implementation of the dual summing block as a single block.

One way to implement the extra subtracting input and the integrator is shown in Fig. 7.57. This DAI is a modification of the DAI shown in Fig. 7.33. The output of this integrator is related to the inputs by

$$v_{out}(z) = o_1(z) \cdot \frac{C_{I2}}{C_{F2}} \cdot \frac{z^{-1/2}}{1-z^{-1}} - v_2(z) \cdot \frac{C_{I2}}{C_{F2}} \cdot \frac{1}{1-z^{-1}} - v_1(z) \cdot \frac{C_{I22}}{C_{F2}} \cdot \frac{1}{1-z^{-1}} \tag{7.116}$$

If we set $C_{I2} = C_{I22} = C_{F2}$ and we realize that the comparator in the second modulator, assuming it is clocked with the rising edge of ϕ_1 (or the falling edge of ϕ_2), adds a half-clock cycle delay in series with the $o_1(z)$ input and a full clock cycle delay in series with $v_1(z)$ and $v_2(z)$, then we can write

$$v_{out}(z) = [o_1(z) - v_1(z) - v_2(z)] \cdot \frac{z^{-1}}{1-z^{-1}} \tag{7.117}$$

Figure 7.58 shows the implementation of a 2-1 modulator.

We could also use the topology shown in Fig. 7.59 to implement the summing block of Fig. 7.56. This topology has the benefit of using a single capacitor for a simpler circuit and no matching differences between C_{I2} and C_{I22}. Unfortunately, the topology is no longer insensitive to the parasitic capacitance on the top plate of the switched capacitor. In the parasitic insensitive topologies, Fig. 7.57 for example, the top plate of

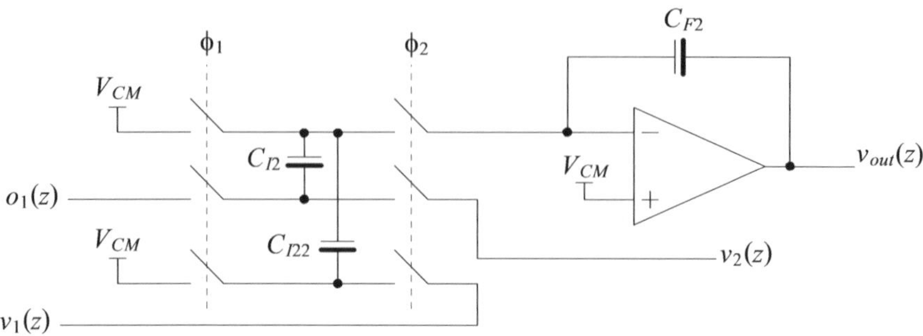

Figure 7.57 Implementing the dual summing block for a cascaded modulator.

the capacitor is always held at the common-mode voltage, V_{CM}. In the topology of Fig. 7.59 the top plate is charged to $v_1(t)$ when the ϕ_1 switches are closed and discharged to V_{CM} when the ϕ_2 switches are closed. The difference between these voltages combined with the value of the unwanted parasitic capacitance to ground on the top plate causes unwanted charge to transfer to the feedback capacitor and a gain error. This by itself isn't too bad. However, the unwanted capacitance can have a large depletion capacitance component, resulting in a voltage-dependent capacitance and thus nonlinear gain. Nevertheless, in some applications this topology may still prove useful.

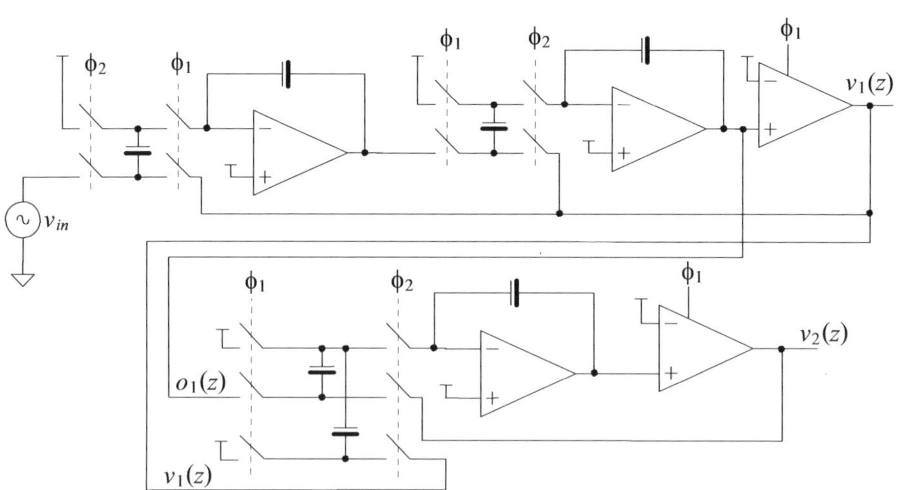

Figure 7.58 Implementation of a 2-1 NS modulator.

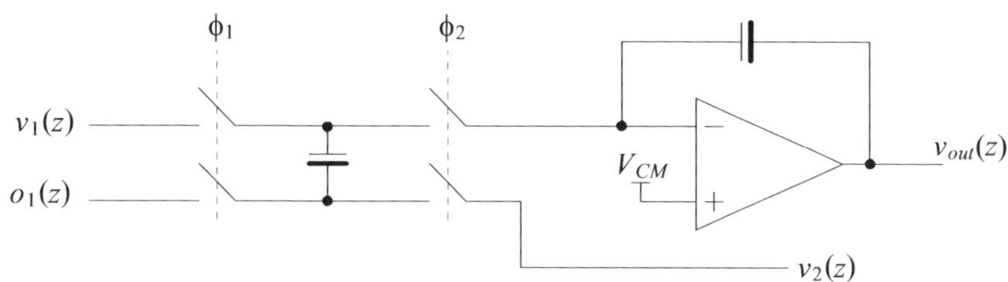

Figure 7.59 Implementing the dual summing block with a single capacitor results in sensitivity to the top plate parasitic capacitance.

ADDITIONAL READING

[1] R. Schreier and G. C. Temes, *Understanding Delta-Sigma Data Converters*, Wiley-IEEE Press, 2005. ISBN 978-0471465850

[2] S. K. Dunlap and T.S. Fiez, "A Noise-Shaped Switching Power Supply using a Delta-Sigma Modulator," *IEEE Transactions on Circuits and Systems I: Regular Papers*, Vol. 51, No. 6, pp. 1051 - 1061, June 2004

[3] A. Eshraghi and T.S. Fiez, "A Comparative Analysis of Parallel Delta-Sigma ADC Architectures," *IEEE Transactions on Circuits and Systems I: Fundamental Theory and Applications*, Vol. 51, No. 3, pp. 450 - 458, March 2004

[4] S. R. Norsworthy, R. Schreier, and G. C. Temes (eds.), *Delta-Sigma Data Converters: Theory, Design, and Simulation*, Wiley-IEEE Press, 1997. ISBN 978-0780310452

[5] R. T. Baird and T. S. Fiez, "Linearity Enhancement of Multibit ΔΣ A/D and D/A Converters using Data Weighted Averaging," *IEEE Transactions on Circuits and Systems II: Analog and Digital Signal Processing*, Vol. 42, No. 12, pp. 753 - 762, December 1995

[6] J. C. Candy and G. C. Temes (eds.), *Oversampling Delta-Sigma Data Converters*, Wiley-IEEE Press, 1992. ISBN 978-0879422851

[7] B. P. Brandt and B. A. Wooley, "A 50-MHz Multibit Sigma-Delta Modulator for 12-b 2-MHz A/D Conversion," *IEEE Journal of Solid-State Circuits*, Vol. 26, No. 12, pp. 1746 - 1756, December 1991

[8] B. E. Boser and B. A. Wooley, "The Design of Sigma-Delta Modulation Analog-to-Digital Converters," *IEEE Journal of Solid-State Circuits*, Vol. 23, No. 12, pp. 1298 - 1308, December 1988

[9] B. E. Boser, K-. P. Karmann, H. Martin, and B. A. Wooley, "Simulating and Testing Oversampled Analog-to-Digital Converters," *IEEE Transactions on Computer-Aided Design*, Vol. 7, pp. 668 - 674, June 1988

[10] J. C. Candy, B. A. Wooley, and O. J. Benjamin, "A Voiceband Codec with Digital Filtering," *IEEE Transactions on Communications*, Vol. 29, pp. 815 - 830, June 1981

QUESTIONS

7.1 Show how to derive Eqs. (7.1) and (7.2) from the block diagram seen in Fig. 7.1.

7.2 After reviewing Sec. 2.2.3, would it be possible to replace the delaying integrator seen in Fig. 7.2 with a non-delaying integrator? If so, what is the *NTF* and *STF* of the modulator? Is the modulator stable?

7.3 Using SPICE simulations, show how passing the digital signal seen in Fig. 7.3 through an RC lowpass filter will reduce the modulation noise in the signal and help to recover the original analog input signal. What happens to the original signal's amplitude if it's filtered, by the added RC filter, too much?

7.4 Show the spectrums (modulator input, digital output, and analog output after filtering) of the signals in question 7.3. Discuss what the spectrums indicate.

7.5 If an extra delay, z^{-1}, was added to the forward path of the modulator in Fig. 7.2 would the resulting topology be stable? Why or why not?

7.6 Show, using timing diagrams, how Eq. (7.3) is correct.

7.7 For the NS modulator shown in Fig. 7.5 used for digital to analog conversion, what component serves as the ADC? What component serves as the DAC?

7.8 Explain how the quantizer in Fig. 7.5 functions.

7.9 What are we assuming about an input signal if the modulation noise follows Eq. (7.5)?

7.10 What is the magnitude of Eq. (7.5) (plot it against frequency)?

7.11 What is the difference between quantization noise and modulation noise?

7.12 Show the steps and assumptions leading to Eq. (7.12).

7.13 Is the statement on page 238 that "every doubling in the oversampling ratio results in 1.5 bits increase in resolution" really true if K is small? Explain.

7.14 Does noise-shaping work for DC input signals? If so, how?

7.15 Show the steps leading up to Eq. (7.22).

7.16 What is the difference between a NS ADC and a Nyquist ADC?

7.17 In your own words, describe ripple in the output of a digital filter connected to an NS modulator.

7.18 Does adding a dither signal to the input of a NS modulator help reduce the peak-to-peak ripple in the digital filter output? Does it help to break up tones in the filter's output?

7.19 Derive Eq. (7.26).

7.20 Repeat Ex. 7.3 if the integrator's gain is set to 0.5.

7.21 Estimate the range of G_c for the quantizer seen in Fig. 7.16. How does this compare to the range of G_c for the 1-bit quantizer seen in Fig. 7.15? Name two benefits of the 1-bit quantizer over multi-bit quantizers.

7.22 Verify that Eq. 7.30 is correct. Use pictures if needed.

7.23 In your own words, and without equations, describe integrator leakage. How would you relate integrator leakage, found in integrators that use an active element as seen in the NS modulators found in this chapter, to the passive integrators used in the NS modulators discussed in the last chapter?

7.24 Would large parasitic op-amp input capacitance affect the settling time of a DAI? Verify your answer using simulations with ideal op-amps (infinite open-loop gain) and non-ideal op-amps (open-loop gains around the oversampling ratio, K).

7.25 In your own words, how does oversampling affect input-referred offset/noise and the effects of a jittery clock on an NS data converter?

7.26 Determine the transfer function of the DAI shown in Fig. 7.20.

7.27 Derive Eq. (7.51).

7.28 Sketch the implementation of the fully-differential second-order NS modulator.

7.29 Derive Eq. (7.61).

7.30 Sketch the fully-differential equivalent of Fig. 7.33.

7.31 Resimulate the modulator in Ex. 7.4 if the gains are set to one. Comment on the stability of the resulting circuit.

7.32 Resimulate the modulator in Ex. 7.4 if the input is only 50 mV. Comment on the stability of the resulting circuit.

7.33 Regenerate Fig. 7.40 by selecting integrator gains so that the maximum output swing of any op-amp is 800 mV peak-to-peak.

7.34 Comment, in your own words, on why the actual SNR of a NS-based data converter can be worse than the ideal values calculated in the chapter.

7.35 Derive Eq. (7.75). Make sure each step of the derivation includes comments.

7.36 Resimulate Fig. 7.44 using two-bit ADC and DAC.

7.37 Sketch a possible implementation of a quantizer for the error feedback modulator shown in Fig. 7.48.

7.38 What transfer function does the following block diagram implement?

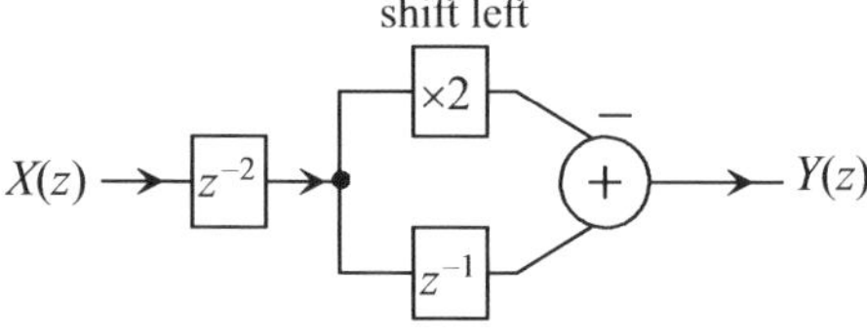

Figure 7.60 Circuit for question 7.38.

7.39 In Fig. 7.54 sketch the block diagram implementation of the circuit in series with the $v_2(z)$ output.

7.40 Derive the transfer function of the topology seen in Fig. 7.61 (show details of your derivation). What is the input common-mode voltage of the op-amp? Is this a concern when not using a negative supply voltage? If the input signals have a common-mode of *VDD*/2, does this affect the common-mode voltage of the circuit's output (remember that the op-amp is part of an integrator). Would it be a good idea, now that the inputs of the op-amp and the top plates of the capacitors are tied to ground or the virtual ground of the op-amp, to swap the bottom and top plates of the capacitors? Why or why not? Use SPICE to support your answers.

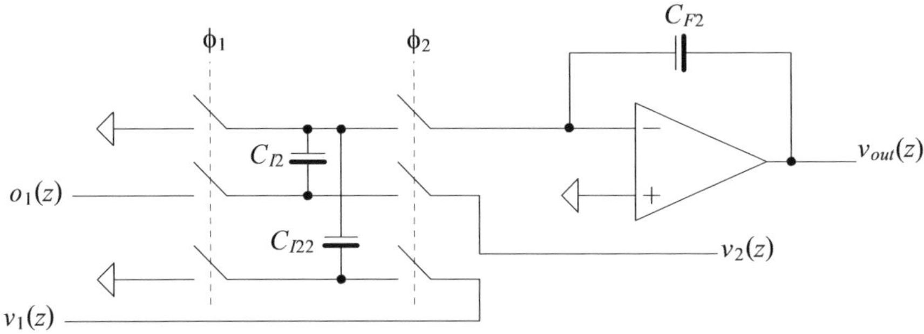

Figure 7.61 Circuit used in question 7.40.

7.41 Repeat question 7.40 for the op-amp circuit seen in Fig. 7.62.

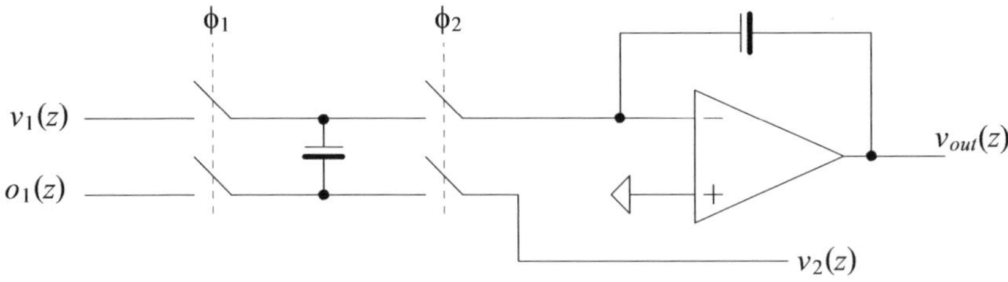

Figure 7.62 Circuit used in question 7.41.

Chapter 8

Bandpass Data Converters

The data converter topologies we've covered in the last few chapters have performed analog-to-digital conversion on a range of frequencies extending from DC to some frequency, B. These topologies are sometimes called lowpass data converters. In this chapter we develop the idea that we can perform data conversion on a bandwidth of frequencies that doesn't include DC. The topologies developed in this chapter are called bandpass data converters. Bandpass data converters are useful in communication circuits, especially wireless communications, where the information is modulated up to some higher frequency and thus contained in some fixed bandwidth in the frequency spectrum.

Before developing the idea of a bandpass data converter, let's review, on pages 4 through 6, why in-phase (I) and quadrature (Q) signals are used in communication systems. In simple terms, I/Q signals are used because we can send more information without increasing the bandwidth of the transmission channel. Note that other schemes that use several phase-shifted sinewaves (e.g., 0, 30, 60, and 90 degrees) may also be employed to further increase the information transmitted without increasing the channel bandwidth. In this chapter, however, we focus only on 0 and 90 degree phase-shifted (I/Q) sinewaves.

As seen in Fig. 1.6 and Eq. (1.11), summing I and Q components results in the I/Q signal. Figure 8.1 shows time-domain, equal amplitude, I and Q signals and their sum (the I/Q signal). The phase shift of the I/Q signal leads the Q component by 45 degrees while lagging the I component by 45 degrees when equal amplitude I and Q signals are used. By varying the amplitude of either (or both) the I or Q signals we vary the phase shift and amplitude of the I/Q signal (this variation in amplitude and phase shift represent the information we are transmitting). Note that this modulation scheme is often called *quadrature amplitude modulation* (*QAM*).

In order to recover the information from the I and Q components in the I/Q signal let's first write, see Eq. (1.10),

$$S_{IQ}(t) = A_I \cdot \cos 2\pi f_c \cdot t + A_Q \cdot \sin 2\pi f_c \cdot t \tag{8.1}$$

where f_c represents the frequency of a carrier signal while A_I and A_Q represent the information we are transmitting. Both A_I and A_Q vary with time and thus have some

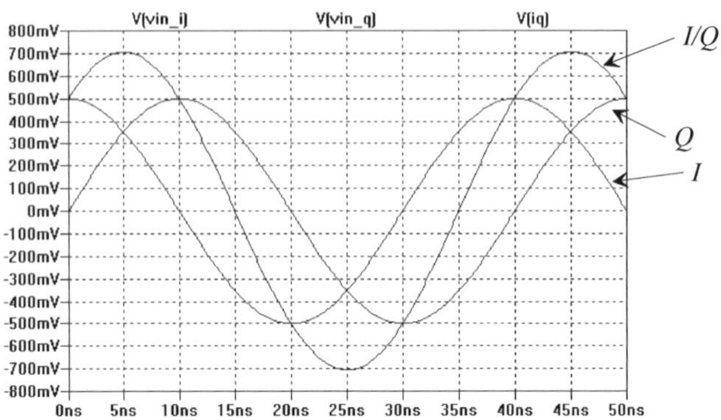

Figure 8.1 In-phase (I), quadrature (Q) signals, and I/Q signals.

spectral representation, $A_I(f)$ and $A_Q(f)$. The A_I component amplitude modulates the cosine term while the A_Q component amplitude modulates the sine term.

In order to recover the A_I and A_Q in the receiver we can multiply the received signal, ideally a scaled version of $S_{IQ}(t)$, by cosine and sine signals. For A_I,

$$S_{IQ}(t) \cdot \cos 2\pi f_c \cdot t = (A_I \cdot \cos 2\pi f_c \cdot t + A_Q \cdot \sin 2\pi f_c \cdot t) \cdot \cos 2\pi f_c \cdot t \tag{8.2}$$

Knowing

$$\cos A \cdot \sin B = \tfrac{1}{2}(\sin[B-A] + \sin[A+B]) \tag{8.3}$$

$$\cos A \cdot \cos B = \tfrac{1}{2}(\cos[B-A] + \cos[A+B]) \tag{8.4}$$

we get

$$S_{IQ}(t) \cdot \cos 2\pi f_c \cdot t = \underbrace{\frac{A_I(f)}{2}}_{\text{Desired signal}} + \underbrace{\frac{A_I}{2} \cdot \cos 2\pi 2 f_c \cdot t + \frac{A_Q}{2} \cdot \sin 2\pi 2 f_c \cdot t}_{\text{Remove by passing through a lowpass filter}} \tag{8.5}$$

The A_I component can be recovered by passing this signal through a lowpass filter to remove the higher frequency components. Note that it's assumed the maximum frequency of interest in $A_I(f)$ is less than f_c.

In order to recover A_Q from the received signal we multiply by a sinewave or

$$S_{IQ}(t) \cdot \sin 2\pi f_c \cdot t = (A_I \cdot \cos 2\pi f_c \cdot t + A_Q \cdot \sin 2\pi f_c \cdot t) \cdot \sin 2\pi f_c \cdot t \tag{8.6}$$

Knowing

$$\sin A \cdot \sin B = \tfrac{1}{2}(\cos[B-A] + \cos[A+B]) \tag{8.7}$$

we get

$$S_{IQ}(t) \cdot \sin 2\pi f_c \cdot t = \underbrace{\frac{A_Q(f)}{2}}_{\text{Desired signal}} + \underbrace{\frac{A_Q}{2} \cdot \cos 2\pi 2 f_c \cdot t + \frac{A_I}{2} \cdot \sin 2\pi 2 f_c \cdot t}_{\text{Remove by passing through a lowpass filter}} \tag{8.8}$$

We'll use these results later in the chapter in Sec. 8.2.4.

8.1 Continuous-Time Bandpass Noise-Shaping

When we developed the (lowpass) passive noise-shaping modulator back in Sec. 6.1 we used a capacitor to sum (sigma) the difference (delta) between the input and the fed back signals. The result was an absence of noise in the modulator's output signal, Fig. 6.5, at DC where the impedance of the capacitor is infinite. In order to implement a bandpass modulator let's replace the capacitor in Fig. 6.4 with an LC tank circuit, Fig. 3.39. Following the same reasoning, the output of our bandpass modulator should have no noise at the resonant frequency of the tank where its impedance is infinite, $f_0 = 1/2\pi\sqrt{LC}$.

8.1.1 Passive-Component Bandpass Modulators

Figure 8.2a shows the implementation of a passive-component bandpass noise-shaping modulator using this approach. To determine the transfer function of this topology let's use Fig. 8.2b. We can write (following what we did in Sec. 6.1 for the lowpass topologies)

$$\left(\frac{v_{in}-v_{int}}{R}+\frac{-v_{out}-v_{int}}{R}\right)\cdot\frac{sL}{1+s^2LC}+V_{Qe}(f)=v_{out} \tag{8.9}$$

and

$$v_{in}\cdot s\frac{L}{R}-v_{int}\cdot 2s\frac{L}{R}-v_{out}\cdot s\frac{L}{R}+V_{Qe}(f)\cdot(1+s^2LC)=v_{out}\cdot(1+s^2LC) \tag{8.10}$$

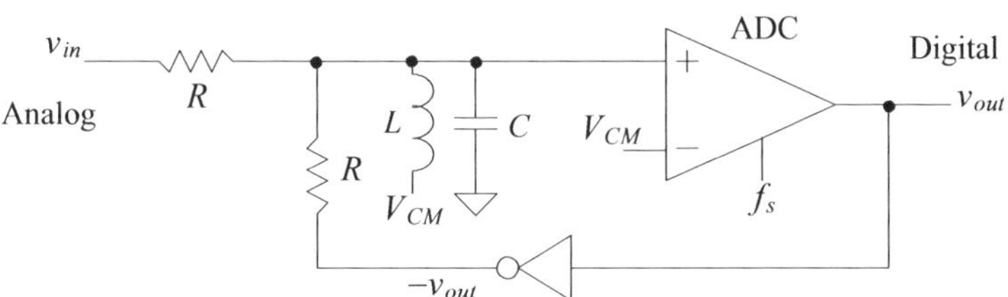

(a) Circuit implementation of a bandpass modulator.

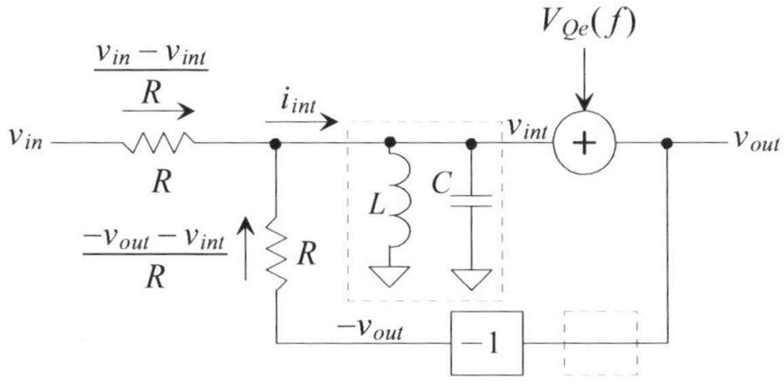

(b) Block diagram

Figure 8.2 A bandpass passive NS modulator.

$$v_{out} = v_{in} \cdot \frac{s\frac{L}{R}}{1 + s\frac{L}{R} + s^2LC} + V_{Qe}(f) \cdot \frac{1 + s^2LC}{1 + s\frac{L}{R} + s^2LC} - v_{int} \cdot \frac{2s\frac{L}{R}}{1 + s\frac{L}{R} + s^2LC} \tag{8.11}$$

and finally (see Fig. 3.39 and then Eq. [6.12] for the analogy with the lowpass passive modulator) we get

$$v_{out} = \overbrace{\frac{s\frac{1}{RC}}{s^2 + s\frac{1}{RC} + \frac{1}{LC}}}^{STF(f)} \cdot v_{in} + \overbrace{\frac{s^2 + \frac{1}{LC}}{s^2 + s\frac{1}{RC} + \frac{1}{LC}}}^{NTF(f)} \cdot V_{Qe}(f) + \underbrace{\frac{-2v_{int} \cdot s\frac{1}{RC}}{s^2 + s\frac{1}{RC} + \frac{1}{LC}C}}_{\text{Extra noise/distortion term}} \tag{8.12}$$

The *STF* shows a bandpass response while the *NTF* shows a band reject response, Fig. 8.3. Like the lowpass modulator we have the extra noise/distortion term that causes the modulator to deviate from the ideal behavior. We know, at this point, that we can drive v_{int} to zero by using an active element (amplifier). This is discussed in the next section.

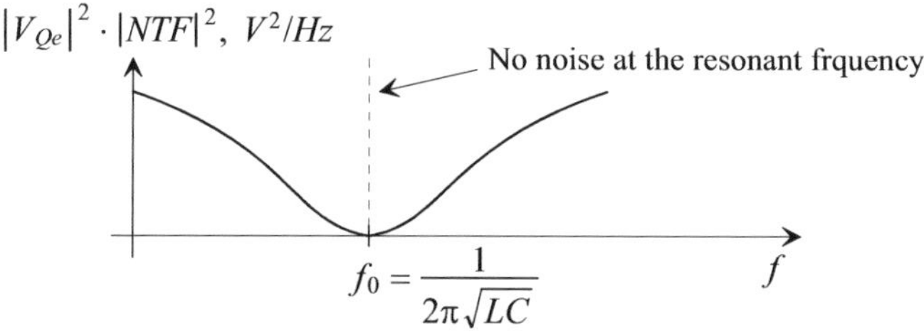

Figure 8.3 Modulation noise spectral density for a bandpass modulator.

Example 8.1

Simulate the operation of the NS modulator seen in Fig. 8.2 when *R* is 1k, *C* is 10 pF, *L* is 4.06 μH, and the clocking frequency is 100 MHz. Comment on the resulting simulation results and the operation/limitations of the circuit.

The conversion is centered around, from Fig. 8.3, 25 MHz. What this means is if our input signal is a sinewave at 25 MHz we should be able recover an exact replica of this input after removing the modulation noise. If a DC signal is applied to the modulator we should get out no signal (actually the modulator should output a sequence, like 101010101 that averages to V_{CM} or, for a mixed-signal system, no signal).

Figure 8.4 shows the simulation results. Note the ripple in the output, after filtering, amplitude. This variation can be removed by more filtering (increasing the *Q* of the simple filter used in the simulation by increasing the resistor) to remove noise. It should be noted that an ideal comparator was used in this simulation. As mentioned on page 207 a practical concern in a CT modulator is the effects of varying comparator delay which result in *amplitude modulation* in the output. Finally, Fig. 8.5 shows the spectrum of the modulator's output when a 25 MHz input tone is applied. Again, as in the passive noise-shaping lowpass topology, the extra, unwanted, term in Eq. (8.12) limits the *SNR*. ■

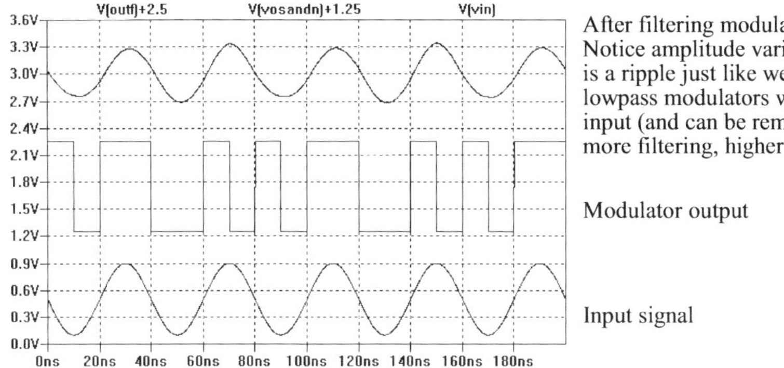

Figure 8.4 Simulation results for the modulator discussed in Ex. 8.1.

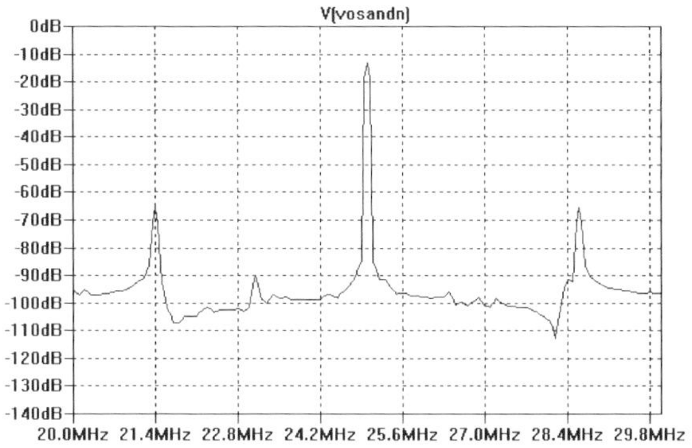

Figure 8.5 Spectrum of the modulator's output in Ex. 8.1.

An Important Note

Notice that the value of the inductor we used in Ex. 8.1 is too big to be integrated with the modulator. If one understands the concepts in this section we see that the *LC* tank can be replaced with an active circuit, like a g_m-C filter, that has a high-impedance over a range of frequencies (the range we want to convert from an analog representation to a digital representation).

8.1.2 Active-Component Bandpass Modulators

To remove the extra noise/distortion term in Eq. (8.12) we can use, as we did back in Sec. 6.2.4, an active circuit. Figure 8.6 shows one such implementation. Noting that the variation in v_{int} is zero when the op-amp's gain is infinite we can rewrite Eq. (8.12) for this topology as

$$v_{out} = \overbrace{\frac{s\frac{1}{RC}}{s^2 + s\frac{1}{RC} + \frac{1}{LC}}}^{STF(f)} \cdot v_{in} + \overbrace{\frac{s^2 + \frac{1}{LC}}{s^2 + s\frac{1}{RC} + \frac{1}{LC}}}^{NTF(f)} \cdot V_{Qe}(f) \tag{8.13}$$

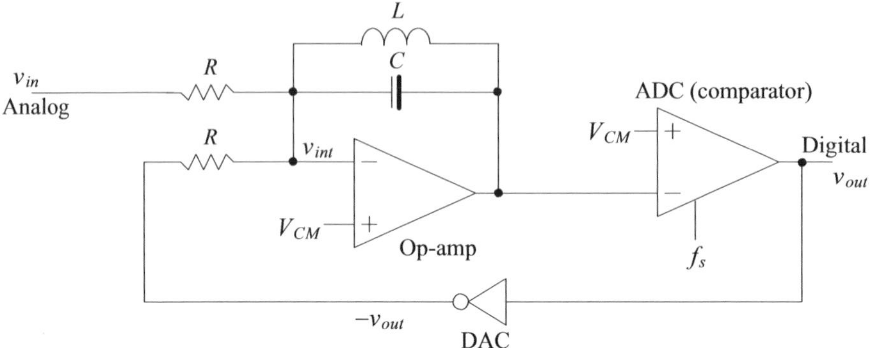

Figure 8.6 An active-integrator bandpass NS modulator.

Simulations demonstrating the operation of this topology are found at CMOSedu.com.

Signal-to-Noise Ratio

The output of the bandpass modulator can be passed through a bandpass filter, Sec. 4.2.3, with a bandwidth $2B$ to remove the modulation noise, Fig. 8.7. The smaller B, the lower the noise in the final digital output word and the larger the *SNR*. Again, the trade-off with using smaller filter bandwidth is that the allowable input signal frequency range shrinks.

Let's use the results in Sec. 6.1.1 (or Sec. 7.1.2) to help with our *SNR* estimate here. When we compare Fig. 6.8 (which doesn't show the noise power contributed from the negative components of the frequency spectrum) to Fig. 8.7, we see exactly the same shape (note that for large Q, as used in Ex. 8.1, this isn't true). Therefore we can use the equations derived earlier for the lowpass modulators to estimate the *SNR* in the bandpass topologies. Notice that if f_0 is 25 MHz and B is 100 kHz then the oversampling ratio, K, is 125.

Notice that the bandpass modulator seen in Fig. 8.6 has a second-order response, two poles in the *NTF*, so it's called a *second-order bandpass modulator* (even though its *SNR* is similar to a first-order lowpass response). A *fourth-order bandpass modulator*, Fig. 8.8 (again see simulations at CMOSedu.com), has behavior (an *SNR)* similar to the second-order noise shaping topology discussed in Secs. 6.2.4 and 7.2.

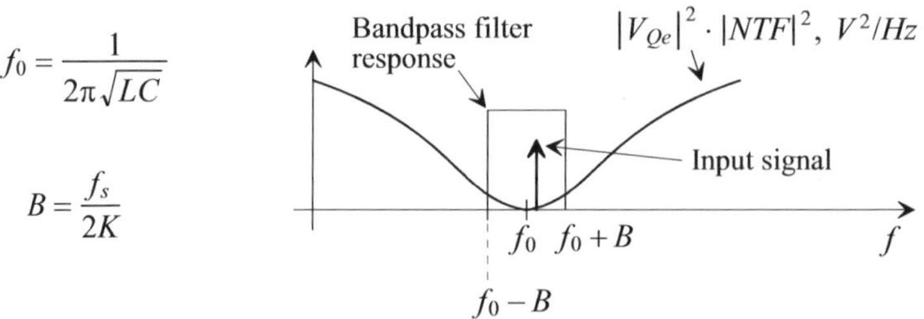

Figure 8.7 Filtering out modulation noise to calculate SNR.

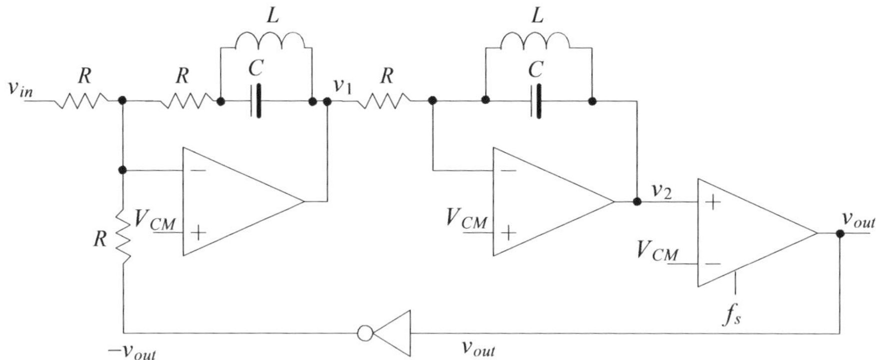

Figure 8.8 Fourth-order bandpass noise-shaping modulator.

8.1.3 Modulators for Conversion at Radio Frequencies

One area, at the time of this writing, that still has substantial room for development is the use of bandpass data converters for wireless (narrowband or radio frequency [RF]) communications. Currently mixing (down converting the transmitted information centered around some carrier frequency) into an intermediate frequency (or to baseband) is performed using a multiplier. A multiplier is an analog circuit (e.g., a Gilbert multiplier) with both continuous-time inputs and output. By using a bandpass converter for mixing we can go directly to digital format (the bandpass converter's input is analog, its output is digital, and it is clocked with a local oscillator). *The key to a successful* modulator design at high operating frequencies is minimizing the delay in both the forward and feedback paths of the converter (to ensure a stable converter). Thus the topology seen in Fig. 8.2 is more likely to be successful at the high conversion rates required in an RF circuit than the topologies seen in Figs. 8.6 and 8.8.

Figure 8.9 shows a topology built from the concepts used in the simple passive modulator in Fig. 8.2. The low-noise amplifier (LNA) isn't part of the modulator and is used to provide gain and isolation for the RF input signal (the noise performance of the receiver is dominated by the performance of this first stage). The output of the LNA is a current, $g_m v_{in}$ (this is the input to the modulator). A portion of the current fed back, I_{FB}, must, on average, equal this input current. The output of the modulator, even though a digital voltage value, can be thought of as the current fed back, i_{out}, to the resonator (which is equal, on average, to $N \cdot I_{FB}$ where N is the number of times the comparator output goes low). Deriving the *STF* and *NTF*, similar to Eq. (8.12) (but with an input current, $i_{in} = g_m v_{in}$, output current, i_{out}, and quantization noise current, $I_{Qe}(f)$) would show that we don't get an extra noise/distortion term. Note that increases in the *SNR* must employ K-path sampling, Fig. 6.24, rather than topologies that result in an increase in the modulator's forward delay. The use of K-path sampling is also required to ensure that an adequate oversampling ratio can be achieved when the carrier frequency is large. In addition, it's used to ensure the inherent lowpass filtering we get (the path filter seen in Fig. 6.24) when combining the comparator's outputs doesn't affect the desired signal (of course we can combine the comparator outputs so that they have a bandpass response, as discussed in Sec. 4.2.3, rather than a lowpass Sinc response). We leave the detailed implementations of these topologies to the refereed literature.

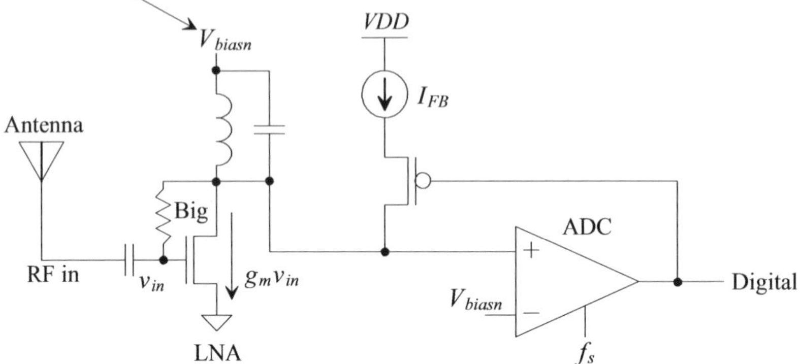

Figure 8.9 Design of a bandpass modulator for data conversion at RF.

8.2 Switched-Capacitor Bandpass Noise-Shaping

In Sec. 4.2.3 we discussed the idea that we can implement a Sinc-shaped averaging bandpass filter centered around $f_s/4$ (or $f_s/6$) having a transfer function, Eq. (4.25), of

$$H(z) = \frac{1-z^{-K}}{1+z^{-2}} \tag{8.14}$$

Comparing this equation to the equation for the equivalent lowpass averaging filter, Eq. (4.10), we see that transforming our lowpass modulator topologies into bandpass modulator topologies with bandpass responses centered at $f_s/4$ can be accomplished by

$$\text{Substituting } z^{-2} \text{ for } -z^{-1} \tag{8.15}$$

The discrete-analog integrator (DAI) discussed in Sec. 2.2.3 is the basic building block used in lowpass, switched-capacitor, modulator implementations. In order to implement a bandpass modulator at $f_s/4$ we need to replace the DAI used in the lowpass topologies with an $f_s/4$ resonator, an analog implementation of Fig. 4.23, or

$$\text{Replace } \frac{1}{1-z^{-1}} \text{ with } \frac{1}{1+z^{-2}} \tag{8.16}$$

or, when we can't avoid a delay in the implementation of the building block,

$$\text{Replace } \frac{z^{-1}}{1-z^{-1}} \text{ with } \frac{z^{-1}}{1+z^{-2}} \tag{8.17}$$

8.2.1 Switched-Capacitor Resonators

In order to move towards implementing a switched-capacitor $f_s/4$ resonator, consider the circuit in Fig. 8.10. The top portion of the circuit is simply the DAI discussed in Sec. 2.2.3 (Fig. 2.54). The bottom portion provides the positive feedback needed for the addition (instead of subtraction) of the delayed output. The transfer function of this circuit is

$$v_{out}(z) = \frac{v_1(z)\cdot z^{-1/2} - v_2(z)}{1+z^{-1}} \cdot \frac{C_I}{C_F} \tag{8.18}$$

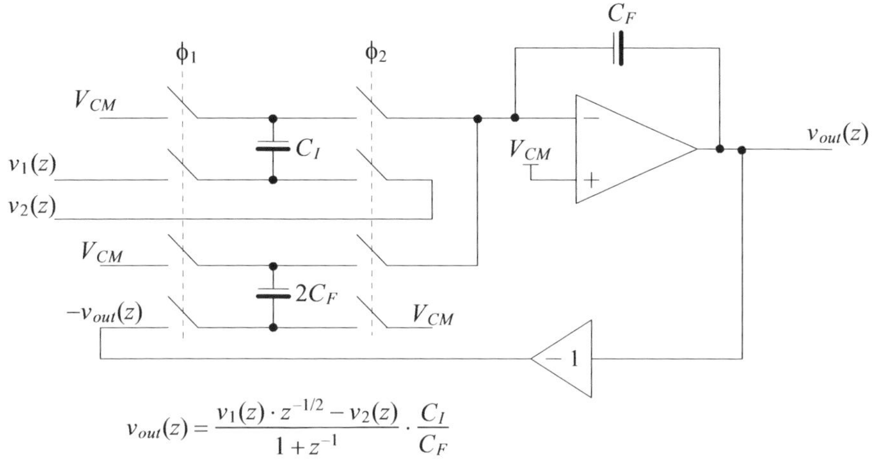

Figure 8.10 Implementing an $f_s/2$ resonator for use in a bandpass modulator.

or an $f_s/2$ resonator (the pole is located at $f_s/2$). In order to implement an $f_s/4$ resonator we'll need to use two delays in the feedback path. Implementing two (analog) delays, with any reasonable level of precision, in the feedback path is challenging.

In order to implement an $f_s/4$ resonator, we'll take two $f_s/2$ resonators, put them in parallel (see Eq. [2.56] in the *K*-path sampling discussion found in Sec. 2.1.6), and switch the phases of the clocks in each topology. So, for example, if we clock two of the resonators in Fig. 8.10 connected in parallel (see Fig. 2.37) at 50 MHz then the output word rate is 100 MHz (the output of the topology changes each time either ϕ_1 or ϕ_2 goes high). Reviewing Eq. (2.56) for two paths, $z \to z^2$, we can then re-write Eq. (8.18) as

$$v_{out}(z) = \frac{v_1(z) \cdot z^{-1} - v_2(z)}{1 + z^{-2}} \cdot \frac{C_I}{C_F} \tag{8.19}$$

or our desired $f_s/4$ resonator transfer function. Note that in this equation the v_1 input is delayed by a full clock cycle (see Fig. 2.55 to see how this is modeled for the DAI). We'll see that this delay means that we can't design a two stage bandpass modulator (a fourth-order modulator) using two non-delaying resonators. This shouldn't be a surprise since we also couldn't design a second-order lowpass modulator, Fig. 7.27c, with two non-delaying DAIs.

Example 8.2
Simulate the operation of the resonator in Fig. 8.10, clocked at 50 MHz, if $C_I = C_F = 1\ pF$. Comment on the stability of the resonator. Verify your comments using a SPICE simulation.

The pole of the resonator is located at $z = -1$, right on the unit circle (located at a frequency of $f_s/2$). We know, from Sec. 4.3.1, that for a discrete-time circuit to be stable its poles must lie inside the unit circle. We, therefore, expect the resonator to oscillate (become unstable), Fig. 8.11. ■

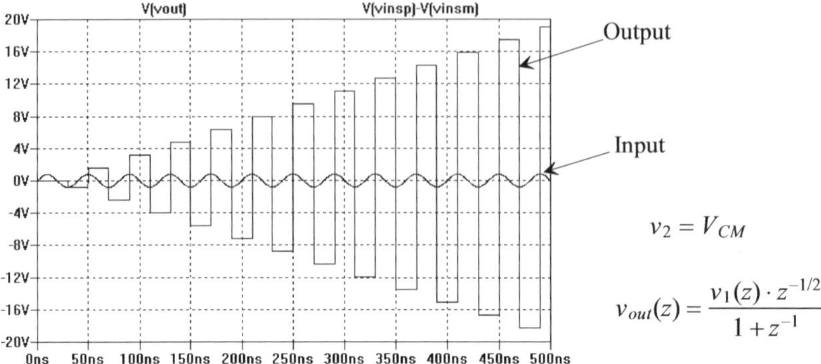

Figure 8.11 Simulating the operation of the resonator in Fig. 8.10.

Example 8.3

Suggest a modification to the resonator in Fig. 8.10 to eliminate the need for the gain of −1 block. Verify your circuit modification with SPICE.

Reviewing Fig. 2.56 we see that if we connect the fed back output signal to the top plate of the $2C_F$ capacitor, Fig. 8.12, then the signal is inverted when it's transferred to the output of the resonator. Simulation files for this circuit verifying correct operation are available at CMOSedu.com. Note that we'll use the topology seen in Fig. 8.10 since it is more tolerant of charge injection errors on the $2C_F$ capacitor than the topology seen in Fig. 8.12. ■

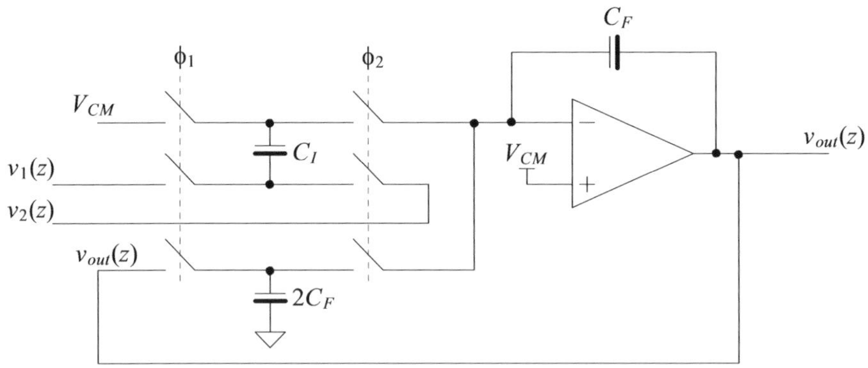

Figure 8.12 Implementing a resonator for use in a bandpass modulator.

8.2.2 Second-Order Modulators

Figure 8.13 shows the implementation of a second-order modulator based on the first-order lowpass topology seen in Fig. 7.1. Remembering, from page 290, that a second-order bandpass modulator's *SNR* is similar to (calculated in the same way as) the first-order lowpass modulator's *SNR* we can write

$$v_{out} = V_{Qe}(z) + (v_{in} - v_{out}) \cdot G_1 G_c \cdot \frac{-z^{-2}}{1+z^{-2}} \tag{8.20}$$

or, assuming $G_1 G_c = 1$ (as discussed in Sec. 7.1.5),

$$v_{out} = v_{in} \cdot \overbrace{(-z^{-2})}^{STF} + V_{Qe}(z) \cdot \overbrace{(1+z^{-2})}^{NTF} \tag{8.21}$$

Note the inversion in the signal transfer function. This is trivial to remove; for an analog signal see Fig. 3.6 and, for a digital signal, see Fig. 6.3. The noise transfer function, $1+z^{-2}$, has the shape seen in Fig. 8.7. At $f_s/4$ the modulation noise is zero increasing as we move away from $f_s/4$. Note that this topology is second-order because the number of poles in the *NTF* is two (a first-order lowpass modulator has one pole in its *NTF*).

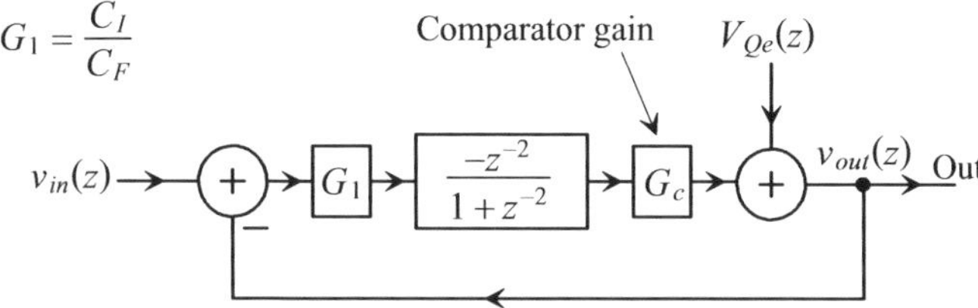

Figure 8.13 Block diagram of a second-order bandpass modulator.

Example 8.4
Simulate the operation of the second-order modulator seen in Fig. 8.13 (show the modulator's output spectrum) with a clocking frequency of 50 MHz (data coming out of the modulator at a rate of 100 MHz because of the 2-paths used) and an input frequency of 25.1 MHz. If the digital output of the modulator is run through an ideal digital bandpass filter with a pass frequency range of 25 $MHz \pm 390$ kHz then estimate the final SNR_{ideal}.

Figure 8.14 shows the modulator's output spectrum. Notice how the modulation noise increases as the frequency moves away from $f_s/4$ (25 MHz here). The oversampling ratio, *K*, can be calculated using, once again,

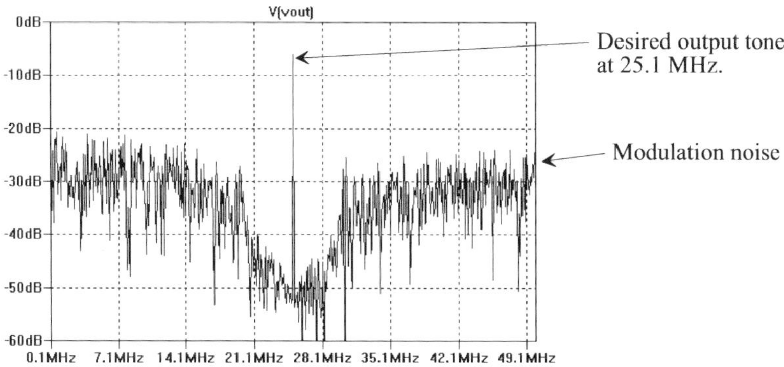

Figure 8.14 Spectrum of the modulator's output discussed in Ex. 8.4.

$$K = \frac{f_s}{2B} \tag{8.22}$$

For the present example f_s = 100 MHz and B = 390 kHz so K = 128. From Eq. (7.14) the SNR_{ideal} is 66 dB. ■

8.2.3 Fourth-Order Modulators

Figure 8.15 shows the block diagram of a fourth-order bandpass modulator formed by transforming the lowpass second-order modulator seen in Fig. 7.26c. Again, the bandpass topology is now called a fourth-order modulator because the number of poles in the *NTF* is four. The transfer function for this, f_s /4, bandpass modulator (sometimes called a *quadrature modulator*) is

$$v_{out}(z) = v_{in}(z) \cdot \overbrace{(-z^{-2})}^{STF} + V_{Qe}(z) \cdot \overbrace{(1 + z^{-2})^2}^{NTF} \tag{8.23}$$

Again, as discussed in Sec. 8.1.2, the fourth-order bandpass modulator's *SNR* can be estimated using the same approach that we used for the second-order lowpass modulator discussed in Sec. 7.2.

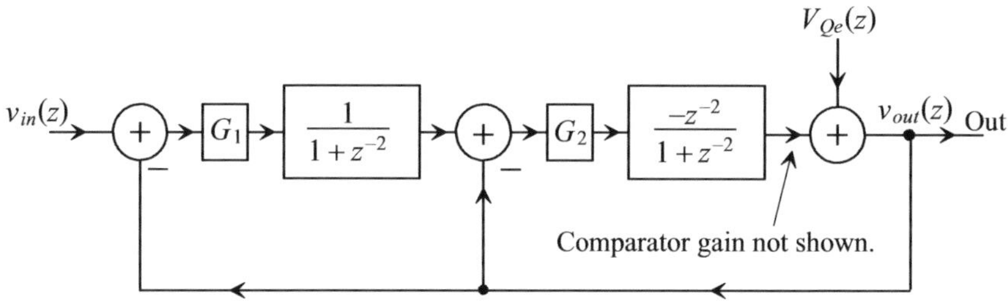

Figure 8.15 A fourth-order bandpass modulator.

Example 8.5
Repeat Ex. 8.4 using the fourth-order topology seen in Fig. 8.15.

Figure 8.16 shows the modulator's output spectrum. The oversampling ratio, K, is again 128 so using Eq. (7.52) the SNR_{ideal} is 100 dB. ■

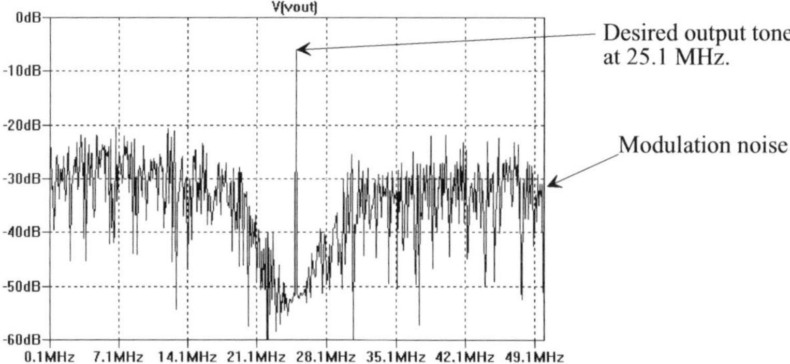

Figure 8.16 Spectrum of the modulator's output discussed in Ex. 8.5.

A Common Error

Back in Fig. 7.39 and the associated text we discussed that adding gratuitous delay to the forward path of the modulator would move it towards instability. Consider using a delaying resonator, Fig. 8.17, for the first stage of the fourth-order modulator. The first stage is made delaying by simply switching the phases of the clock signals used in this circuit. Unfortunately, the added two clock cycle delay in the forward path, for a total of four clock cycles of delay, makes the topology unstable. Simulation examples at CMOSedu.com, for Fig. 8.17, can be used to help show that this modulator *is always unstable*. The output of the modulator oscillates at $f_s/4$ regardless of the input signal.

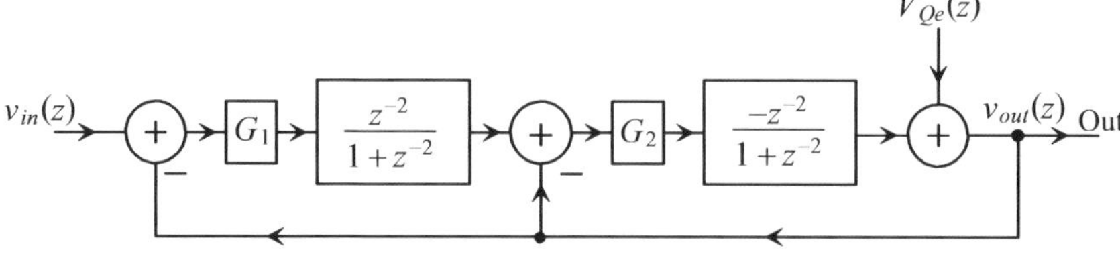

Figure 8.17 A fourth-order bandpass modulator using two delaying resonators (bad).

A Comment about $1/f$ Noise

Notice that in the bandpass modulators we've discussed $1/f$ noise isn't a concern. The $1/f$ noise doesn't have a significant effect on the signal at $f_s/4$ (assuming $f_s/4$ is a relatively large frequency). At lower frequencies the digital filter connected to the modulator's output removes both the modulation noise and the $1/f$ noise.

8.2.4 Digital *I/Q* Extraction to Baseband

One of the benefits of digitizing the signal around $f_s/4$ is the ease with which we can extract the *I* and *Q* components in the transmitted signal (see the first few pages of this chapter). The multiplication by the sine and cosine terms, see Eqs (8.2) and (8.6), used to recover the original signals simplifies to multiplying by either +1, −1, or 0, Fig. 8.18. Extracting the *I/Q* components digitally eliminates the *I* and *Q* channel phase response mismatch (difference in delay through each path) encountered in typical baseband demodulators. Figure 8.19 shows the simulation results showing how the 100 kHz (offset from 25 MHz) signal seen in Fig. 8.16 is demodulated to baseband. Note that the input in the simulation is simply a 25.1 MHz sinewave, not an *I/Q* generated signal (see simulation example at CMOSedu.com using an *I/Q* signal).

After looking at Fig. 8.18 we might wonder how we multiply the modulator output, a 1-bit word, by 1, 0, and −1. As seen in Fig. 6.3, a 1 coming out of the modulator corresponds to 01 (+1) in two's complement while a 0 corresponds to 11 (−1). Multiplying these outputs by 0 results in an output of 00 (0) in two's complement. Multiplication by +1 or −1 results in 01 × (+1) = 01, 01 × (−1) = 11, 11 × (+1) = 11, and 11 × (−1) = 01. A data selector and a multiplexer with some logic for output selection can be used to implement the multiplier (see the simulation example used to generate Fig. 8.19 at CMOSedu.com for one example).

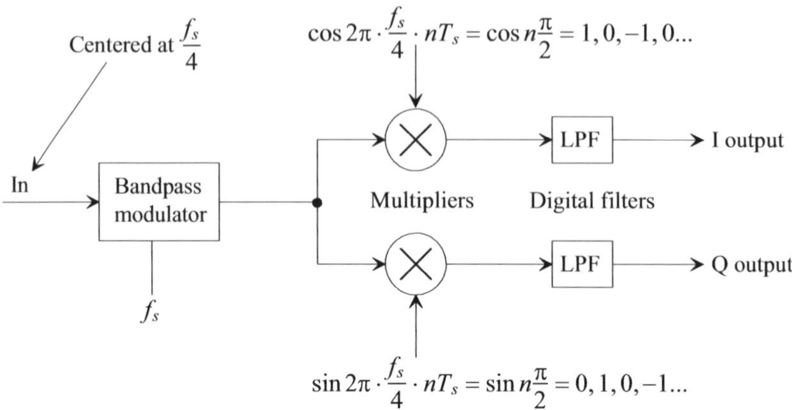

Figure 8.18 Digital I/Q demodulation.

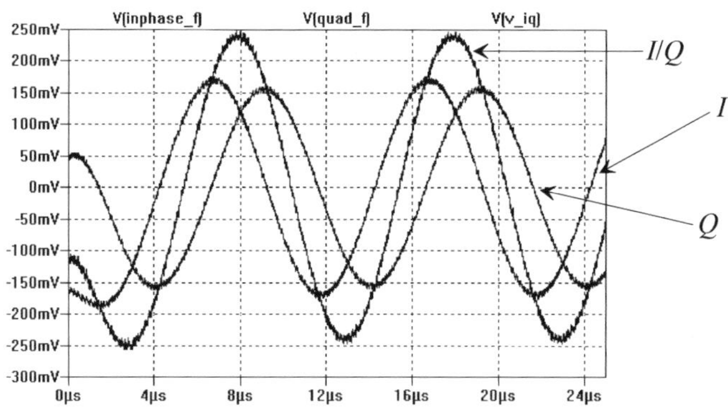

Figure 8.19 Showing how I and Q components of a signal can be extracted digitally.

ADDITIONAL READING

[1] R. Schreier and G. C. Temes, *Understanding Delta-Sigma Data Converters*, Wiley-IEEE Press, 2005. ISBN 978-0471465850

[2] A. I. Hussein and W.B. Kuhn, "Bandpass ΣΔ Modulator Employing Undersampling of RF Signals for Wireless Communication," *IEEE Transactions on Circuits and Systems II: Analog and Digital Signal Processing*, Vol. 47, No. 7, pp. 614 - 620, July 2000

[3] A. Jayaraman, P.F. Chen, G. Hanington, L. Larson, and P. Asbeck, "Linear High-Efficiency Microwave Power Amplifiers using Bandpass Delta-Sigma Modulators," *IEEE Microwave and Guided Wave Letters*, Vol. 8, No. 3, pp. 121 - 123 March 1998

[4] A. K. Ong and B. A. Wooley, "A Two-Path Bandpass SD Modulator for Digital IF Extraction at 20 MHz," *IEEE Journal of Solid-State Circuits*, Vol. 32, No. 12, pp. 1920-1934, December 1997

[5] S. R. Norsworthy, R. Schreier, and G. C. Temes (eds.), *Delta-Sigma Data Converters: Theory, Design, and Simulation*, Wiley-IEEE Press, 1997. ISBN 978-0780310452

[6] B.-S. Song, "A 4th-Order Bandpass ΔΣ Modulator with Reduced Number of Op-Amps," *IEEE Solid-State Circuits Conference*, pp. 204-205, 367, February 1995

[7] B.-S. Song, "A Fourth-Order Bandpass Delta-Sigma Modulator with Reduced Numbers of Op-Amps," *IEEE Journal of Solid-State Circuits*, Vol. 30, No. 12, pp. 1309-315, December 1995

[8] D. B. Ribner, "Multistage Bandpass Delta Sigma Modulators," *IEEE Circuits and Systems II: Analog and Digital Signal Processing*, Vol. 41, No. 6, pp. 402 - 405, June 1994

[9] S.A. Jantzi, W.M. Snelgrove, and P.F. Ferguson Jr., "A Fourth-Order Bandpass Sigma-Delta Modulator," *IEEE Journal of Solid-State Circuits*, Vol. 28, No. 3, pp. 282 - 291, March 1993

[10] J. C. Candy and G. C. Temes (eds.), *Oversampling Delta-Sigma Data Converters*, Wiley-IEEE Press, 1992. ISBN 978-0879422851

QUESTIONS

8.1 Show, using SPICE, how to adjust the phase and amplitude of the I and Q signals discussed in the beginning of the chapter to modulate the amplitude and phase of the resulting I/Q to construct a constellation diagram for 8-level rectangular QAM.

8.2 Suggest a topology for the bandpass passive-integrator NS modulator where the input and fed back signals are currents. Derive a transfer function for your design. Does your topology have the extra noise/distortion term seen in Eq. (8.12)? Why or why not? Simulate the operation of your design.

8.3 Show the details of deriving the transfer function for the modulator in Fig. 8.6.

8.4 Repeat question 8.3 for the modulator seen in Fig. 8.8.

8.5 Derive the transfer function for the modulator seen in Fig. 8.9.

8.6 Sketch the implementation of a modulator, based on the topology seen in Fig. 8.9, but using a multi-bit quantizer and feedback DAC.

8.7 Show the details of how Eq. (8.18) is derived.

8.8 Derive the transfer function of the modulator seen in Fig. 8.12.

8.9 Using the modulator topology in Ex. 8.4, show that if we apply a 25 MHz input sinusoid to the modulator we can recover this input signal by passing the output digital data through a bandpass filter with a very small bandwidth (show that the input and output signal amplitudes are equal).

8.10 Derive the transfer function of the topology seen in Fig. 8.17. Verify that the topology is unstable by determining the location of the topology's poles.

8.11 Using a bandpass modulator and digital demodulation (sketch the schematic of your design) show how to recover a 10 kHz sinewave that is amplitude modulated with a carrier frequency of 1 MHz. Use SPICE to verify the operation of your design.

Chapter

9

A High-Speed Data Converter

The majority of the data converter topologies we've discussed up to this point have traded off time for resolution (so the signal bandwidth is generally much smaller than the clocking frequency). In other words, to get higher signal-to-noise ratios, SNRs, (wider output word size, N) we've averaged (filtered) the output of a noise-shaping modulator to remove the modulation noise. The averaging (again, filtering) has the undesired effect of reducing the noise-shaping modulator's allowable signal bandwidth. In this chapter we turn our attention towards high-speed topologies. Our approach is based on the material covered in Sec. 6.2.3 where K-paths are used but with switched-capacitor, SC, circuits used for the required feedback subtraction from the input (delta) and integration (sigma).

9.1 The Topology

Our high-speed topology, Fig. 6.24, consists of a single integrator (or resonator for a bandpass topology) that sums the difference between the input signal and the signals from the K feedback paths. Let's start out this section by discussing the clock signals.

9.1.1 Clock Signals

Examine the clock signals seen in Fig. 9.1. Notice that the last four clock signals in this figure can be generated by simply inverting the first four clock signals (so for K-paths we need $K/2$ phases of a clock signal at f_s). By sampling an input waveform on the rising (or falling) edges of these clock signals we get an effective sampling frequency of

$$f_{s,new} = K_{path} \cdot f_s \tag{9.1}$$

The key thing to note is that the frequency of the clock signal, f_s *, doesn't set the new sampling frequency.* Rather the delay between rising edges sets the effective sampling frequency, $f_{s,new}$. Why is this important? If we generate these clock signals using a ring oscillator, for example, made with inverters having delays of 10 ps then the new, or effective, sampling frequency is 100 GHz! If our desired signal bandwidth is 100 MHz then we are oversampling 1000 to 1. The frequency f_s is selected to allow the circuits in each path to have enough time to respond (settle). Using the ring oscillator example, this consideration (path settling time) tells us how many inverters we need, the number of paths in the converter, and thus the ring oscillator's oscillation frequency f_s.

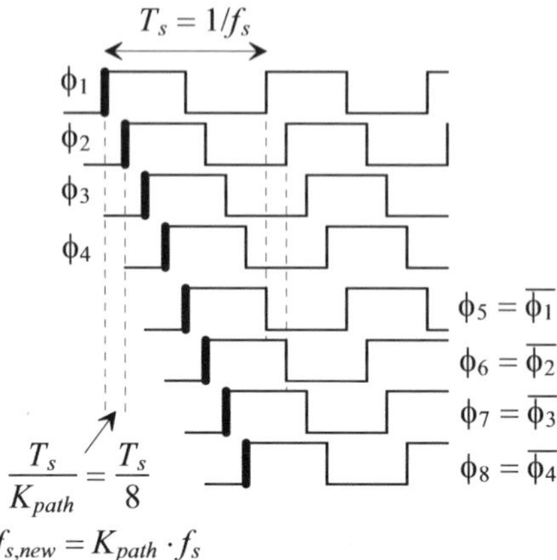

Figure 9.1 Showing the clock signals used for time-interleaved sampling and the high-speed topologies discussed in this chapter.

Path Settling Time

Figure 2.37 details the clock signals used in traditional multi-rate signal processing. These signals should be compared to the clock signals seen in Fig. 9.1. When processing digital signals that are clocked, for example, on the rising edge of the clock signal, the signals in Figs. 2.37 and 9.1 are equivalent. In an analog circuit, however, the signals seen in Fig. 9.1 allow the capacitors more time to be charged or discharged. This is illustrated in Fig. 9.2. *We need to be careful here.* When sampling the input signal we must ensure that the input signal can charge the sampling capacitors, using either clock signals, in T_s/K_{path}. If this isn't the case then we effectively filter our input signal reducing its amplitude (the benefit of this inherent filtering is a built-in anti-aliasing filter, AAF). The reward for using the wider clock signals is that we can adjust the feedback signal's timing to help keep the topology stable. It's also critically important for a precision feedback signal from the DAC. The fact that the rate the capacitors charge/discharge is highest right after the switches close can be used to ensure that a single path pushes the summing circuit in the right direction. However, if the "push" isn't significant enough (doesn't occur quickly enough before clocking the next path) the topology will oscillate. The summing circuit's signal, say the voltage across C in Fig. 6.24, will move around confused because of the varying, or conflicting, information fed back from each path in the converter (see Sec. 9.1.5 for further discussions).

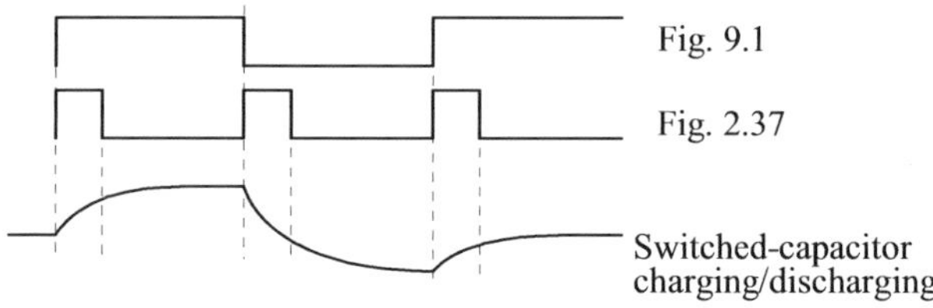

Figure 9.2 Charging discharging a switched-capacitor.

9.1.2 Implementation

Figure 9.4 shows how we would implement a high-speed analog-to-digital topology, *a single integrator, 1-sigma, with K feedback paths, K-deltas*, (see Fig. 6.24) using switched-capacitors, Fig. 7.2, building blocks. The amplifier used in the active integrator has to respond very quickly, as described in the last chapter. Further note that the integration capacitor, $4C_i$, is selected based on an integrator gain of 0.25, again, discussed in detail in the last chapter. Also seen in this figure are the clock signals and the way the outputs from each path can be combined together. The outputs of the K paths can be summed to generate an output code ranging from 0000 (all outputs are low) to 1000 (all outputs are high). Summing the outputs together results in Sinc filtering with a transfer function of

$$H(z) = \frac{1 - z^{-8}}{1 - z^{-1}} \text{ with } f_{s,new} = K_{path} \cdot f_s \tag{9.2}$$

as seen in Fig. 9.3. Note that this is exactly the same response that we get when using the sample-and-hold, S/H, Fig. 2.17, clocked at f_s. While there are other ways to combine the path outputs let's do some examples.

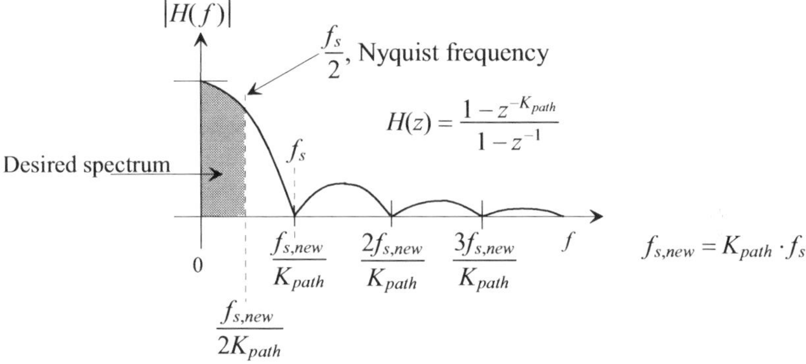

Figure 9.3 Frequency response of the summing circuit (path filter) seen in Fig. 9.4.

Example 9.1

Simulate the operation of the data converter seen in Fig. 9.4 if C_i = 100 fF and f_s = 100 MHz. Show that an input frequency of 50 MHz results in, as seen in Figs. 9.3 and 2.17, attenuating the input by 3.9 dB (= 0.64).

Figure 9.5 shows the simulation results when the input is a 7 MHz sinusoid centered around 500 mV with an amplitude of 400 mV. Notice that the rate at which the data comes out of the summing circuit is 800 MHz (the output changes every 1.25 ns). Also notice that the change in the output is the reference voltage, here 1 V, divided by 8 or 125 mV.

Figure 9.6 shows the input and output if the frequency is increased to 50 MHz (the Nyquist rate if we were to decimate down to a clocking frequency of 100 MHz). We expect the amplitude of the output to be $0.64 \cdot 0.4 + 0.5$ or 756 mV. The simulation shows an output of 750 mV (remember the quantization noise). ■

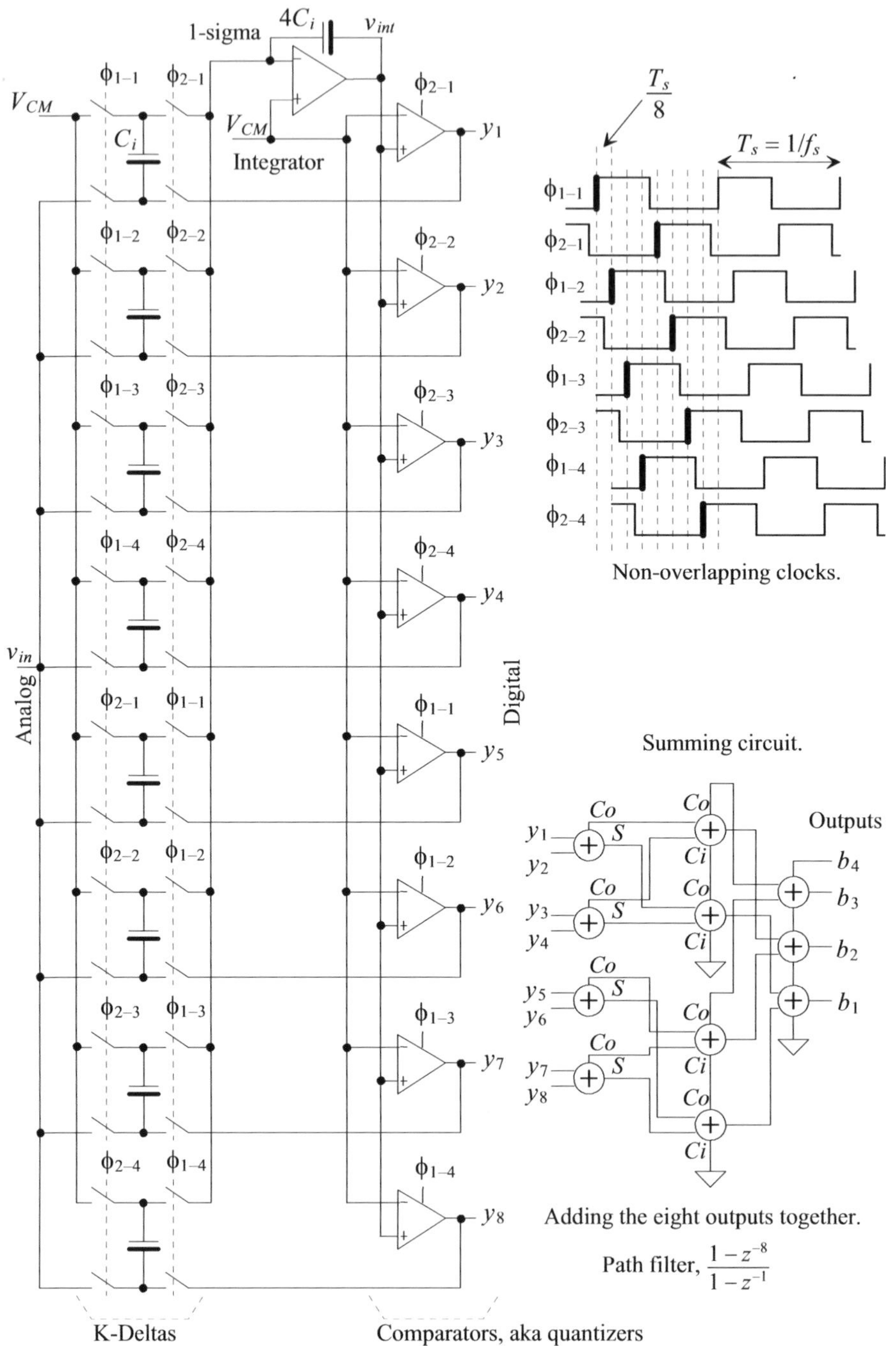

Figure 9.4 A topology for high-speed data conversion using mixed-signal techniques, the K-delta-1-sigma topology.

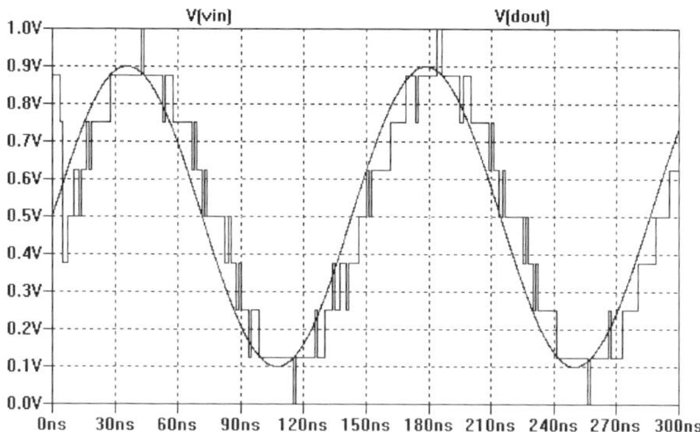

Figure 9.5 Simulating the operation of the data converter in Fig. 9.4 with a 7 MHz input signal.

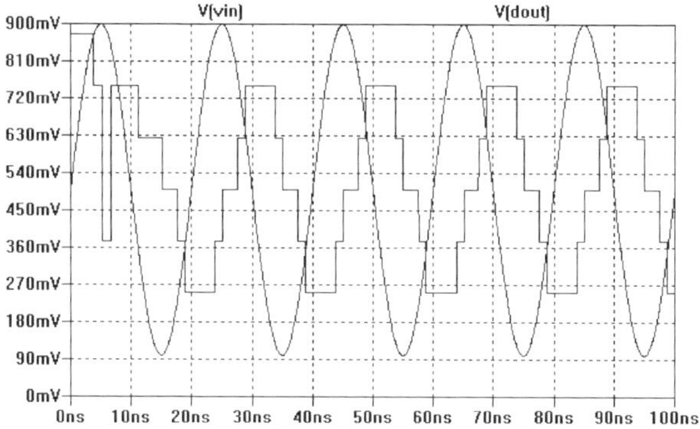

Figure 9.6 Again, simulating the operation of the data converter in Fig. 9.4 but now with a 50 MHz input signal.

Example 9.2
Repeat Ex. 9.1 if the output is decimated down to the 100 MHz clock rate.

Figure 9.7a shows that adding a register in series with the K outputs can be used to decimate the word rate down to f_s. A K-bit register can be placed in front of the summation circuit or a $\log_2 K$ wide register can be placed on the output of the summing circuit (not shown in Fig. 9.7). Timing mistakes in this digital part of the circuit are an important practical concern. The digital signals are moving quickly so eliminating the skew introduced by the summing circuit, by placing the register on the input side (Fig. 9.7a), is the approach we take here.

The simulation results are seen in Fig. 9.8. Notice that, when the input signal frequency is at the Nyquist rate in Fig. 9.8b, that the sampling instances repeat. Avoiding decimation (down-sampling), as can be seen when comparing Figs. 9.5, 9.6, and 9.8, and discussed in detail earlier in the book, helps keep the original signal intact.

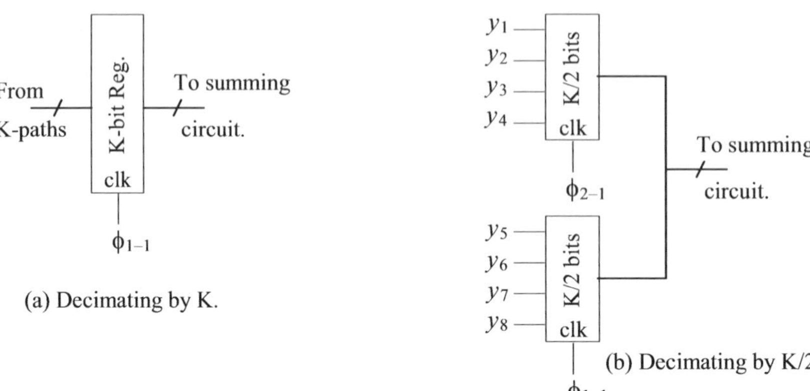

(a) Decimating by K.

(b) Decimating by K/2.

Figure 9.7 Adding a register to decimate the outputs of the high-speed modulator.

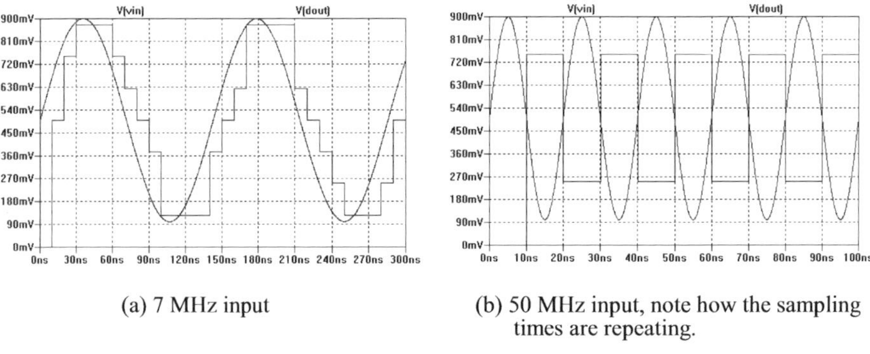

(a) 7 MHz input

(b) 50 MHz input, note how the sampling times are repeating.

Figure 9.8 Repeating Ex. 9.1 but with a decimation of 8.

In order to move towards reducing the unwanted effects of decimation (aliasing) and timing errors, examine the topology seen in Fig. 9.7b. Here we've split the word up into two paths (though we could also use 4 paths with the cost of tighter timing concerns). Summing the words together, we still get the filtering specified by Eq. (9.2). However, our decimated word rate is now $2f_s$. Figure 9.9 shows the simulation results using these techniques. The benefits should be obvious. ■

9.1.3 Filtering

Thinking about what we've done in this section we might come to the conclusion that the topology seen in Fig. 9.4 is nothing more than a flash ADC implemented with 8-comparators. We are only getting, effectively, 3-bits out of the topology (see previous examples) so why not simply use a 3-bit flash ADC? The simple answer to this is that averaging (read filtering) has little, Fig. 5.31, to no (for a DC input) effect on the resolution of a Nyquist-rate ADC while averaging can be used in a noise-shaping topology to increase the resolution, Fig. 7.7. We are repeating material presented earlier in the book so let's just do some examples to provide additional discussions.

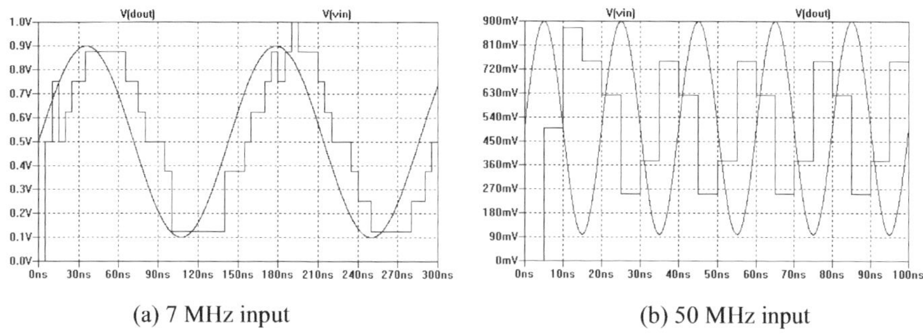

(a) 7 MHz input (b) 50 MHz input

Figure 9.9 Repeating Ex. 9.1 but with a decimation of 4.

Examples

Let's say that we want to design an ADC that is clocked with a 100 MHz clock and that outputs 8-bit words at least at a 100 MWord/s rate. In other words, we want to use mixed-signal circuit techniques to design the equivalent of a Nyquist-rate ADC (e.g., pipeline or flash) replacement (so the signal bandwidth ranges from DC to 50 MHz).

Let's use the topology seen in Fig. 9.4 as the basis for discussing this design. Generating the phases of the clocks is something we'll discuss later. Further we assume that the reader understands that selecting the size of the capacitors used in the topology is based on *kT/C* noise, Table 2.1, considerations. We'll continue to use 100 fF capacitors in our topology with eight paths for an effective capacitance, for calculating input-referred noise, of 800 fF.

As mentioned at the beginning of the section, we must be able to charge the 100 fF capacitors with the input signal in T_s/K seconds. For the present example $T_s/K = 1.25$ ns so the switch resistance should be much less than 12.5 kΩ (very easy to ensure).

What this design comes down to is selection of both the appropriate decimation rate on the output of the *K*-path topology and the digital filter. Examine the block diagram seen in Fig. 9.10. Here we use the 8-path topology seen in Fig. 9.4 clocked at 100 MHz. Directly on the output of the modulator, which changes at a rate of 800 MHz, we decimate by 8 back down to 100 MHz, with the register, and add the outputs together. As indicated by Eq. (9.2) this lowpass filters the data. If our input is a sinusoid at 6.25 MHz centered around 500 mV with a peak amplitude of 400 mV, then we can estimate the attenuation of our signal from this addition (filtering) using

$$\frac{\sin \pi \cdot \frac{6.25}{100}}{\pi \cdot \frac{6.25}{100}} = 0.993 \tag{9.3}$$

The output of the summing circuit is a 4-bit word that ranges from 0000 (0 in decimal) to 1000 (8 in decimal). We pass the 4-bit output through two Sinc filters which increase the word size by 3-bits each for a total output word size of 10-bits. However, the word is not centered around the common-mode voltage so we can shift the word left to do a multiply by 2 and center the word (as discussed earlier). This reduces the word size to 9-bits, still more than required. The issue is that we also reduce the signal bandwidth. For our 6.25

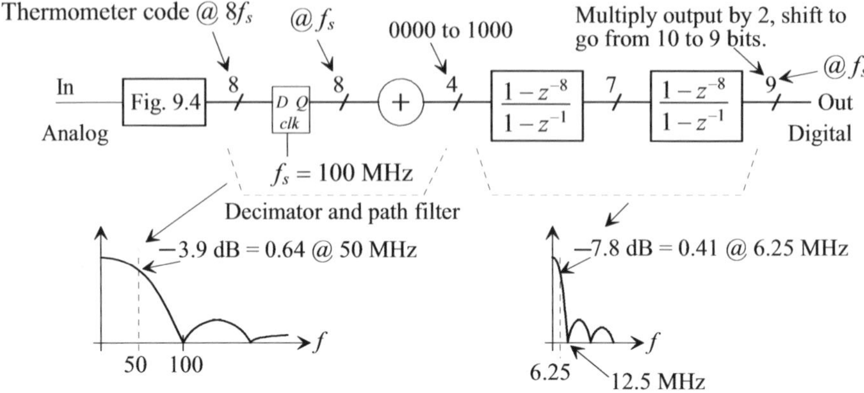

Figure 9.10 Decimating and filtering the output of the K-path modulator, an approach that won't work for Nyquist-rate conversion.

MHz input signal, as seen in Fig. 9.10, this means that we attenuate the signal by 0.41. The total attenuation of our 6.25 MHz input signal is then 0.407 so the peak amplitude out of the filter is 663 mV (ideally it's 900 mV). Figure 9.11 shows simulation results verifying our hand calculations. Note the delay through the filter when starting.

Studying Fig. 9.10, we see that the decimation right at the beginning of the digital filter is limiting our bandwidth. Consider spreading the decimation out through the filter as seen in Fig. 9.12 (see also Sec. 4.2.5). Here we've put a register on the output of the summing circuit instead of on its input, Fig. 9.7a. For ideal signals we may not need this register; however, as mentioned earlier, eliminating skew and timing errors is important for proper data converter operation. Simulation results are seen in Fig. 9.13. The input signal undergoes less attenuation but the final resolution is 7-bits instead of 9-bits (indicating that there is less filtering in Fig. 9.12 than 9.10). In order to move towards our design requirements, an 8-bit converter with a signal bandwidth of 50 MHz that is clocked at 100 MHz, our final output clock rate will be increased to 200 MWords/s (instead of 100 MWords/s as in Figs. 9.10 and 9.12).

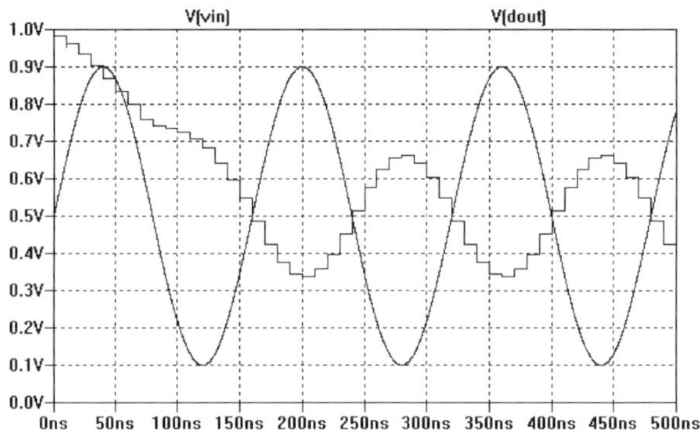

Figure 9.11 Simulating Fig. 9.10 with a 6.25 MHz input signal.

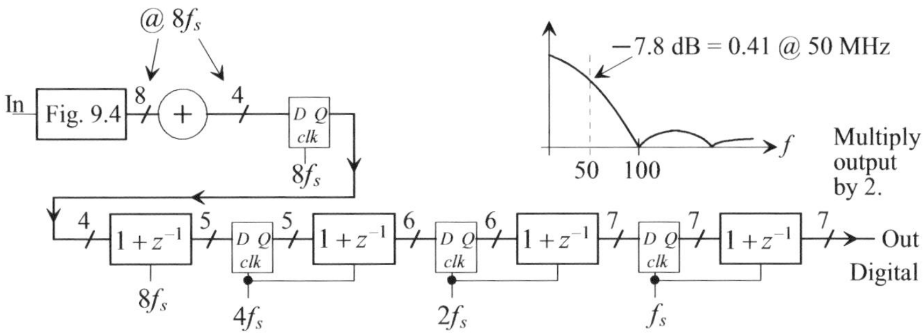

Figure 9.12 Decimating through the filter, see Sec. 4.2.5.

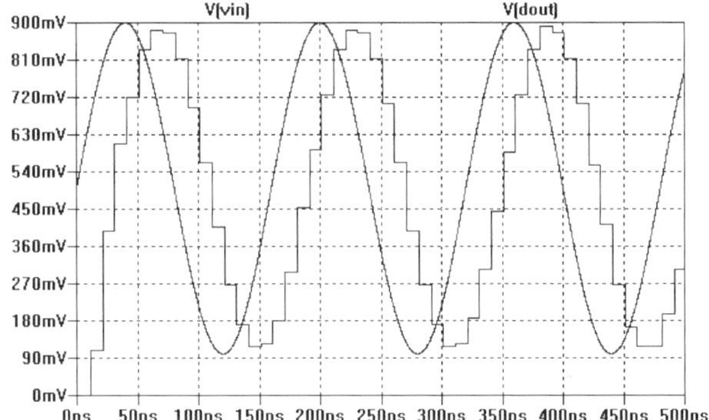

Figure 9.13 Simulating Fig. 9.12 with a 6.25 MHz input signal.

Figure 9.14 shows our next attempt at filtering and decimation. The simulation results, with a 6.25 MHz input, are seen in Fig. 9.15. By adding together the 8 outputs in Fig. 9.4 we are still attenuating our input signal with a Sinc response. Note that the output of the adder is changing at 200 MWords/s. We don't employ decimation at any other place in the filter. The chain of averaging filters, on the adder's output, increases the word size by 1-bit per stage. The attenuation from this digital filter at 50 MHz is $(0.707)^5 = 0.177$, see Fig. 1.17. This, combined with the attenuation from the addition, results in an overall attenuation for a 50 MHz input signal of 0.113 (change the input signal frequency and simulate to verify).

Figure 9.16 shows yet another filter. This filter doesn't employ decimation so a 50 MHz input should see the –3.9 dB attenuation from adding the 8 outputs together, Fig. 9.10. In addition, as seen in Fig. 1.17 (but with frequencies on the x-axis scaled by 8 so the first zero point occurs at 400 MHz), the averaging stages will provide some additional, minor, attenuation, Fig. 9.17. The question we need to ask is "Does this filter remove the modulation noise in addition to increasing the word size?" If the answer is "No, the filter doesn't remove the modulation noise," then how do we meet our design goals? Further, how do we determine if the filter removes the modulation noise?

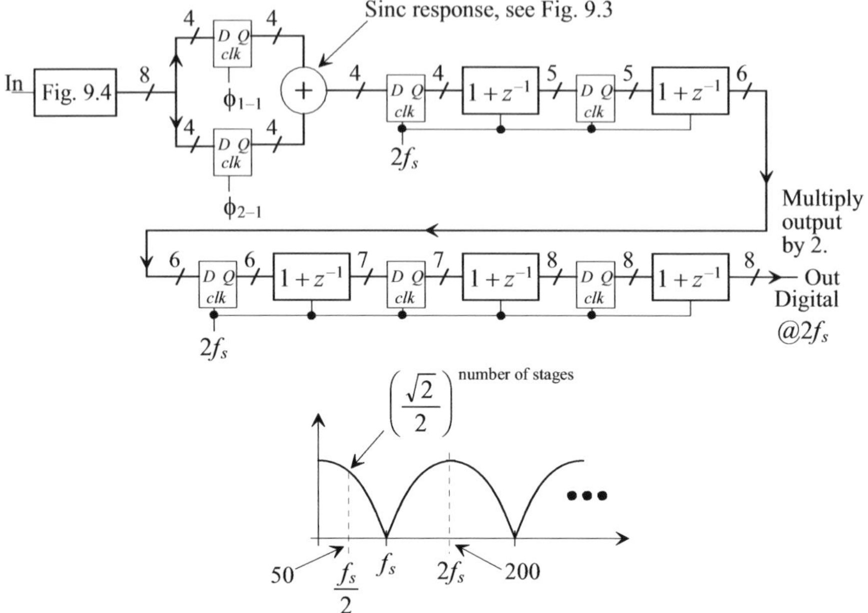

Figure 9.14 Another filter; however, this topology has problems removing the modulation noise.

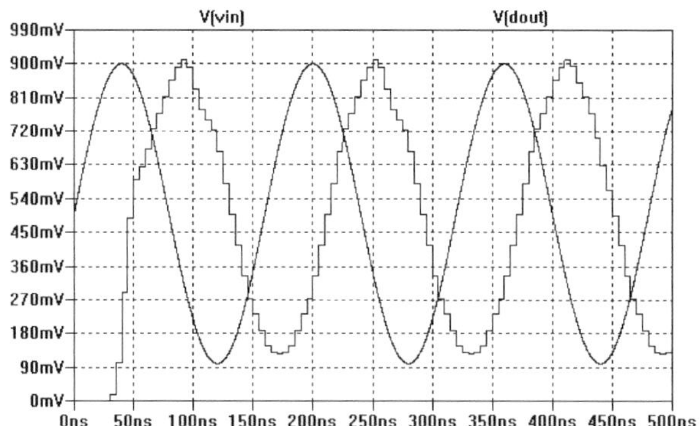

Figure 9.15 Simulating Fig. 9.14 with a 6.25 MHz input signal. Note the the nonlinearity in the output.

In order to move towards answering these questions, notice the nonlinearity in both Figs. 9.15 and 9.17. We implemented the filter by focusing on only increasing the output word size to 8-bits. We didn't concern ourselves with removing the modulation noise and thus we see noise and nonlinearity in our output signal. Reviewing Sec. 7.1.3, we can remove the modulation noise by using the filter seen in Fig. 9.10 but without the decimation (noting that if $L = 2$ then only the path filter and one Sinc filter are used for a 7-bit output). This also helps us move closer towards the design goals. Figure 9.18 shows the simulation results when a 50 MHz input is applied to the topology seen in Fig. 9.10 without decimation. The input signal undergoes 3-Sinc filter responses $(0.64)^3 = 0.262$.

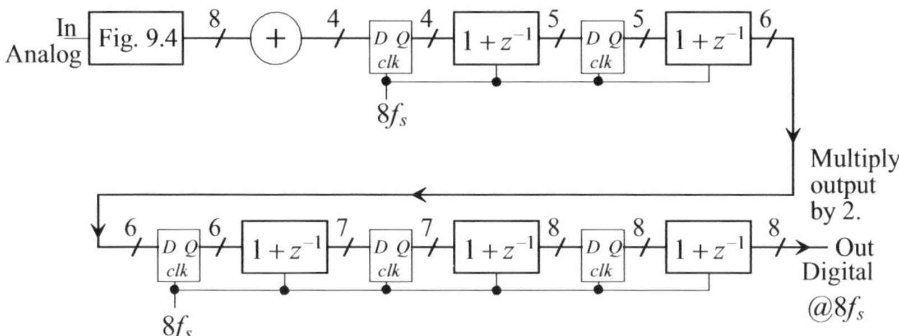

Figure 9.16 Filtering without decimation. This filter won't do a good job removing the modulation noise but it does increase the output word size to 8-bits.

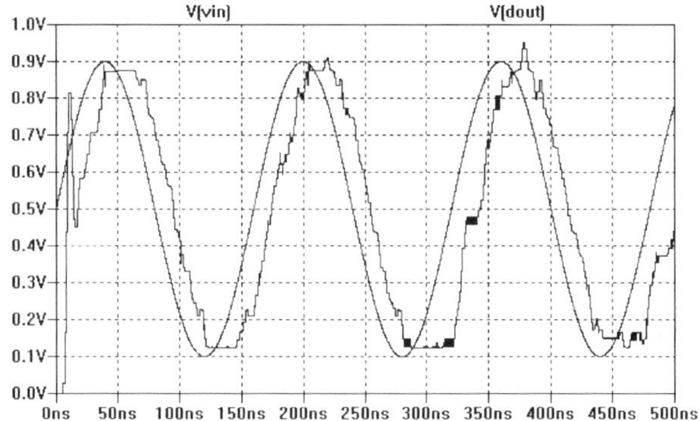

Figure 9.17 Simulating Fig. 9.16 with a 6.25 MHz input signal. Again, note the the nonlinearity in the output.

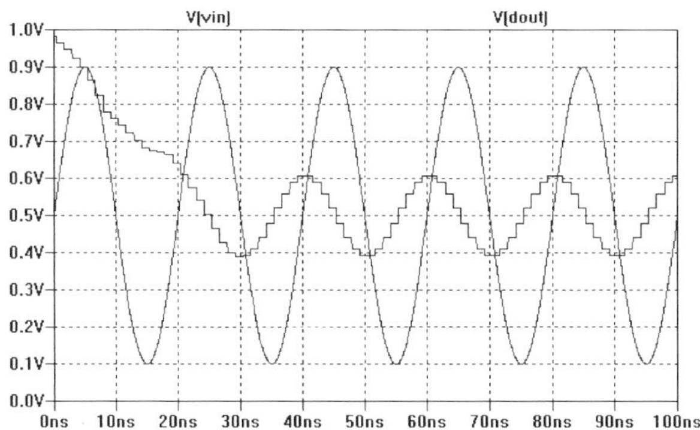

Figure 9.18 Using the filter seen in Fig. 9.10 without decimation. Attenuation is due to 3-Sinc stages. The input signal frequency is 50 MHz.

Direction

In order to move towards our design goals we can use the filter seen in Fig. 9.12 but with an additional averaging, $1+z^{-1}$, stage on the filter's output. This increases the word size to 8-bits. The penalty is an extra 3 dB attenuation at 50 MHz (so the total attenuation at 50 MHz is 10.8 dB or 0.288). The practical issue with this topology, for very high-speed operation, is avoiding timing errors in the stages clocked at $8f_s$. For a general design, we'll use the topology seen in Fig. 9.10. Timing errors are straightforward to minimize. The drawback is the reduction in signal bandwidth.

An area that should be investigated further is the use of biquad filters for removing the modulation noise, as discussed in Sec. 4.3.4.

9.1.4 Discussion

If the NS modulator seen in Fig. 7.2 is clocked at 100 MHz we get the noise transfer function, *NTF*, shape seen in Fig. 9.19a (out to 800 MHz or $8f_s$). While we are used to focusing on the spectral content between DC and f_s (= 100 MHz here), we know that any discrete-time system's frequency response repeats with the sampling frequency, f_s. The signal bandwidth, as seen in the figure, is 6.25 MHz, assuming K_{avg} = 8. Note that we are not showing the frequency response of the digital filter used for removing the modulation noise (which is where K_{avg} is used, Eq. [6.47] and Fig. 7.8). Figure 9.19b shows the assumed spectrum of the quantization noise while Fig. 9.19c shows the single path clocked at 100 MHz (noting we are not showing the non-overlapping clock signals).

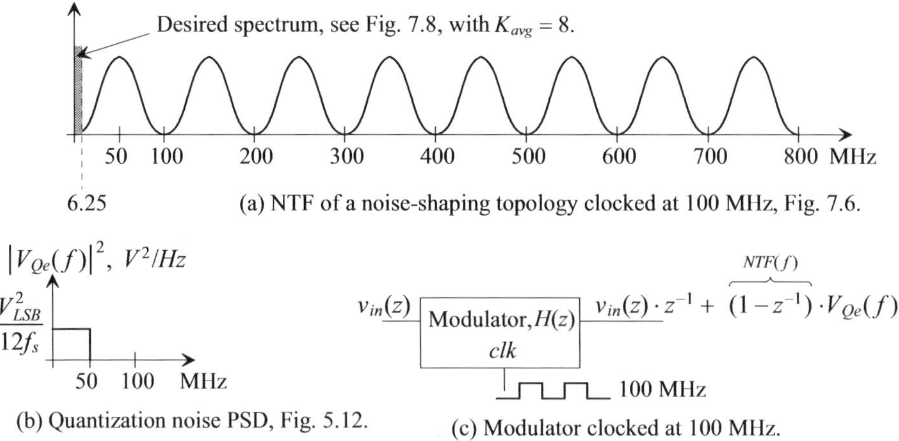

Figure 9.19 NTF and quantization noise spectrums of a first-order NS modulator clocked at 100 MHz with K_{avg} = 8.

Next, examine the 8-path topology of modulators seen in Fig. 9.20 (modified from Fig. 2.37 for this discussion). In (a) of the figure each modulator is clocked at f_s, the same as in Fig. 9.19c, but with time-shifted clock signals. In (b) we show the equivalent circuit, see discussion on page 224 for details, noting the equivalent circuit's output changes with a frequency of $K_{path}\cdot f_s$. By using an effectively higher clock frequency we spread the

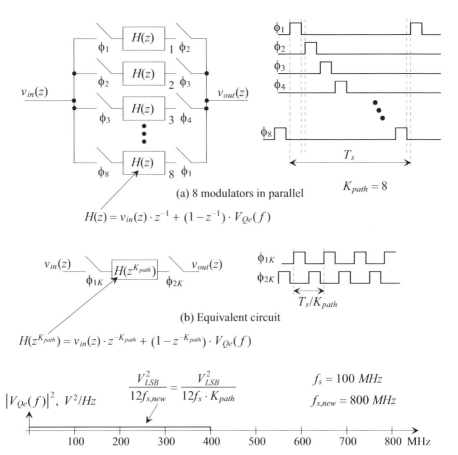

$$H(z) = v_{in}(z) \cdot z^{-1} + (1 - z^{-1}) \cdot V_{Qe}(f)$$

(a) 8 modulators in parallel $K_{path} = 8$

$$H(z^{K_{path}}) = v_{in}(z) \cdot z^{-K_{path}} + (1 - z^{-K_{path}}) \cdot V_{Qe}(f)$$

(b) Equivalent circuit

$|V_{Qe}(f)|^2$, V^2/Hz $\qquad \dfrac{V_{LSB}^2}{12 f_{s,new}} = \dfrac{V_{LSB}^2}{12 f_s \cdot K_{path}}$ $\qquad f_s = 100\ MHz$, $f_{s,new} = 800\ MHz$

(c) Quantization noise for an 8-path topology clocked 100 MHz, compare to Fig. 9.19b.

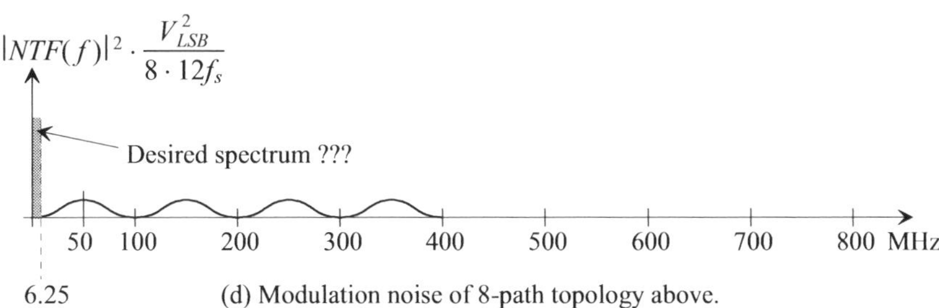

(d) Modulation noise of 8-path topology above.

Figure 9.20 Eight modulators in parallel (a time-interleaved topology).

quantization noise out over a wider range of frequencies, (c). This reduces the modulation noise added to our signal in the bandwidth from DC to 6.25 MHz as seen in Fig. 9.20d. If our desired signal bandwidth remains at 6.25 MHz then we will clearly get an improvement in SNR. Note that the output of the topology is 1-bit changing at a rate of 800 MHz.

Suppose we want to use this topology for high-speed data conversion where we use the fact that our effective output clock frequency is 800 MHz (and so the Nyquist frequency is 400 MHz). Is this possible or does it make sense? This is a good point to **remember the fundamentals**. As mentioned on page 201, using noise-shaping does not result in a reduction in quantization noise. Rather, noise-shaping topologies push the quantization noise to frequencies outside the region of interest. Reviewing Fig. 9.20d we see that (all of) our quantization noise is still present in the spectrum from DC to 400 MHz (see example below). It's impossible to reduce this noise with a digital filter without effecting the desired signal in this same frequency range.

Example 9.3

Estimate the RMS value of the quantization noise for the 8-path topology clocked at 100 MHz seen in Fig. 9.20d.

$$V_{Qe,RMS}^2 = \int_0^{4f_s} |NTF(f)|^2 \cdot \frac{V_{LSB}^2}{8 \cdot 12 f_s} \cdot df = \int_0^{4f_s} 4 \cdot \sin^2\left(\pi \cdot \frac{f}{f_s}\right) \cdot \frac{V_{LSB}^2}{8 \cdot 12 f_s} \cdot df$$

$$= \frac{V_{LSB}^2}{4 \cdot 12 f_s} \cdot \int_0^{4f_s} \left[1 - \cos\left(2\pi \cdot \frac{f}{f_s}\right)\right] \cdot df = \frac{V_{LSB}^2}{12}$$

or $V_{Qe,RMS} = V_{LSB}/\sqrt{12}$, no reduction in quantization noise. While this result was derived for the case $K_{path} = 8$ the reader should see that the result is valid for any value of K_{path}. ■

Example 9.4

Estimate SNR_{ideal} for the topology seen in Fig. 9.20a.

The procedure for calculating SNR was given in Sec. 5.2. Since we just calculated the quantization noise, all that's left is to calculate the desired output signal power. If we apply a sinusoid to the topology with a peak amplitude of V_p, then adding the output signal powers from each path results in $K_{path} \cdot V_p^2/2$. We can then write

$$SNR_{ideal} = 20 \cdot \log \frac{\sqrt{K_{path}} \cdot V_p/\sqrt{2}}{V_{LSB}/\sqrt{12}}$$

or (see Fig. 5.31)

$$SNR_{ideal} = 6.02N + 1.76 + 10 \log K_{path} \tag{9.4}$$

For every doubling in the number of paths we get a 3-dB (0.5 bits) increase in the SNR. Such a modest increase in SNR is, generally, too insignificant for the hardware costs so this approach isn't used. ■

The virtues of the topology seen in Fig. 9.4 should be clearer. Using a single integrator (1-sigma) with K feedback paths (K-deltas) we can attain conversion bandwidths approaching $f_s/2$ with reasonable resolutions. Since a single integrator is used, common information is applied to the input of each quantizer. This allows the quantization noise to be pushed to very high frequencies. In other terms, we get fast sampling by using K feedback paths while we are able to push the quantization noise to higher frequencies using a single integrator common to all quantizers.

9.1.5 Understanding the Clock Signals

Notice that in Fig. 7.2 we used ϕ_1 to clock both the input switches and the comparator. In our high-speed topology seen in Fig. 9.4, however, we used an earlier clock signal to strobe the comparator. For example, examining the top path in Fig. 9.4, the input switches are clocked with ϕ_{1-1} while the comparator is clocked with ϕ_{2-1}. The input signal is captured on C_i at the falling edge time of ϕ_{1-1}. A very short time later, T_s/K_{path}, we clock the comparator. Since *near-ideal components* are used in our simulations the delay associated with transferring the charge on C_i to the integrator's output is nearly zero. When the comparator makes a decision it is immediately fed back to the integrator, again with minimal delay, so that the next path's decision is directly dependent on the previous path's decision. The result is the topology behaves like a single, first-order, noise-shaping topology clocked at $K_{path} \cdot f_s$ with an input/output relationship given by Eq. (7.2).

In a practical implementation of the modulator, however, the delay through the integrator and comparator is finite and so the information fed back experiences a delay. This delay causes *a limit-cycle oscillation* (a self-sustained oscillation). Figure 9.21 shows an example of this oscillation. In this figure the input signal is the DC common-mode voltage, 500 mV here, and the clock signals are selected so that, as in Fig. 7.2, the same clock signal is used for both the input pair of switches and the comparator. The average value of the output signal does equal the input signal; however, we have the unwanted oscillation seen in the figure. Figure 9.22 shows another example but with an AC input signal. Since this oscillation always occurs at a relatively high frequency the digital filter on the output of the modulator reduces it.

The practical problem with using the clock phases seen in Fig. 9.4 with non-ideal elements is that the comparator delay varies making it appear as though the clock is jittering, see Secs. 5.2.1 and 5.2.2. The result is a reduction in *SNR* and *amplitude modulation* in the data converter's output, Fig. 5.22. Adding an amplifier in series with the output of the integrator to ensure that the comparators make a fast decision can help with this problem. In the design presented in the next section we use the same clock phase for both the input switches and the comparator. This reduces the requirements placed on the comparator allowing a decision to be made in nearly an entire (slow) clock cycle T_s. The drawback is that the quantization noise is no longer centered on $K \cdot f_s/2$ and so the *SNR* suffers.

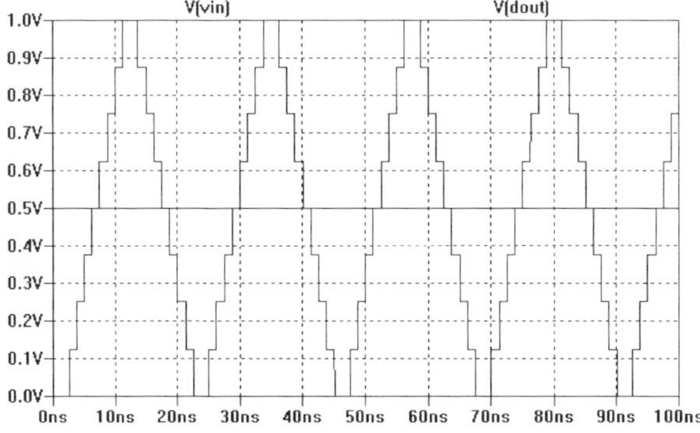

Figure 9.21 Showing limit-cycle oscillations in the high-speed modulator.

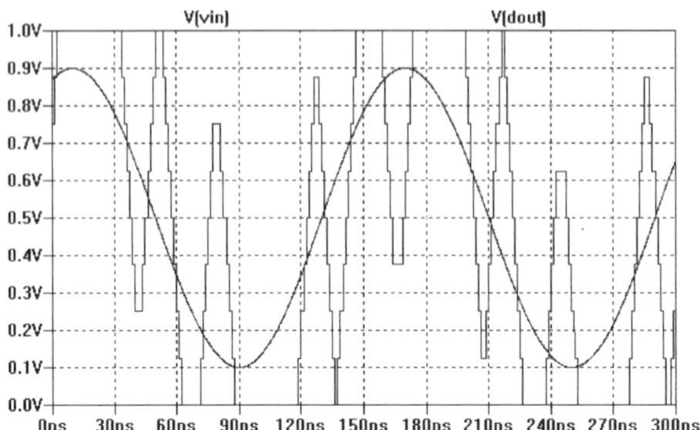

Figure 9.22 Showing limit-cycle oscillations with a sinewave input signal.

9.2 Practical Implementation

In this section we'll discuss the practical implementation of the high-speed topology proposed in this book. The goal of this section is to provide discussions and provoke thought that should prove helpful when designing a converter using this topology. The goal is not to provide definitive solutions for specific applications. This endeavor is left for discovery by the engineers and researchers doing mixed-signal circuit design.

9.2.1 Generating the Clock Signals

Generating the 16 clock signals needed for the topology seen in Fig. 9.4 can be challenging. We could use a delay-locked loop (DLL) that takes an input clock signal and generates the 16 clock signals (but that adds complexity). Here we use a ring oscillator that runs asynchronously with an external clock signal. Figure 9.23 shows the basic delay stage schematic and icon used in the oscillator. Figure 9.24 shows the complete ring oscillator while Fig. 9.25 shows some simulation results in a 500 nm, 5-V, CMOS

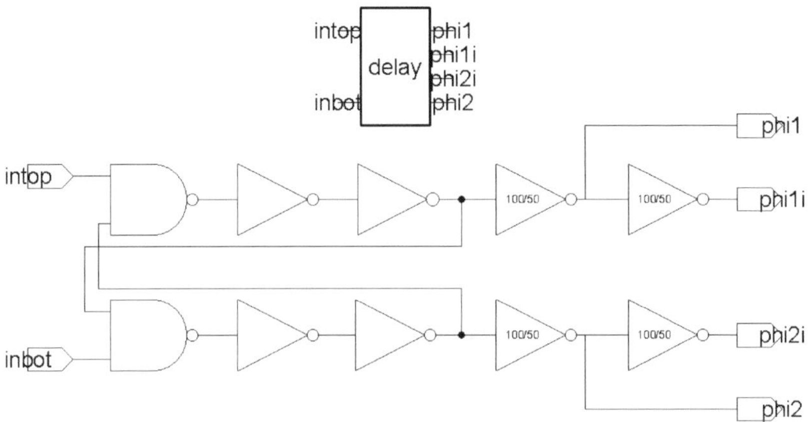

Figure 9.23 Delay stage used in the ring oscillator, schematic and icon.

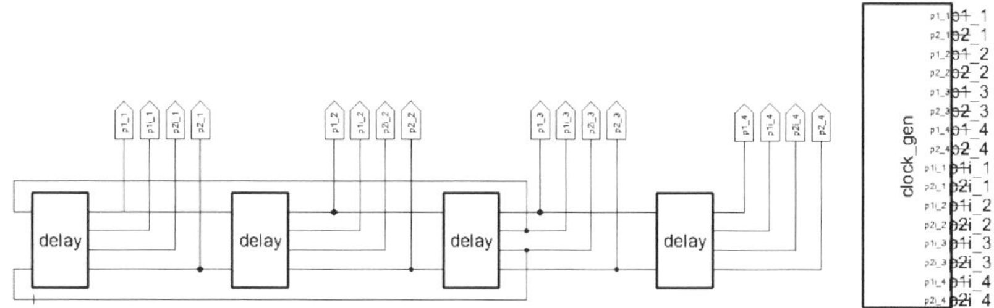

Figure 9.24 Ring oscillator, schematic and icon, for use with the data converter.

process. Note that the frequency of the oscillator can be adjusted by adding or removing inverters in the delay stage. Also notice that the simulated oscillation frequency is around 200 MHz (about half of this for a simulation with layout parasitics). The analog input signal is sampled at a rate of $f_{s,new}$ or 1.6 GHz. The 8 path outputs are added together, Fig. 9.4, and then decimated, Fig. 9.7a.

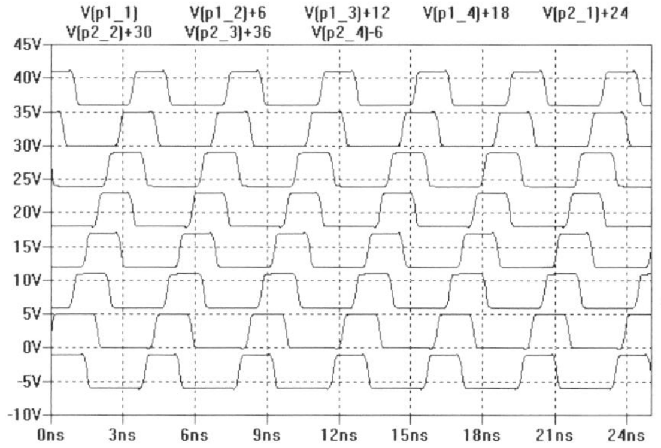

Figure 9.25 Simulating the oscillator in a 500 nm process.

Figure 9.26 shows how the external (synchronous) and internal (asynchronous) clocks can be interfaced using a synchronizer. We are assuming that the internal clock is running faster than the external clock (so the output is decimated). Note that the synchronizer doesn't introduce aperture jitter as discussed in Sec. 5.2.1 since it only processes digital signals. After reviewing this section, however, we might wonder *if the ring oscillator is a practical choice for providing the clock signals to the data converter.* Ring oscillators are certainly not as stable as crystal-controlled oscillators. However, notice that by combining the K_{path} outputs together we reduce the variance of the aperture jitter by K_{path}, see, for example, Eq. (5.30). Further filtering reduces the effects of a jittery clock signal. Slow variations in *VDD*, ground, or temperature will also have essentially no affect on the data converter's performance (via the ring oscillator) since these changes simply vary the sampling rate (the internal clock frequency). Note that for large differences in the internal and external clock frequencies aliasing concerns, when decimating, should be taken into consideration in the synchronizer (as should metastability concerns, e.g., the external clock (not) going low just after the internal clock goes high resulting in a glitch on the output of the AND gate in Fig. 9.26).

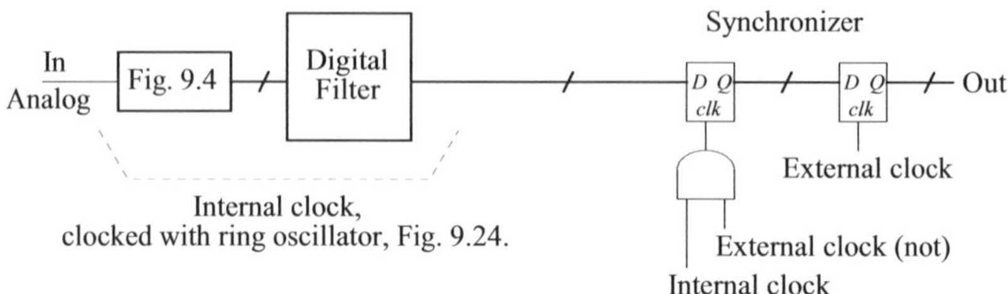

Figure 9.26 Synchronizing the external and internal clock signals.

9.2.2 The Components

The Switched-Capacitors

Figure 9.27 shows the schematic and the icon for the switched-capacitors used in the modulator. The concerns, when selecting the widths of the MOSFETs, as discussed on pages 302 and 307, are that the capacitors can be charged/discharged within T_s/K_{path}. The effective resistances of each MOSFET in Fig. 9.27 is 500 Ω. Since two MOSFETs are in parallel with switches on each side of the 100 fF capacitor, the time constant is 50 ps. Finally, the size of the capacitors used in this circuit, as discussed on page 307, is set by thermal noise considerations.

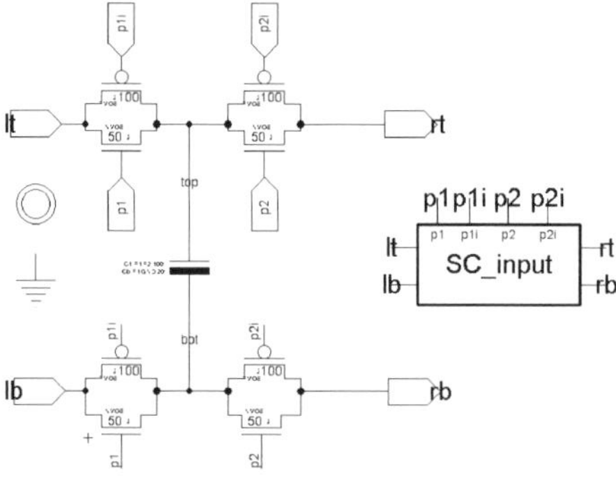

Figure 9.27 Schematic and icon of the switched-capacitors used in the modulator.

The Amplifier

Figure 9.28 shows the schematic and icon for a self-biased amplifier. This topology was picked because it's simple and fast. Further, for the comparator discussed next, the outputs are complementary and generated from NMOS devices (which interface nicely with the comparator's inputs). Simulation results are seen in Fig. 9.29. The non-inverting amplifier input is held at 2.5 V while the inverting input is pulsed from 2.45 to 2.55 V (100 mV change). The output changes from roughly 2.8 to 0.8 (2 V change) so the gain is 20 (which should be larger than K_{path} or 8 here). Both outputs of the amplifier are loaded, in the simulation, with 400 fF capacitors. Our concern, Sec. 7.1.8, is that we don't see slewing in the response. The amplifier sizes picked here are larger than required for our final design (so power is wasted).

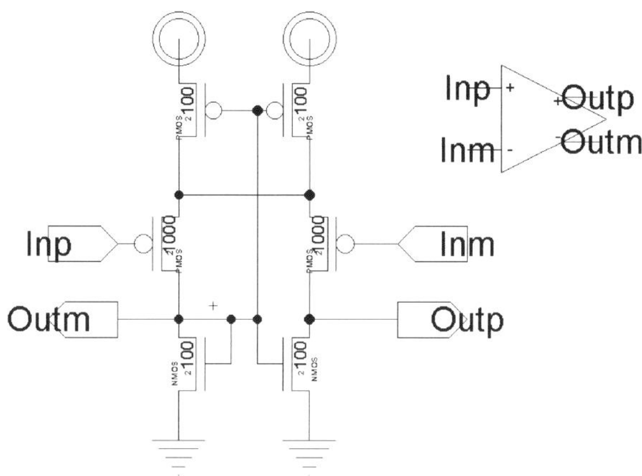

Figure 9.28 Self-biased amplifier and icon.

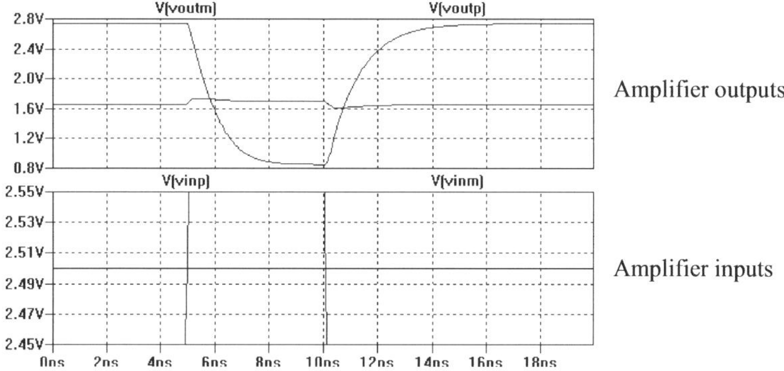

Figure 9.29 Simulating the operation of the amplifier in Fig. 9.28 with 400 fF loads.

The Clocked Comparator

The schematic and icon of the clocked comparator used in the design is seen in Fig. 9.30. The input clock signal is sharpened and buffered up so that the input capacitance of the comparator is reduced and the edges in the comparator are fast (so that a good, reliable, decision can be made). The basic latch is formed with cross-coupled inverters (as usual). No DC current flows in the comparator. Current is pulled from *VDD* when the clock signal changes states. The NAND gates are used to ensure that the comparator output changes states (only) on the rising edge of the input clock signal. The outputs of the NAND gates are buffered up to ensure that they can drive the switched-capacitors. If these outputs don't fully charge the switched-capacitors or the *VDD*/ground connected to these gates droops or bounces, the feedback signal will be affected causing noise and nonlinearity. Finally, the inputs to the comparator should be above the threshold voltage of the NMOS device but not so large that the input devices are pushed into deep triode. Note, as discussed already, that we need to be careful to ensure minimal forward delay through the comparator (and, of course, the integrator).

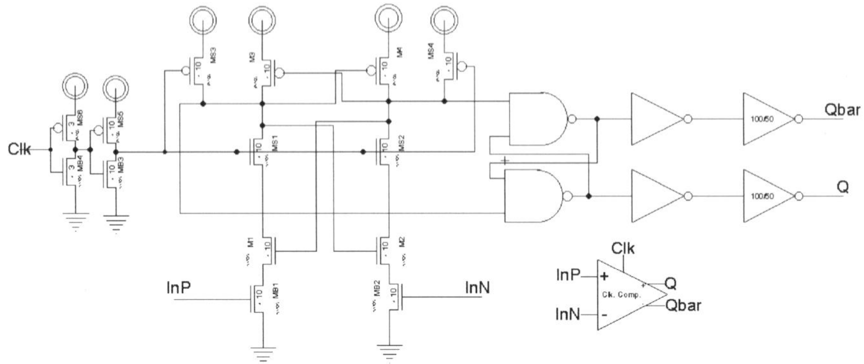

Figure 9.30 Clocked comparator and icon.

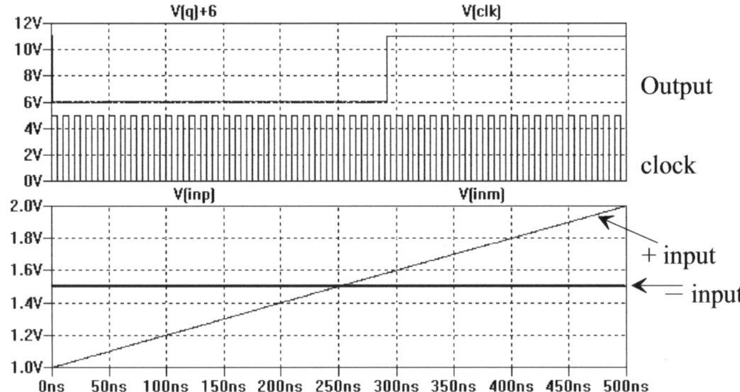

Figure 9.31 Simulating the operation of the comparator in Fig. 9.30.

9.2.3 The ADC

Figure 9.32 shows the schematic of the high-speed ADC (minus the digital filter) we designed in this section using a 500 nm CMOS process with a *VDD* of 5 V. The resistors are used (for SPICE simulations) to provide a simple method of summing the digital outputs for ease of viewing. While they are useful for speeding up the simulation, they are not useful for looking at the performance of the converter. The finite transition times, glitches, etc. in the digital data limit their usefulness when viewing analog data. Figure 9.33 shows the results of simulating the ADC with a 10 MHz input signal.

At this point, in order to characterize the behavior of the ADC, we need to add the digital filter to the modulator's output. Ideally the *K*-delta-1-sigma topology seen in Fig. 9.4 behaves like a single, first-order, noise-shaping modulator clocked at $f_{s,new}$ ($=f_s \cdot K_{path}$ or 1.6 GHz here) with a 1-bit output (we selectively connect each path's output to the overall topology's output every T_s/K_{path} seconds). Let's use the decimator and filter seen in Fig. 9.10, but with, as discussed in Sec. 7.1.3, $L = 2$ (the adder and only one Sinc filter). The final output word size is 7-bits, the effective 1-bit output by the modulator and then 3-bits (each) for the addition and the Sinc filter. We won't

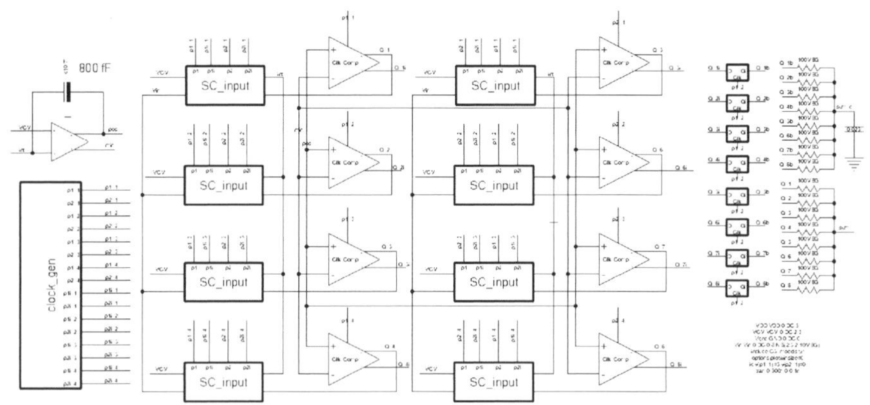

Figure 9.32 K-delta-1-sigma modulator.

synchronize the output data to an external clock. This means that the output will change at the rate of the ring oscillator (roughly every 5 ns or 200 MHz). An input signal frequency of (roughly) 200 MHz/16 or 12.5 MHz will be attenuated by –3.9 dB. Figure 9.34 shows the simulation results for the ADC (modulator and filter) when the input signal is 10 MHz (same as Fig. 9.33 but showing the output of the digital filter). These conversion rates are comparable to any high-speed converter topology implemented in a 500 nm CMOS process (even faster if we use less decimation early in the digital filter).

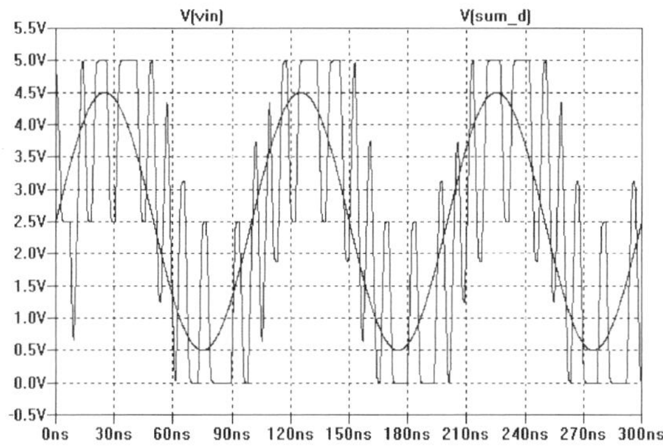

Figure 9.33 Simulating the ADC in Fig. 9.32 with a 10 MHz input signal.

Figure 9.35 shows a simulation where the ADC's input is a ramp signal that transitions from 0 to *VDD* in 1 μs. The delay through the filter is 50 ns so the input signal should be shifted 50 ns later in time when comparing the ADC's input to its output. Of course, the ADC's output will then be valid only after this delay (so we neglect the start-up transient). We can estimate the minimum step size in the output of the ADC as $VDD/2^6$ or 78 mV. We use 2^6 instead of 2^7 (the exponent is the number of bits coming out of the filter) because in order for the output signal to swing rail-to-rail we have to

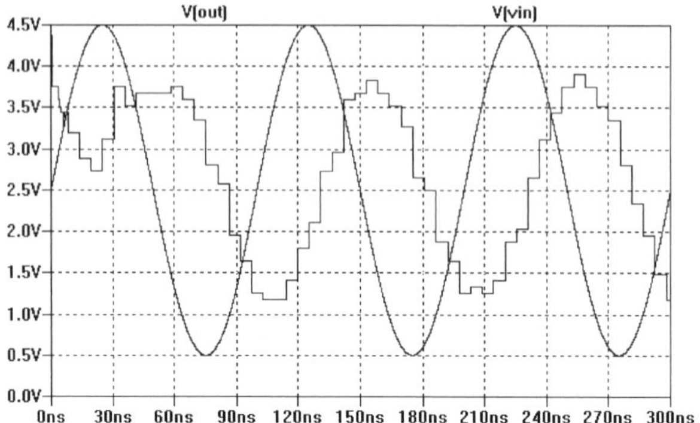

Figure 9.34 Same as Fig. 9.33 (10 MHz input) but looking at the output of the digital filter.

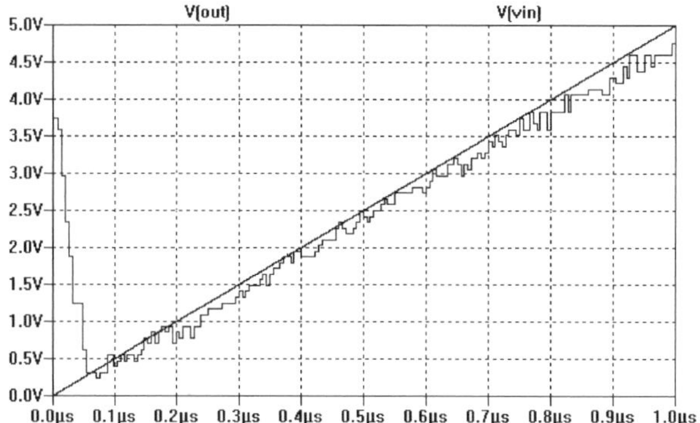

Figure 9.35 Simulating the modulator with a ramp input signal. Shift input later in time by 50 ns to compare to output.

multiply it by two before applying it to the ideal DAC used to display the digital data as an analog output voltage (as discussed earlier).

Remember that increasing the output word size (increasing resolution) requires additional filtering. Also, again, recall that to increase the signal conversion bandwidth we need to reduce the amount of decimation used immediately following the modulator and prior to the adder that sums the K_{path} outputs together (see Fig. 9.7b for the decimate by $K_{path}/2$ example).

9.3 Conclusion

This chapter has proposed a new topology for high-speed analog-to-digital conversion using the mixed-signal circuit design techniques presented in this book. The topology should prove useful when designing high-speed ADCs in nanometer CMOS. Future work can be focused in many areas including, but not limited to: fully-differential signal paths, the integrator (amplifier) design, higher-order (perhaps passive, Sec. 6.2.1) topologies, bandpass converters, segmenting feedback paths, digital calibration (e.g. for offset, Sec. 7.1.9) and the design of the digital filter (including the decimating stage).

ADDITIONAL READING

[1] F. Borghetti, C.D. Fiore, P. Malcovati, and F. Maloberti, "Synthesis of the Noise Transfer Function in N-path Sigma-Delta Modulators," *5th IEE International Conference on Advanced A/D and D/A Conversion Techniques and their Applications*, Limerick, Ireland, pp. 171-176, July 25-27, 2005.

[2] A. Eshraghi and T. S. Fiez, "A Comparative Analysis of Parallel Delta–Sigma ADC Architectures," *IEEE Trans. on Circuits and Systems - I: Regular Papers*, Vol. 51, No. 3, pp. 450-458, March 2004

[3] A. Eshraghi and T. S. Fiez, "A Time-Interleaved Parallel DS A/D Converter," *IEEE Trans. on Circuits and Systems - II: Analog and Digital Signal Processing*, Vol. 50, No. 3, pp. 118-129, March 2003

[4] V.-T. Nguyen, P. Loumeau, and J. F. Naviner, "Advantages of High-Pass DS Modulators in Interleaved DS Analog-to-Digital Converter," *Proceedings of the 45th Mid-West Symposium on Circuits and Systems*, August 4-7, Tulsa, OK, pp. I-136-I-139, 2002

[5] M. Kozak, and I. Kale, "Novel Topologies for Time-Interleaved Delta–Sigma Modulators," *IEEE Trans. on Circuits and Systems - II: Analog and Digital Signal Processing*, Vol. 47, No. 7, pp. 639-654, July 2000

[6] E. T. King, A. Eshraghi, I. Galton, and T. S. Fiez, "A Nyquist-Rate Delta–Sigma A/D Converter," *IEEE J. Solid-State Circuits*, Vol. 33, No. 1, pp. 45-52, Jan. 1998

[7] R. Khoini-Poorfard, L. B. Lim, and D. A. Johns, "Time-Interleaved Oversampling A/D Converters: Theory and Practice," *IEEE Trans. on Circuits and Systems - II: Analog and Digital Signal Processing,* Vol. 45, No. 8, pp. 634-645, August 1997

[8] I. Galton and H. T. Jensen, "Oversampling Parallel Delta-Sigma Modulator A/D Conversion," *IEEE Trans. on Circuits and Systems - II: Analog and Digital Signal Processing*, Vol. 43, No. 12, pp. 801-810, Dec. 1996

[9] R. Schreier, G. C. Temes, A. G. Yesilyurt, Z. X. Zhang, Z. Czarnul and A. Hairapetian, "Multibit Bandpass Delta-Sigma Modulators using N-path Structures," *Proceedings of the IEEE International Symposium on Circuits and Systems*, pp. 593-596, May 10-13 1992

QUESTIONS

9.1 What is a time-interleaved data converter? Why is a time-interleaved converter different from the *K*-Delta-1-Sigma converter seen in Fig. 9.4? Sketch the implementation of a time-interleaved data converter implemented with Delta-Sigma modulators. Also sketch the clock signals used in the topology.

9.2 Using the modulator from Ex. 9.1 show that capacitor matching isn't important in the *K*-Delta-1-Sigma topology.

9.3 Repeat question 9.2 but show that the open-loop gain of the amplifier used in the integrator isn't critical (compare, using simulations, the performance of the converter using different gains).

9.4 Show the details of (derive from the time-domain outputs of the *K*-Delta-1-Sigma modulator) how the path filter seen in Fig. 9.4 has a *z*-domain transfer function of

$$\frac{1-z^{-8}}{1-z^{-1}}$$

Explain how this filter performs a moving-average filtering of the modulator's outputs. Does this filter decimate the *K*-Delta-1-Sigma outputs? Why or why not?

9.5 Sketch the decimate by *K*/4 topology similar to the topologies seen in Fig. 9.7. Ensure the proper clock signals are used in your sketch.

9.6 Explain, in your own words, why oversampling (averaging the outputs) using a 3-bit Flash converter (eight comparators), won't result in as significant improvement in SNR as the *K*-Delta-1-Sigma topology.

9.7 What is the frequency response (an equation) of the filter seen in Fig. 9.10?

9.8 What is the frequency response (an equation) of the filter seen in Fig. 9.12?

9.9 What is the frequency response (an equation) of the filter seen in Fig. 9.14?

9.10 What is the frequency response (an equation) of the filter seen in Fig. 9.16?

9.11 Show how the switches on the inputs and outputs of the 8 modulators in parallel seen in Fig. 9.20 can be described using the unit matrix and delays or

$$\begin{bmatrix} 1&0&0&0&0&0&0&0\\ 0&1&0&0&0&0&0&0\\ 0&0&1&0&0&0&0&0\\ 0&0&0&1&0&0&0&0\\ 0&0&0&0&1&0&0&0\\ 0&0&0&0&0&1&0&0\\ 0&0&0&0&0&0&1&0\\ 0&0&0&0&0&0&0&1 \end{bmatrix} \cdot \begin{bmatrix} z^{-0}\\ z^{-1/8}\\ z^{-2/8}\\ z^{-3/8}\\ z^{-4/8}\\ z^{-5/8}\\ z^{-6/8}\\ z^{-7/8} \end{bmatrix}$$

where $z = e^{j2\pi f \cdot T_s}$. Using these relationships show how to relate the inputs of the *K* paths in parallel, Fig. 9.20, to the transfer function *H*(*z*) and the resulting topology's outputs.

9.12 The effective sampling frequency of the *K*-Delta-1-Sigma ADC discussed in Sec. 9.2 is roughly 1.6 GHz. Can any component of the ADC operate at, or be clocked at, 1.6 GHz? Verify your answers using SPICE and the 500 nm, 5 V, CMOS models used to generate these figures. What is the most critical component then, from a timing perspective, (the DFF used to capture the eight bits coming out of the modulators) and what is critical in that component (the DFF's setup and hold times)? What happens if an error is made in the most critical component? (The wrong count is captured. For example, we should capture 6 logic 1s but we actually capture 5 or 7 logic 1s. Since we have a significant amount of averaging in the digital filter the effects should be small. If an equal number of positive and negative errors are made the errors average to zero and don't affect the converter's performance.)

Index

A

B

C

D

E

F

G

H

I

J

K

L

M

N

O

P

Q

R

S

T

U

V

W

Z